Dyeing of Cellulose Fibres

Society of Dyers and Colourists

Society of Dyers and Colourists (SDC) is the world's leading independent educational charity dedicated to advancing the science and technology of colour. Our mission is to educate the changing world in the science of colour.

SDC was established in 1884 and became a registered educational charity in 1962. SDC was granted a Royal Charter in 1963 and is the only organisation in the world that can award the Chartered Colourist status, which remains the pinnacle of achievement for coloration professionals.

We are a global organisation. With our head office and trading company based in Bradford, United Kingdom, we have members worldwide and regions in the United Kingdom, China, Hong Kong, India and Pakistan.

Membership: To become a member of the leading educational charity dedicated to colour, please email members@sdc.org.uk for details.

Coloration Qualifications: SDC's accredited qualifications are recognised worldwide.

Please email edu@sdc.org.uk for further information.

Colour Index: The unique and definitive classification system for dyes and pigments used globally by manufacturers, researchers and users of dyes and pigments (www.colour-index.com).

Publications: SDC is a global provider of content, helping people to become more effective in the workplace and in their careers by educating them about colour. This includes textbooks covering a range of dyeing and finishing topics with an ongoing programme of new titles. In addition, we publish *Coloration Technology*, the world's leading peer-reviewed journal dealing with the application of colour, providing access to the latest coloration research globally.

For further information please email info@sdc.org.uk or visit www.sdc.org.uk.

Dyeing of Cellulose Fibres

Rıza Atav

Tekirdağ Namık Kemal University
TU

WILEY

Registered Office(s)
John Wiley & Sons, Inc., 111 River Street, Hoboken, NJ 07030, USA
John Wiley & Sons Ltd. New Era House, 8 Oldlands Way, Bognor Regis, West Sussex, PO22 0NO, UK
Wiley-VCH GmbH, Boschstr. 12, 69469 Weinheim, Germany
John Wiley & Sons Singapore Pte Ltd, 134 Jurong Gateway Road, #04-307H, Singapore 600134

For details of our global editorial offices, customer services, and more information about Wiley products visit us at www.wiley.com.

The manufacturer's authorized representative according to the EU General Product Safety Regulation is Wiley-VCH GmbH, Boschstr. 12, 69469 Weinheim, Germany, e-mail: Product_Safety@wiley.com.
Wiley also publishes its books in a variety of electronic formats and by print-on-demand. Some content that appears in standard print versions of this book may not be available in other formats.

Library of Congress Cataloging-in-Publication Data has been applied for:

Hardback: 9781394366828
ePDF: 9781394366842
epub: 9781394366835
oBook: 9781394366859

Cover Design: Wiley
Cover Image: © Mykola Ivashchenko/Shutterstock

Set in 9.5/12.5pt STIXTwoText by Lumina Datamatics

Contents

Biography

Rıza ATAV was born on 21.09.1981 in Izmir, Türkiye. He graduated from Izmir Private Turkish High School in 1998 as a highest ranked student. He started his university education in the same year and graduated from Ege University Faculty of Engineering, Textile Engineering Department with the title of Textile Engineer as the top student of the department in 2002. In the same year, he started to work as a research assistant and began his postgraduate education at Ege University. He received MSc degree in 2005 and PhD degree in 2009. Afterwards, he started to work as assistant professor in Tekirdağ Namık Kemal University in August of the same year. He was awarded the title of associate professor in 2012. Since December 2017, he has been a professor at Tekirdağ Namık Kemal University Textile Engineering Department.

He speaks fluent English, good Italian, intermediate Spanish and novice French. In 2008, he won the Italian Language Education Scholarship at the Universita' per stranieri di Siena, ranking first in the exam held by the İzmir Italian Cultural Centre.

His main research and interest fields are textile dyeing-printing, functional-intelligent textiles and luxury animal fibres. For more than 15 years, he has been giving undergraduate courses namely Dyeing Technology I, Dyeing Technology II, Textile Printing, Textile Chemistry, Finishing of Denim Fabrics, Professional English and Italian for Engineers; master courses namely Functional Textiles, Luxury Hair Fibres and Their Finishing Processes, Luxury Secretion Fibres and Their Finishing Processes, and New Technologies and Methods in Textile Dyeing; and PhD courses namely Scientific Basis of Textile Dyeing and Applications of Enzymes and Nanotechnological Products in Textile Dyeing.

He has over 140 articles published in various national and international journals, over 140 papers presented at national and international congresses, 2 national books, 1 international book and 5 international book chapters. In addition, he has more than 100 completed and ongoing projects in total. He carries out tasks such as referee, scientific advisory board member and editorial board member in more than 30 national and international journals. He has also been guest editor for the special issue of "Angora Fibre" of the "American Journal of Materials Engineering and Technology" in 2014 and special issue of "3rd International Congress of Innovative Textiles & 2nd International Congress on Wool and Luxury Fibres" published by "Colouration Technology" in 2023. He has graduated 17 MSc and 1 PhD students so far, and he is currently consulting MSc and PhD students. His scientific studies were cited over 1,000 times in various journals.

He carried out various administrative duties such as Vice-Head of Department, Vice Dean, Head of Department, Vocational School Directorate and various commission memberships. He took part in the congress organisation committee of the International Innovative Textiles Congress (ICONTEX) held in 2011, 2019 and 2022. In 2019 and 2022, he held the International Wool and Luxury Fibres Congress (ICONWOOLF).

He is currently working at Tekirdağ Namık Kemal University Textile Engineering Department as a professor. At the same time, he carries out consultancy activities in the R&D centre of various companies within the scope of university-industry cooperation. His R&D studies in the field of textile dyeing are as follows, but not limited to:

- improvement of conventional dyeing processes in an economical and ecological way,
- saving water, time, energy, dye and chemicals in dyeing,
- adaptation of new dyeing methods (layer-by-layer, etc.),
- the use of new technologies (ozone, ultrasound, plasma, etc.) in the dyeing,
- improving the dyeability of fibres by enzymatic treatments,
- industrialisation of natural dyes,
- applications of biotechnological products (dendrimer, liposome, etc.) in dyeing,
- obtaining various colour effects in dyeing,
- applications of dyes in the field of functional and smart textiles,
- digitalisation of textile dyeing,
- obtaining colour without using dyes,
- dye synthesis from structures obtained by recycling from waste,
- obtaining coloured nanosurface via electrospinning,
- reuse of the dyeing wastewater and
- decolourisation of dyeing wastewater.

Preface

It was the afternoon of an ordinary March day. In the background, from the television that was on, the discussion about humanity going through a new challenge and the frequently repeated word like "cobbit" were playing in his ears. At first glance, it was like a little quirk of March who loves bad surprises; there was nothing to be afraid of. While the days, wrapped in webs of obscurity, passed by slowly, everyone, young and old, almost rewinded to Piaget's pre-operational period and stuck there. Where was this virus? You know, it did not exist. The minds that thought that they could not be defeated by fighting seemed to take shelter in denial. At that moment, the words of Albert Camus vaguely came to his mind from The Plague: "Indeed, disasters are a common thing, but when they happen to you, it is difficult. There have been plagues as well as wars in the world. Plagues and wars also catch people unprepared." This is exactly how it happened, the plague of the age, COVID-19. Look, he even learnt its name. The simplest things suddenly became the hardest for humanity. Life didn't seem to flow. Business had stopped; people were confined to their homes. Everyone was in shock; the obscurity turned into despair, and even science and technology remained helpless at the beginning. Humanity, which was planning to go to other planets, could not even get in front of their door. He, on the other hand, was looking for a way out of the wreckage of the negative news that followed. According to Wilhelm Reich, since love, work and knowledge are the inexhaustible resources of our lives, they should rule our lives. Maybe that was the exit!

That's exactly how it happened, dear readers. Although the pandemic had restricted the movement of feet, there was no limit to the hands and brain, and that was enough to write a book. In the first days of the pandemic, I thought of a locomotive spark to bring together the books, which I had been preparing for many years, but could not find the opportunity to put together and put on display, one by one. It was literally turning crisis into opportunity. In this way, my books that you are reading came out. Of course, do not get it wrong; the writing of these books was not something that happened during the pandemic period. It was just opening the puzzle box and pouring it in front of you in order to assemble the puzzle pieces that were built with the meticulousness of a diamond craftsmanship over a period of about 20 years. In my opinion, the duty of a scientist is like a three-legged trivet: to contribute to the literature, even if small, by making permanent scientific researches; to train scientists and to write permanent books that will be masterpieces in their own field, and no matter how solid the other two of these legs are, if the third one is missing, it would make no sense. In this context, I wanted to crown the first two, which I have tried to do as much as I can for many years, with these books that I hope will be permanent. By the way, it is not possible to distinguish one from the other, but I would like to state that I think, raising scientists is one step ahead when it comes to human beings. Just as the sustainable one of dyeing is acceptable, the sustainable one of science is preferable. Because no matter how valuable the work you do, the important thing is to train someone who can keep your studies up and carry them to even higher levels. As in all areas of life, the main thing in business life is to touch someone's life. As Voltaire put it beautifully, "Every man is guilty of all the good he did not do."

Colouring processes carried out to increase the attractiveness and added value of a textile product by changing its appearance can be obtained by two different techniques, dyeing and printing. The purpose of the dyeing process is to colour the entire surface of the material. The phenomenon of colouration actually already exists in nature by itself. For example, essential oil-based organic compounds called carotene in orange peel and limonene in lemon peel give these fruits colour as well as scent. Many beings that we can count around us are self-coloured. To give an example from the textile field, the wool of some sheep is in various colours such as grey, brown or black due to the coloured pigments inside the cortical cells that make up the cortex layer. Colouring of non-coloured materials for various reasons is only possible with dyeing and printing processes.

Since the colour and pattern have an important effect on the selection of a textile product, the majority of textile materials such as clothes and home textiles are produced in different colours and patterns. It will be enough for you to understand what I'm talking about, simply by putting your head up and looking at the clothes and textile materials of the people around. Throughout history, colour has been used for a wide variety of purposes. For example, American Indians coloured their faces both to determine their status in the tribe and to intimidate their enemies. Later, people were influenced by the colours in nature and wanted to colour their goods and clothes. The brown linen woven fabrics found wrapped around mummies in the tombs of the pharaohs in Egypt are proof of this. Colour has also functioned as an indicator of social status in history. An example of this is the practice of Julius Caesar. In those times, the purple-coloured natural dye called "purpura" obtained from the secretion of the Murex snail was very valuable. Because it required thousands of snails to obtain 1 gram of dye. Due to the rarity and cost of these dyes, purple gradually became a colour symbolising power and high position in society.

Although I have thought of collecting it in a single book from time to time, I believe that presenting textile dyeing, which I consider to be a bottomless pit, in separate sections, will be an easier source for you, the beneficiaries, to digest, so the "Textile Dyeing" book series emerged.

Incidentally, in this preface I would like to explain why I have chosen the word "dye" rather than "dyestuff," given that the two terms are semantically identical. It can be said that using the word "dye" would be a more environmentally friendly approach, especially considering the fact that the use of the term "dye" instead of "dyestuff" in written works will save six characters and lead to less paper consumption and indirectly less trees to be cut. Maybe there will be those who say, "There is no need to get hung up on the details, they both fulfil the function of explaining what is meant," but I would like to reply to this with Gülten Akın's lines "Oh, nobody has time to stop and figure out the finer things."

In this book, I have dealt with dyeing of cellulose fibres. Although the names change over time, the dye groups and dyeing methods are permanent, and for this reason, I did not specifically include commercial names in my books. Instead, I have tried to explain in detail the dyes and dyeing methods that have been put on the market in the past; some of them have been discontinued and some of them are still in use, as well as recently developed dyes and dyeing methods in recent years. I just wanted to draw attention to the firsts by giving the trade names of the dye and/or dyeing methods that were first developed in that field or that became iconic in the field. However, by the way, I would like to underline that I do not intend to highlight the products and methods of a particular manufacturer where trade names are used, albeit rarely, within the framework described. Because the fact that its name is included is not due to its superiority over the others, but only because it is the first or the only one in its field.

While talking about the dyeing of each fibre, I discussed not only the dye classes used to dye those fibres today, but also the alternative dye classes that can be used in dyeing those fibres, even if they are not of significant use today. Because sometimes a dye class that has lost its importance can find a use again with the developments in dye technology, auxiliaries and dyeing processes. Sulphur dyes, which once lost their importance due to ecological problems in the dyeing method and have found a place for reuse in obtaining the "vintage effect" with the developments provided, are a good example of this. While explaining how to dye with each dye class, I took care to include the latest developments in the light of economy and ecology criteria, as well as how a more environmentally friendly dyeing can be done at a lower cost, in addition to conventional methods. I have ended my checks that I have been making over and over again to see whether or not something is left incomplete in my books or if there is anything that could have been expressed better, by saying "perfect is the enemy of good."

Throughout my professional life, the biggest shortcoming I saw in my field of expertise was that sometimes textile dyeing was seen as just an application, and the principles behind the events were ignored by focusing on the carrying out of the processes without questioning how and why. This is the biggest problem that leaves textile engineering, which is an applied science, at the level of technicians and hinders progress. The key to highlighting the engineering part of textile dyeing, which is an applied expertise, is to rely on natural sciences (mathematics, biology, physics and especially chemistry). For this reason, my students who take my courses know well that it has been my greatest goal to make textile dyeing not a verbal course and to make the engineering of the business talk. I tried to reflect this point of view in my books as much as I could. Although I have bombarded you with intense chemistry from time to time, I hope I have been able to instil in you the questioning that enables our minds to think systematically in a certain algorithmic order, also for textile dyeing. Although there is only one correct answer, the questions that will lead you to that answer are as wide as your imagination. At this point, being able to ask the right questions is sometimes more important than the answers. Because the source of the answers is the questions. How beautifully Sadi Shirazi said, "He doesn't know since he doesn't ask, if he knew, he would have asked. He doesn't ask since he doesn't know, if he asked, he would have known."

For the reasons explained above, I tried to give detailed information about the scientific foundations of the subjects in my books. On the other hand, I tried to keep physics, biology, chemistry and mathematics at a level that can be understood by those who have taken the undergraduate level courses, by not ignoring the fact that the main target of these books is undergraduate level courses. Taking into account the learning requirements of the modern age and the new generation, I have presented the experimental plans and related research questions that will enable the practical application of the subjects covered in the content of each book in a laboratory environment in the last section of the relevant book.

In recent years, textile dyeing has also had its share of developments in technology, environment and ethical awareness, and for that reason, not only what you produce but also how you produce it has become questionable. While, on one hand, the serious limitations imposed have increased the pressure to produce more ecologically in terms of environment and human health by consuming less water, less energy, less dye and chemicals, on the other hand, to be able to respond quickly to customer demands in a way that keeps up with the pace of change in fashion, and to offer innovative products and effects to the market have become very important. In addition to this, textile dyeing, the foundations of which were built on small scales in simple boilers with natural dyeing, and then started to be carried out under large-scale industrial production conditions in automatically controlled devices with the development of synthetic dyes and dyeing devices, today has changed its crust and gone beyond being a simple means of gaining colour and visual appeal and has evolved to a completely different point. Colour can add a wide variety of functions as well as providing visual appeal and being distinguished from others. For example, the chameleon changes colour to provide chemical camouflage. It does this with the help of melanin pigment, which also helps to browning of human skin in the sun. In the event of fear or anger, the melanin-producing cells of the chameleon become active and spread the melanin particles, which can produce various colours ranging from yellow to brown and black, thus protecting the chameleon from its enemy by changing to a different colour. Biomimetics, that is, imitation from nature, is a subject that has gained great importance in the field of textile dyeing as in many fields of science, starting from the idea that much needed information is hidden in nature. This ability to change colour, which exists in creatures like chameleons and provides various functions, can now be added to fabrics, thanks to textile dyeing, and fabrics dyed with chromic dyes that change colour with various factors (heat, light, pH, etc.), especially for camouflage in the military field, can be produced. Furthermore, smart bandages that change colour to inform that the wound is healed, smart (notifying by changing colour when the individual has a fever, an infection, etc.) clothes for individuals who cannot express themselves (babies, disabled people, etc.) or dye-sensitised batteries with higher efficiency in solar cells can be listed among the functional applications of textile dyeing.

In recent years, while it has been aimed to achieve colour in textile materials in a shorter time by using less water, less energy, less dye and less auxiliary, nowadays it has been started to be examined how to obtain colour without using any dye. Rainbow, which is a meteorological phenomenon in which the colours of the light spectrum appear in the form of an arc, which occurs when raindrops or fog clouds reflect and refract the sun's rays by showing optical prism properties, can be given as an example of the formation of colour without dye. Although there are subjects such as the use of naturally coloured fibres, structural colouring with nanocrystals, microbial colouring with bacteria and fungi, and colouring through heating in textile dyeing, which are currently being studied in terms of obtaining colour without using dye in textile dyeing, many different techniques will be developed that we cannot even imagine today, and since humanity will not give up on colour, it will come to the point where there will be no need for dyeing processes that pollute the environment more or less. Therefore, what we need is an idea and a breathtaking move that will shatter all our classical knowledge and experience, as stated by M.T. in his book "The Delusions of Ikaros, The Story of a 'Destructive Intelligence'." Considering the cumulative progress of science, it can be said that the realisation of this can be achieved by starting from the existing knowledge and especially by multidisciplinary studies. However, it is useful to keep in mind that this may not always be the case, and that an invention may emerge from an unexpected area. Because, as the writer M.T. said, "The invention of the Internet was not the result of the continuous development of the telegraph. Likewise, electricity was not found as a result of the continuous development of the candle."

I would like to point out that I have discussed all these more advanced topics in the *"Advanced Textile Dyeing"* series of books. The subjects of these books are actually a good proof of textile dyeing to be transformed into a multidisciplinary science. Because to understand the use and effects of ultrasound in textile dyeing, and structural colouring a good physics and physical chemistry knowledge; to understand the enzymatic or microbial colouring a good biology and biochemistry knowledge; to understand the adsorption isotherms, diffusion theory, redox reactions, bonds and many more subjects in dyeing a good chemistry knowledge; to understand the various modellings and optimisations related to dyeing a good mathematics knowledge is obligatory. It is obvious that a natural scientist alone cannot realise scientific research that can carry these issues forward, and that a textile engineer alone will not be sufficient for these issues. For this reason, I can say that I believe that multidisciplinary studies in the field of textile dyeing can contribute to the development of this field even more.

Textile engineering is a branch of science that is very diverse, and each requires a separate expertise, starting from the production/cultivation of the fibres, firstly the transformation of these fibres into yarn, then the surface (woven fabric, knitted fabric or non-woven surface) and finally textile product and clothing, and subjecting them in various forms (loose fibre, yarn, fabric, piece garment, etc.) to finishing processes such as pretreatment, colouring (dyeing or printing) and finishing. At this point, I would like to underline that specialisation in industrial business life, and especially in academic life, is critical both to contribute to the field and to benefit oneself. A textile engineer should have a general and broad vision that can look down on all of the above-mentioned issues, but on the other hand, he/she should specialise in a coast he/she chooses in order not to drown in such a wide textile sea. In other words, in my opinion, what is acceptable in textiles, as in every other field, is a T-type person who specialises in one area of textiles but is basically competent in other areas. There is no mistake in metaphor; if you take a little bit of this and a little bit of that without deepening in any of them, then you'll witness an emerge of an ostrich (in Turkish "ostrich" is called as "camel-bird"), which is neither bird nor camel...

As a single author, I would like to say that in addition to the obligations of my full-time professional business life, I found the strength necessary to write all these books from my great love and admiration for textile dyeing. Inspired by Marshall McLuhan's quote "The quality of your life is determined by two things: the books you read and the people you talk to," I can say that the quality of your business life is determined by the books you read and the people you work with. I would like to state that it is my greatest desire for this series of books to contribute to your business life in this sense.

I think I would have done something wrong if I had been content with just squeezing the names of the owners of the books, each of which is a masterpiece, between the lines in the reference lists without mentioning them here. I would like to thank the author of the books "Textile Dyeing-I" and "Textile Dyeing-II" **Prof Dr Işık Tarakçıoğlu**; the co-author of the book "Fundamentals of Dyeing & Printing" and also my MSc and PhD supervisor **Prof Dr Abbas Yurdakul**; the editor of the book "The Theory of Colouration of Textiles" **Prof Dr Alan Johnson** and all chapter authors of the book; The author of the book "Blends Dyeing" and the editor of the book "Cellulosic Dyeing" **John Shore** and all chapter authors of the book "Cellulosic Dyeing"; the editor of the book "Wool Dyeing" **Prof Dr David M. Lewis** and all chapter authors of the book; the author of the book "Chemical Principles of Synthetic Fibre Dyeing" **Prof Dr Stephen M. Burkinshaw**, and the authors of the book "Giles' Laboratory Course in Dyeing" **Prof Dr Charles Hugh Giles, David G. Duff** and **Roy S. Sinclair**, and all other scientists who contributed to the formation of my knowledge in textile dyeing field with their studies and books.

My appreciation is also extended to the **Society of Dyers and Colourists (SDC)** and **Wiley-VCH publishing company** for their interest to publish this book and their excellent collaboration. I would especially like to thank **Dr Andrew Towns** and **Andrew Filarowski** from SDC for their support and motivation. I would also like to present my acknowledgements to **Vishali Chandra Mohan, Felix Bloeck** and **Dhanapriya Velumurthy** from Wiley for their help and disciplined work throughout the project.

With the hopeful expectations of these books to be useful to anyone interested in textile dyeing, especially dear students...

Prof Dr Rıza ATAV
Tekirdağ/29.10.2023

1

General Information About Cellulose Fibres

1.1 Classification of Cellulose Fibres and Their Production in the World

Fibres composed entirely or mostly of cellulose macromolecules are called "cellulose fibres".[1] Cellulose fibres are divided into two groups: natural and regenerated (Figure 1.1).

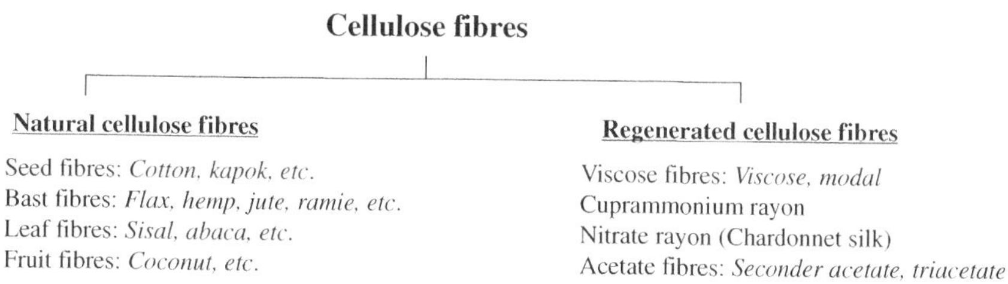

Figure 1.1 Classification of cellulose fibres.

Cellulose is the most abundant of all naturally occurring organic polymers, and thousands of millions of tons of cotton are produced worldwide each year through photosynthesis.[2] Within the global fibre production of approximately 132,000 million kg in 2024, cotton became the second most important fibre after polyester with a share of approximately 19%. Other plant fibres apart from cotton include various fibres such as kapok, flax, hemp, jute, ramie, sisal, agave, abaca and coir fibre. The market share of plant fibres other than cotton, which have a global production volume of approximately 6,900 million kg, was approximately 5% of the total global fibre production in 2024.[3]

Cotton, the purest form of cellulose found in nature, is a seed fibre obtained from plants of the genus *Gossypium*. The most important cotton-growing countries are the United States, China, India, Pakistan and the Central Asian Republics. Other cotton-producing countries include Mexico, Turkey, Brazil, Egypt and Sudan.[2] The cottons of commerce are almost all white (creamy yellow to bright white). In recent years, there has been a renewed interest in naturally pigmented and coloured cottons. The naturally coloured cottons (in brown and green colours) available today are usually shorter, weaker, and finer than regular Upland cottons, but they can be spun successfully into ring and rotor yarns by blending with normal white cottons or among themselves. With the use of naturally coloured cotton, for a limited number of colours, the use of dyes and other chemicals can be completely omitted in textile finishing; this is an important advantage in terms of environmentally friendly production.[4]

Cotton, as a natural textile fibre, is one of the most widely used fibres in the world due to its softness, breathability, biodegradability, comfort and other good properties.[5] However, cotton requires intensive use of water and chemicals.[6] The water consumed in cotton cultivation is approximately 3800 L/kg. The pesticides and synthetic fertilizers used affect many living ecosystems and cause pollution of ground and surface waters, and as a result, they have a great negative impact on global health.[7] Therefore, the development of sustainable and economical alternative natural fibres and their adaptation to conventional production processes are required.[6] Research on alternative raw materials that will reduce the rate of cotton use in textile products mainly point to flax and hemp fibres with the increase in the popular use of bast fibres (flax, hemp, jute, ramie, etc.) in recent years.[8] Jute, coir and flax are the most produced with market shares of approximately 54, 26 and 5% of all non-cotton, plant-based fibres, respectively. Jute is used for home textiles such as curtains and carpets, as well as for making twine, rope, mats and packaging materials. While coir fibre is used in the production of home textiles such as floor mats, doormats, brushes and mattresses; processed flax fibres are used in various products such as home textiles and clothing.[3] Ramie is used in high-quality fabrics. On the other hand, leaf fibres are coarse and are mainly used in ropes and cords.[9]

Flax and hemp are hardy crops and can grow in nutrient-poor soils while needing much less water, chemicals and pesticides for cultivation than cotton. It is also considered a more sustainable way of producing textiles, as hemp grows faster (3–4 months) than trees and other crops.[7] Compared to cotton, which is the most consumed natural fibre, flax and hemp fibres stand out with their superior ecological properties and organic textile production potential.[10] Fabrics produced from hemp fibre exhibit superior distinguishing features such as high moisture content and moisture absorbency, breathability, UV resistance, high temperature resistance and tensile strength.[11] Like hemp, fabrics produced from flax fibres are preferred by consumers for their many properties, such as rapid moisture absorption, antibacterial, UV resistance, antistatic, unique style and vintage appearance, and are widely used in summer clothing and ready-to-wear.[12]

Regarding regenerated cellulose fibres, it can be said that they constitute approximately 6% of the total fibre production volume with an annual production of 8,400 million kg, according to 2024 data. The market shares of viscose, modal and Lyocell fibres among regenerated cellulose fibres are approximately 80, 3 and 5%, respectively. The market share of cupro fibres is approximately 0.2%. Acetate fibres are the second most produced fibre after viscose, with a share of approximately 13% among regenerated cellulose.[3]

In the light of this information, it can be said that cotton fibres are the most important among cellulose fibres. Since the basis of all the fibres given in the diagram above is cellulose, their dyeing principles are basically similar to each other. For this reason, in this section, firstly, general information about the structure of cellulose fibres will be given, and then the dyeing of cotton fibres will be discussed in detail in the following sections. Finally, in a separate section, information will be given about the differences and the special points to be considered in dyeing of vegetable fibres other than cotton and regenerated cellulose fibres. Considering their use in the field of textile clothing in the world today, leaf and fruit fibres will not be mentioned both in this section and in the following sections, since vegetable fibres other than seed and bast fibres do not occupy an important place. In addition, since cellulose nitrate and acetate (secondary acetate and triacetate) fibres are obtained by chemical modification of cellulose, their dyeing behaviour is not similar to cotton fibres but to other fibres (e.g. dyeing acetate fibres is similar to polyester). Therefore, nitrate and acetate fibres are also excluded from the scope of this book.

1.2 Chemical Structure of Cellulose Fibres

Cellulose is a polysaccharide formed as a result of the combination of *n* units of β-glucose, a monosaccharide.[13] Cellulose macromolecules are formed by connecting *n* units of β-D-glucopyranose rings via oxygen bridges over the first and fourth carbon atoms of one another.[14] (Figure 1.2). There are 10,000–15,000 glucose units in the cellulose macromolecule. The glucose building blocks in the cellulose chain are arranged along a helical line.[13]

There are three hydroxyl (alcohol) groups in each glucose building block that makes up the cellulose macromolecule.[14] Therefore, there are approximately 30,000–45,000 functional alcohol groups in a cellulose macromolecule.[13] The three —OH groups play an important part in contributing to the ability of cotton cellulose to form highly crystalline regions through hydrogen bonding.[15] Of these, those attached to carbon number 6 are primary alcohol groups, while those attached to second and third carbon atoms are secondary alcohol groups. The reactivity of the primary and secondary alcohol groups is different from one another, and the reactivity of the two secondary alcohol groups is also different from each other. While primary hydroxyls determine the dye sorption of the cellulosic material and the dyeability of the fibre, the reactivity of the hydroxyl groups adjacent to the glycosidic bond at the α position is greater against alkalis and oxidative agents.[14]

Cotton fibres have a very high cellulose content of 94%. In this respect, it surpasses all other natural cellulosic fibres. Apart from cellulose, cotton fibres contain 1.3% protein,

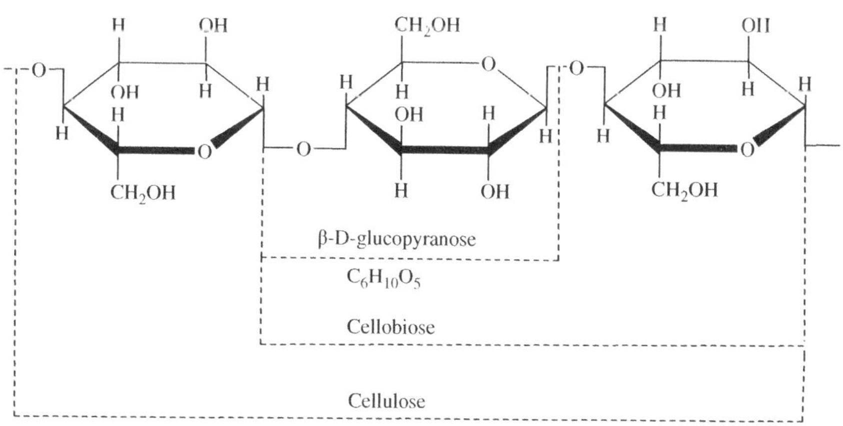

Figure 1.2 Formation of cellulose from β-D-glucopyranose. Ref. [14]/Ege Üniversitesi Tekstil ve Konfeksiyon Araştırma-Uygulama Merkezi Yayınları.

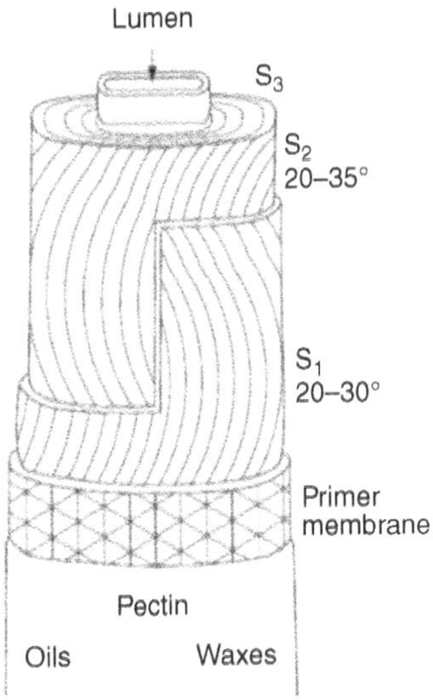

Figure 1.3 Morphology of cotton fibre. Ref. [2]/with permission of Society of Dyers and Colourists.

1.2% pectin, 0.6% wax, 1.2% ash and 1.7% other substances (usually water-soluble organic acids and sugars). When the morphology of cotton fibres, shown schematically in Figure 1.3, is examined, it is understood that they consist of three layers: primary wall, secondary wall and lumen. Primary wall consists of a network of cellulose fibrils covered with an outer layer (or cuticle) of pectin, protein, mineral matter and wax. Although this layer makes up only 5% by weight of the fibre, most of the non-cellulosic components are present here. Wax prevents the fibre from permeating water and aqueous solutions unless there is a wetting agent. The secondary wall forms the bulk of a mature fibre and consists almost entirely of cellulose fibrils spirally arranged around the fibre axis. The secondary wall consists of several layers, and the spiral angle of the fibrils varies from one layer to the next ranging from about 20° to 35°.[2]

Bast fibres are obtained from the stems of various plants, and the most important is the fibre obtained from the *Linum usitatissimum* plant known as flax. These fibres form a protective layer around the woody middle part of the stem and are protected by the outermost cuticle, which contains waxes and other substances. The fibres are composed of thick-walled bundles of cells held together by adhesive substances, and the sizes of the fibres vary from one species to another. Bast fibres contain much less cellulose than cotton. After cellulose, the most abundant component in their structure is the intercellular material consisting of pectins, hemicelluloses and lignin. In addition, they contain wax, ash and other substances.[2]

The chemical compositions of various plant fibres are given in Table 1.1. In this table, fibres are coded as seed (s), bast (b), leaf (l) or fruit (f) fibres according to their origin.

As seen in Table 1.1, the main components of all plant fibres are cellulose, hemicellulose, lignin and pectin.[17]

Considering the regenerated cellulose fibres, it can be said that viscose fibres are the most important. In the production of these fibres, linters or cellulose material are treated with caustic soda, and the sodium cellulose obtained is converted into cellulose xanthogenate by

Table 1.1 Chemical composition of various plant fibres.

Plant fibres	Cellulose (%)	Hemicellulose (%)	Lignin (%)	Pectin (%)
Cotton[s]	82–96	2–6.4	0–5	<1–7
Kapok[s]	13	—	—	—
Flax[b]	60–81	14–20.6	2.2–5	1–4
Hemp[b]	70–92	18–22	3–5	1
Jute[b]	51–84	12–20	5–13	0.2
Kenaf[b]	44–87	22	15–19	2
Ramie[b]	68–76	13–15	0.6–1	2
Banana[l]	60–65	6–19	5–12	3–5
Pineapple[l]	70–82	16–19	5–12	2–3
Sisal[l]	43–78	10–24	4–12	0.8–2
Coconut[f]	43–46	0.25	45–46	3–4

Ref. [16].
s. Seed
b. Bast
l. Leaf
f. Fruit fibres.

treating with carbon sulphide (CS_2). The cellulose xanthogenate is then dissolved in dilute caustic soda, passed through nozzles and solidified in an acidic bath. Since the resulting fibres are obtained by only physical modification of cellulose, the chemical structures of these fibres are similar to cotton.[18] The most important drawback of viscose fibres is their low wet strength. For this reason, second-generation modal and third-generation Lyocell fibres with high wet strength were produced later on. There are two types of modal fibres: "polynosic", which has a higher alkali resistance and wet strength than viscose, and "High Wet Modulus (HWM)", which has a slightly lower strength than polynosic fibres and is not resistant to alkalis. The chemical structure of Modal and Lyocell fibres consists of cellulose.[19]

In the production of cuprammonium rayon, after the linters or cellulose material is dissolved in copper tetramine hydroxide and the prepared solution is passed through the nozzles, the fibres are solidified by contacting them with water and dilute acid. Since the resulting fibres are obtained by only physical modification of cellulose, the chemical structures of these fibres are similar to cellulose.[18]

Cellulose nitrate (or nitrocellulose) is produced by nitration of natural cellulose from cotton linter or wood pulp.[20] Nitrate rayon is obtained by treating cellulose with a mixture of 40% HNO_3/40% H_2SO_4/20% H_2O at 45 °C for 2 hours.[18]

Cellulose acetate is produced by acetylation with acetic acid and acetic anhydride with sulphuric acid as catalyst. Cellulose acetate crystallises best when all hydroxyl groups at positions 2, 3 and 6 are substituted by acetyl ($COCH_3$) groups. This fibre is called cellulose triacetate or primary cellulose acetate. On the other hand, fibres spun from partially saponified cellulose triacetate are called secondary or two-and-a-half acetate.[21]

1.3 Physical and Chemical Properties of Cellulose Fibres

The physical and chemical properties of the fibres must always be taken into consideration in dyeing processes.[17] The physical properties of plant fibres are given comparatively in Table 1.2.

Table 1.3 provides comparative information on various physical-technological properties of cotton and regenerated cellulose fibres.

Table 1.2 Physical properties of various plant fibres.

Plant fibres	Length (mm)	Diameter (μm)	Moisture Content (%)
Cotton[s]	20–64	11.5–17	8.5
Kapok[s]	8–32	15–35	10.9
Flax[b]	27–36	17.8–21.6	12
Hemp[b]	8.3–14	17–23	12
Jute[b]	1.9–3.2	15.9–20.7	12
Kenaf[b]	2–61	17.7–21.9	17
Ramie[b]	60–250	28.1–35	8.5
Banana[l]	2–3.8	—	—
Pineapple[l]	—	20–80	—
Sisal[l]	1.8–3.1	18.3–23.7	11
Coconut[f]	0.9–1.2	16.2–19.5	13

Ref. [16].
s. Seed
b. Bast
l. Leaf
f. Fruit fibres.

Table 1.3 Various physical-technological properties of cotton and regenerated cellulose fibres.

Physical-technological property	Cotton	Viscose	Polynosic	HWM	Lyocell	Cupro fibre
Breaking strength (dry) (cN/tex)	4–28	20–25	35–40	34–36	40–44	15–20
Breaking strength (wet) (cN/tex)	25–30	10–15	27–30	19–21	34–38	9–12
Elongation at break (dry) (%)	7–9	20–25	10–15	13–15	14–16	7–23
Elongation at break (wet) (%)	12–14	25–30	10–15	13–15	16–18	16–43
Average degree of polymerisation	2000–3000	250–300	500–600	300–600	550–600	450–550

Ref. [22].

The effects of various factors on cellulose fibres are summarised below.

- **Water:** As a result of water being retained by the hydroxyl groups in the intermicellar region and on the outer surface of the crystallites, the fibres swell. The swelling is mostly in the direction perpendicular (transverse) to the fibre axis, and the increase in the longitudinal direction is very small.[14] The hygroscopic moisture content of the fibres is cotton: 8.5%, raw linen: 12%, bleached linen: 8.75%, jute: 13.75%, rayon: 11%.[1]
- **Acids:** Cellulose fibres are very sensitive to acids since the oxygen bridges (acetal linkages) connecting the building blocks forming the fibres are not resistant to acids. Cellulose damaged by acids is called hydrocellulose. Since organic acids (acetic acid, formic acid, etc.), which are weaker than inorganic acids, have a less risk of damaging the fibres, organic acids can be used when necessary in the finishing processes of these fibres.[14]
- **Alkalis:** Normally, natural cellulose fibres have low solubility in alkalis. However, the effect of alkaline solutions on cellulose is greater, especially in the presence of air. Since products made of cellulose fibres during textile finishing can be processed several times in an alkaline environment, special care should be taken during these processes, especially for products made from regenerated cellulose fibres.[14]
- **Salts:** Salts have a swelling and dissolving effect on cellulose fibres, depending on the type of salt. Normal, undamaged natural cellulose fibres have low solubility in salts.[14]
- **Oxidising agents:** Cellulose fibres are generally sensitive to oxidising agents. Cellulose damaged by oxidising substances is called oxycellulose. If the products made from these fibres are treated with oxidising agents carefully and moderately, the fibres are not damaged. However, if the treatment with oxidising agents is done uncontrollably and quickly, it is possible for the fibres to be easily damaged as a result of the rupture of the oxygen bridges.[14]
- **Reducing agents:** Cellulose fibres are generally resistant to reducing agents.
- **Organic solvents:** Cellulose fibres are insoluble in known organic solvents (alcohol, ether, benzene, etc.). However, the reactivity of fibres treated with some organic solvents increases.[13]
- **Heat:** Cellulose, as an organic substance, shows limited resistance to high temperatures. When the cellulose fibres are heated up to 150 °C for a short time, no significant change occurs, while pyrolysis (thermal decomposition) starts above 200 °C, the pyrolysis rate increases well above 350 °C and above 400 °C the resulting pyrolysis gas mixture is self-igniting.[14]
- **Light:** Under the influence of light, cellulose fibres are photo-oxidised by atmospheric oxygen and turn into oxycellulose. Under these conditions, fibre strength and degree of polymerisation decrease. The resistance to light of various cellulose fibres differs among themselves. The time required for 50% strength loss to occur due to sunlight is 900 hours for viscose, 940 hours for cotton and 999 hours for linen.[13]
- **Microorganisms and harmful insects:** The microorganisms responsible for microbial damage are always present, but moisture content of around 9% (although microbial damage has been detected at lower levels) is needed for their growth. Microbial

damage, caused by either bacteria or fungi, can occur after boll opening, prior to or after harvesting, during storage, and in cotton fabrics.[4] As a result, effects such as the appearance of stains of different colours on the material and strength losses occur.[13] Cellulose fibres are not affected by moths and other harmful insects.

References

1 Tarakçıoğlu, I. Tekstil terbiyesi ve makinaları, Cilt 1: Tekstil terbiyesinde temel işlemler ve selüloz liflerinin terbiyesi. İzmir, Türkiye: Ege Üniversitesi Tekstil Fakültesi Yayınları, 1979.

2 Nevell, T.P. Cellulose: Structure, properties and behaviour in the dyeing process. In Cellulosics Dyeing (Ed. J. Shore, Chapter 1, 1–73). Bradford, West Yorkshire, UK: Society of Dyers and Colourists, 1995.

3 Preferred Fiber & Materials Market Report 2025. Link of access: https://textileexchange.org/knowledge-center/reports/materials-market-report-2025/, [Date of access: 28.11.2025].

4 Wakelyn, P.J., Bertoniere, N.R., French, A.D., Thibodeaux, D.P., Triplett, B.A., Rousselle, M.-A., Goynes, W.R., Edwards, J.V., Hunter, L., McAlister, D.D., Gamble, G.R. Cotton Fiber Chemistry and Technology, International Fiber Science and Technology Series (Ed. M. Lewin), CRC Press, Taylor & Francis Group, 2006.

5 Ling, C., Guo, L. Preparation of a flame-retardant coating based on solvent-free synthesis with high efficiency and durability on cotton fabric. Carbohydrate Polymers, 2019; 115648.

6 Schumacher, A.G.D., Pequito, S., Pazour, J. Industrial hemp fiber: a sustainable and economical alternative to cotton. Journal of Cleaner Production, 2020; 268, 122180.

7 Ahirwar, M., Behera, B.K. Development of hemp-blended cotton fabrics and analysis on handle behavior, low-stress mechanical and aesthetic properties. The Journal of the Textile Institute, 2022; 113(5), 934–942.

8 Ebskamp, M.J. Engineering flax and hemp for an alternative to cotton. Trends in Biotechnology, 2002; 20(6), 229–230.

9 Ciechanska, D., Wesolowska, E., Wawro, D. An introduction to cellulosic fibres. In Handbook of textile fibre structure Volume 2: Natural, regenerated, inorganic and specialist fibres (Eds. S.J. Eichhorn, J.W.S. Hearle, M. Jaffe, T. Kikutani, Chapter 1, 3–61). Woodhead Publishing in Textiles, UK, 2009.

10 Gedik, G. Kenevir liflerinden üretilen kumaşların optimum ağartma koşullarının ve yöntemlerinin belirlenmesi, MSc Thesis, Pamukkale Üniversitesi Fen Bilimleri Enstitüsü, Denizli-Türkiye, 2010.

11 Kabir, S.M.M., Karim, R., Islam, K. A comparative study on dyeing properties of hemp and cotton fiber. European Scientific Journal, 2017; 13(33), 378.

12 Liu, S., Dai, J., Jia, H., Liu, X., Xu, B. Effect of siro-spun processing parameters on properties of 55/45 flax/cotton blended yarn. Advanced Materials Research, 2011; 331, 502–511.

13 İskender, M.A. Tekstil kimyası I ders notları. Uludağ Üniversitesi, 1989.

14 Seventekin, N. Tekstil kimyası. İzmir, Türkiye: Ege Üniversitesi Tekstil ve Konfeksiyon Araştırma-Uygulama Merkezi Yayınları, 2004.

15 Mather, R.R., Wardman, R.H. The Chemistry of Textile Fibres, The Royal Society of Chemistry, UK, 2011.

16 Mwaikambo, L. Review of the history, properties and application of plant fibres. African Journal of Science and Technology, 2006; 7(2), 121.

17 Ansell, M.P., Mwaikambo, L.Y. The structure of cotton and other plant fibres. Handbook of Textile Fibre Structure, 2009; 62–94.

18 Harmancıoğlu, M. Tekstil maddeleri II: Rejenere ve sentetik lifler. Bornova, İzmir, Türkiye: Ege Üniversitesi Matbaası, 1981.

19 Atav, R., Durak, G., Öktem, T., Seventekin, N. Modal liflerinin üretimi, özellikleri ve kullanım alanları. Tekstil ve Konfeksiyon, 2004; 14(2), 81–86.

20 El Nemr, A., Eleryan, A., Mashaly, M., Khaled, A. Rapid synthesis of cellulose propionate and its conversion to cellulose nitrate propionate. Polymer Bulletin, 2021; 78, 4149–4182.

21 Ganster, J., Fink, H.-P. The structure of man-made cellulosic fibres, In Handbook of textile fibre structure Volume 2: Natural, regenerated, inorganic and specialist fibres (Eds. S.J. Eichhorn, J.W.S. Hearle, M. Jaffe, T. Kikutani, Chapter 6, 201–233). Woodhead Publishing in Textiles, UK, 2009.

22 Özgüney, A.T., Körlü, A., Bahtiyari, M.İ., Bahar, M. Viskon liflerinin fiziksel özellikleri ve makromolekülerüstü yapısı. Tekstil ve Konfeksiyon, 2006; 16, 100–104.

2

Processes of Cellulose Fibres Before Dyeing

2.1 Conventional Pretreatment Processes Applied to Cellulose Fibres

In order to obtain a uniform dyeing, cotton fibres must be pretreated and thoroughly cleaned before dyeing. Due to the oils and resins it contains, raw cotton can hardly be wetted by the dyeing liquor, resulting in an uneven dyeing that has not penetrated well. Although in some cases a uniform dyeing of cotton can be achieved without any pretreatment (e.g. in dyeing with sulphur dyes, since the dyeing bath is a boiling, alkali bath, pretreatment can be omitted), in most cases this is not possible. Good wetting of the cotton material by the dye liquor can be achieved by previously boiling the material in an alkali solution. Especially since the dyeing time is short in continuous dyeing methods, it is essential to ensure that the material gains good wetting and absorbing properties with careful pretreatment. Sometimes, the mentioned drawback can be reduced by treating the material to be dyed with solutions containing wetting agent or by adding wetting agent to the dyebath.[1]

Cotton fabrics are subjected to singeing, desizing, scouring and bleaching processes before dyeing. Proper singeing is important in order to remove the fibre ends on the surface and thus obtain more uniform and clearer colours in dyeing. Desizing is required in woven fabrics, because the sizing layer will give the fabric a hydrophobic feature, making it difficult for the liquor to be absorbed by the fibres, in other words, the fibres from wetting and swelling. Alkali treatment is required to remove substances such as fat, wax and pectin that give hydrophobic properties to cotton fibre. Nowadays, this is not usually applied as a separate process but is done in one step along with bleaching. If the material will be dyed in dark shades, bleaching may not be necessary. However, they are generally subjected to a mild peroxide bleaching process (bleaching using less peroxide). Effective bleaching must be done on materials that will be dyed in light and pastel shades.

Another pretreatment process that can be applied to cotton fibres is mercerisation. As a result of mercerisation, the cross-section of the fibres swells well and the crystal structure is changed. So, the dye uptake rate and amount of cotton fibres, which gain a more open structure, increase. In addition, since mercerised fibres with smoother surfaces reflect light more uniformly and the length of the path through which the rays pass through the swollen cross-section fibres increases, the mercerised fibres dyed with the same amount of dye appear darker compared to non-mercerised fibres.[2] As the stretching applied to the material during the mercerisation process decreases, the dye uptake increases further. Therefore, if the main purpose of mercerisation is to increase the dye uptake, causticisation, i.e., tension-free mercerisation, should be performed.[3]

In recent years, the use of enzymes in the finishing of various textile products has been increasing. The reason for this increase is that enzymatic processes and their wastes do not cause environmental pollution.[4] Cotton materials can be enzymatically treated with cellulase. Cellulases consist of complex mixtures of "Endo-Gluconases", "Exo-Gluconases", "Cellobiohydrolases" and "β-Glucosidases". The following effects are obtained as a result of enzymatic treatment with cellulase:

- Strong reduction in pilling tendency.
- Less fibre debris, therefore a smoother fabric surface structure.
- Improvement of the fabric's handle and draping ability.
- Improvement of the ability of the fibres to absorb water into their structure.[5]
- Obtaining deeper shades due to cleaning the fibre surface during enzyme treatment.[6]

Since the amount of foreign matter contained in flax fibres is high, removing it before dyeing is essential for obtaining a dyeing with good fastness values. This can be done by boiling the linen with soda ash solution before dyeing it and even then bleaching it.[1] However, since alkaline treatments remove non-cellulosic materials and therefore cause undesirable weight loss, linen materials are subjected to pretreatments such as scouring and bleaching when dyeing only pastel shades. This is because dyeing at 60–80 °C causes a 5–10% weight loss, when an alkaline treatment followed by a bleaching is done under high-temperature conditions, a 20% weight loss occurs.[7] After proper pretreatment, linen yarns and linen fabrics can be dyed using conventional methods developed for cotton. If lignin residues remain, they take up more dye from the cellulosic component and result in a mottled appearance and sometimes a deviation in hue. The colour of these darker dyed spots can be problematic, although the colour may diminish as a result of desorption of the poorly bound dye during subsequent washing operations. Therefore, it is important to remove as much impurities as possible before dyeing.[8]

Ramie has a low impurity content compared to other bast fibres and can be processed very similar to cotton. Other fibres, however, have large amounts of non-cellulosic components and are usually slightly brown in colour. Impurities include lignin, which is responsible for the brown colour, as well as hemicelluloses and pectins (and their salts).[9] While jute is usually used unpurified, flax and ramie are usually scoured and sometimes bleached. Most of the non-cellulosic material is removed to the required level depending on the quality desired. Jute is bleached before dyeing, but often significant amounts of lignin remain, resulting in poor light fastness.[3]

2.2 Pretreatments Applicable to Improve and Change the Dyeing Properties of Cellulose Fibres

Beyond the conventional pretreatments described so far, various pretreatments are sometimes used to improve or alter the dyeing properties of cellulose fibres. For example, as can be seen from Figure 2.1, if cellulose is treated with p-toluene sulphochloride, the cellulose derivative obtained is not dyed by dyes that dye cellulose, so it is called "immune yarn".[10]

Figure 2.1 Obtaining immune thread from cellulose. Ref. [1]/Ege Üniversitesi Mühendislik Fakültesi Çoğaltma Yayınları.

Figure 2.2 Obtaining amine thread from immune thread. Ref. [1]/Ege Üniversitesi Mühendislik Fakültesi Çoğaltma Yayınları.

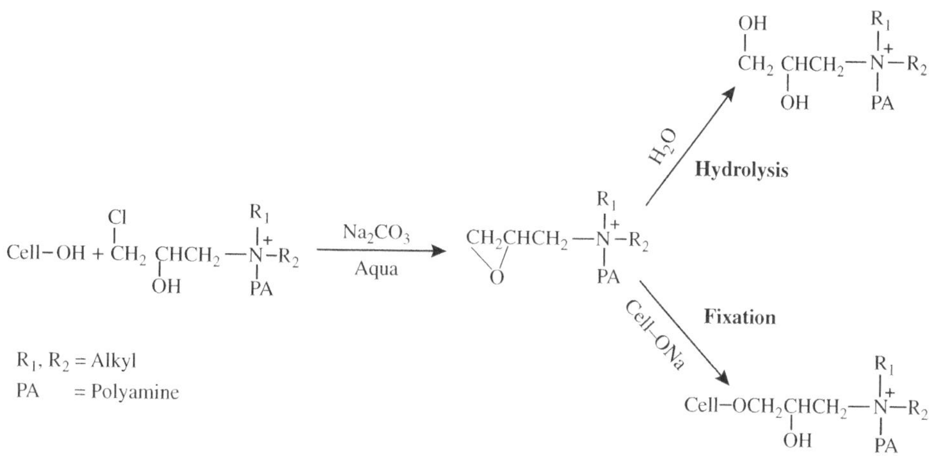

Figure 2.3 Hydrolysis and fixation mechanism of the product based on polyaminochlorhydrin quaternary ammonium compound containing epoxy group. Ref. [12]/DergiPark.

As a result of the treatment of the immune thread with ammonia, the p-toluene sulphonic acid group is broken off and replaced by the amino group, and the cellulose fibres obtained in this way are called "amine thread" (Figure 2.2). Since they have a basic amino group in their structure, they can be dyed well with acid dyes used in wool dyeing.[10]

Similarly, if cellulose fibres are treated with substances that impart acidic properties (e.g. with tannin), they gain affinity for basic dyes.[1]

Studies are also carried out on the addition of amine groups to cotton fibres by pretreatment with cationic substances and thus on the modification of the dyeing properties of the fibres. Currently, amines or quaternary cationic substances containing reactive groups (epoxy, active halogen, ethoxylate or amino) are used as cationising agents.[11] In Figure 2.3, the hydrolysis and fixation mechanism of the polyaminochlorhydrin quaternary ammonium compound-based product containing the epoxy group is given.[12] The aforementioned cationisation processes find application especially in cotton garment dyeing. After the pretreatment made in this way, the products are generally dyed with acid/1:2 metal complex dyes.

Dendrimer technology is one of the subjects that have just begun to be studied in order to improve the dyeability of cellulose fibres by chemical modification. Dendrimer is derived from the Greek words “dendron” (tree) and “meros” (part) and refers to tree-like branched compounds.[13] The term dendrimer was first proposed by Tomalia in 1985.[14] At the same time, the Newkome group independently synthesised macromolecules with similar structures.[15] Dendrimers consist of a core branching units around the core and surface groups, also called branched functional groups. The diversity of dendrimers is provided by functional groups. Branching units, on the other hand, allow dendrimers to grow repeatedly.[16] Dendrimers, which emerged as a product of the developments in polymer technology, have found a wide variety of applications due to their superior properties. One of the new application areas for dendrimers is textile.[17]

In the dendrimer with an amine end group, the amine groups react and bond more easily with the aldehyde groups formed by the oxidation of the alcohol groups, rather than with the alcohol groups in the cotton fibres[18] (Figure 2.4). It is stated that as a result of this pretreatment, less salt can be used in dyeing cotton fibres with reactive dyes, and the use of salt may even be eliminated.

With the re-evaluation of wastes on the agenda in recent years, instead of being left to rot, shellfish waste is re-evaluated by chemical or biological methods and new products are obtained. Chitin and its main derivative, chitosan (Figure 2.5), are the leading products obtained in this way.[20] Chitin and chitosan, a non-toxic and natural biopolymer, are mainly obtained from crab and shrimp shells. In addition, insect shells are also very rich in terms of chitin.[21] Chitin is the second most common biopolymer in the world, after cellulose.[20] If a pretreatment with chitosan is applied to the cellulose fibres before dyeing and chitosan molecules are attached to the structure of the fibres, an increase in the anionic dye uptake ability of the fibres occurs in the subsequent dyeing process to be performed with anionic dyes.

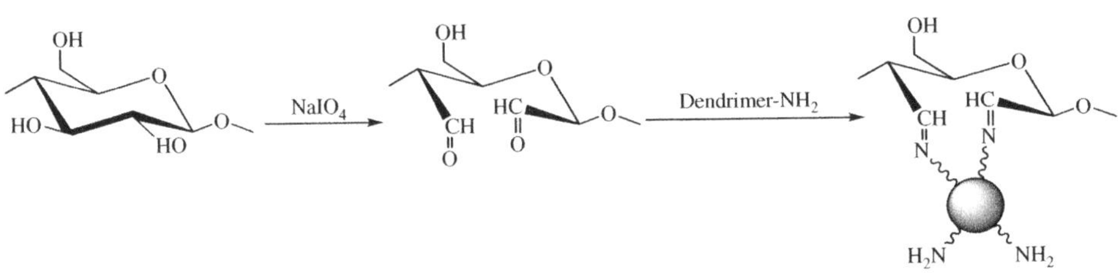

Figure 2.4 Mechanism of action of m-periodate on cellulose fibres. Ref. [19]/Elsevier.

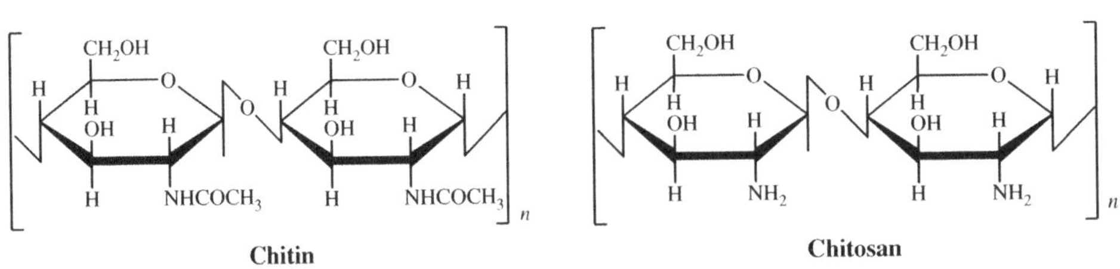

Figure 2.5 Chemical structure of chitin and chitosan. Ref. [20]/DergiPark.

It is worth remembering that the aforementioned processes that change or improve the dyeing properties of the fibres can also be used to obtain various colour effects in dyeing. By blending the cationised and untreated fibres and dyeing them in a single bath, a mélange effect can be obtained. In the case that untreated and different degrees of cationised yarns are used on the same fabric in a certain order, stripe or different pattern effects can be obtained after dyeing. If cellulose fibres with hydroxyl groups removed (immune yarn) and normal or cationised cellulose fibres are used together, colour-white contrasts can be achieved in the same bath.

Before closing the discussion of the processes of cellulose fibres before dyeing, it is useful to briefly mention the effects of various pretreatments on the dyeability of cellulose materials, which have gained importance in recent years and are still being investigated.

Although ***ultraviolet technology*** has a wide area of use in the surface treatment of various polymers, it has not been used widely in textiles until now. UV rays are divided into three groups according to their wavelengths, from the longest to the shortest: UVA, UVB and UVC. UVA and UVB rays, which have lower energy, are not very effective in chemical modification to improve the dyeability of textile fibres. Therefore, UVC rays are used when aiming to improve dyeability. UVC rays provide the formation of new groups on the surface of the fibres. In this way, it increases the properties of materials such as wettability and adhesion. The chemical changes that occur with UVC are limited only to the fibre surface, so there is no significant change in strength.[22] The UV irradiation of the fabric is a factor affecting the colour strength of the fabric. Previous studies show that UV irradiation adds value to colouring and also increases the dye uptake ability of cotton fabrics through oxidation of the surface of cellulose fibres. UV irradiation affects the colour strength values of the dyed fabric and the resulting hue of the colour.[23]

Another technology that can be used in the modification of fibres is ***ozone technology***. Ozone (O_3) is the strongest oxidant with an oxidising potential of 2.07 volts. It is 25 times more effective than hypochlorous acid (HOCl), 2500 times more effective than hypochlorite (NaOCl) and 5000 times more effective than chloramine (NH_2Cl). However, it does not produce waste or harmful substances like them.[24] The ozonation process generally leads to an increase in the degree of whiteness and a decrease in strength in cellulosic fibres. In a study conducted in the literature, it is stated that after ozonation, a decrease in the reactive dye uptake ability of the fibres occurs, and this is due to the conversion of hydroxyl groups in the fibres to aldehyde and carboxyl groups. However, in the experiments, it was determined that the ozonation process did not have a significant effect on the dyeing of cotton fibres with vat dyes. The reason for this is based on the fact that vat dyes are a class of dyes that do not react chemically with the functional groups of the fibres, even if changes occur in their functional groups as a result of damage to the fibres.[25]

Plasma consists of radicals, electrons, ions, neutral atoms, excited particles and UV rays.[26] When the plasma interacts with the material, the species forming the plasma give their energy to the material as a result of collisions, some bonds on the surface are broken, radicals are formed, and accordingly, surface modifications occur. It has been reported by many researchers that plasma treatment increases the hydrophilicity of cotton fabrics. In addition to the oxidation of waxes and oils on the surface of the raw fabric by the effect of plasma and their conversion to compounds with lower molecular weights that can evaporate during the process, the formation of new channels and polar groups on the surface also causes an increase in hydrophilicity.[27] In general, when the literature is examined, it is seen that there is an increase in the rate and amount of dye uptake of plasma-treated cotton fibres. Contributing factors to increased dye uptake may include:

- Change in fabric surface area per unit volume due to surface abrasion.
- Abrasion of the fibre surface and removal of impurities (e.g. cotton wax, etc.) on the surface.
- Change in surface properties due to the formation of free radicals on the cellulosic chains of the cotton.[28]

Modification of the material surface with laser is one of the technologies that have been studied in recent years. ***Laser*** is an acronym of the definition of "light amplification by stimulated emission of radiation". Since laser beams are of high frequency, they have the characteristics of sunlight. The basis of the laser is the electron transitions between atomic or molecular energy levels. Since atoms whose energy level is increased tend to be in a low energy state, they emit a certain amount of photons when returning to this level.[26] Adequate power levels for a given application are essential in surface modification processes because excessive amounts of energy can damage polymers. Infrared lasers such as CO_2 are the most powerful lasers. Serious thermal damage can occur if the applied power level is not appropriate. However, this problem can be overcome by using pulsed-mode CO_2 lasers, which are easier to control than lasers operating in continuous wave mode.[29] Possible applications of laser technology in the textile industry include removing the indigo dye from denim, creating patterns on textile materials to change the dyeability of the fibres, creating surface roughness, and cutting textile surfaces.[30] Although laser technology can be used to increase dyeability, the use of laser energy in textiles is still rare and is generally limited to denim garment finishing.[31] In the literature, it has been observed that laser treatment with different resolutions and pixel times leads to the formation of anionic carboxylic acid groups in cotton fabrics, thereby reducing dye uptake.[30, 32]

Studies on the interaction of high-energy radiation with polymers have attracted the attention of many researchers.[33] Gamma radiation is also a type of high-energy radiation.[34] Cross-linking as a result of irradiation changes the crystal structure of cellulose. In addition, gamma radiation causes photomodification of surface fibres due to oxidative degradation. In the literature, it is stated that irradiated modified fabrics will provide advantages such as fixing more dyes or pigments, producing darker shades, and achieving faster fixation of dyes at low temperature.[23]

In addition to ultraviolet, ozone, plasma, laser and gamma radiation technologies, there are many more new technologies such as electron bombardment, ion implantation. However, studies on the use of these technologies on cellulose fibres are either absent or very limited.

References

1 Tarakçıoğlu, I. Tekstil Boyacılığı I Teksiri. İzmir, Türkiye: Ege Üniversitesi Mühendislik Fakültesi Çoğaltma Yayınları, 1982.

2 Tarakçıoğlu, I. Tekstil terbiyesi ve makinaları. Ege Üniversitesi Emel Akın Meslek Yüksek Okulu, Bornova, İzmir, 1996.

3 Nevell, T.P. Cellulose: structure, properties and behaviour in the dyeing process. In Cellulosics Dyeing (Ed. J. Shore, Chapter 1, 1–73). Bradford, West Yorkshire, UK: Society of Dyers and Colourists, 1995.
4 Korkmaz, A., Öktem, T. Enzimatik işlem görmüş yünlü mamüllerin çeşitli özelliklerinin araştırılması. Tekstil Teknolojisi ve Kimyasındaki Son Gelişmeler Sempozyumu IX Kitapçığı, 2003; 13–28.
5 Duran, K., Ayaz, Ö.Y. Selülazların rejenere selüloz liflerinde kullanımıyla alternatif terbiye prosesleri. Tekstil ve Konfeksiyon, 1999; 5, 390–396.
6 Cavaco-Paulo, A. Mechanism of cellulase action in textile processes. Carbohydrate Polymers, 1998; 37, 273–277.
7 Latham, F.R. Dyeing with vat dyes. In Cellulosics Dyeing (Ed. J. Shore, Chapter 5, 246–278). Bradford, West Yorkshire, UK: Society of Dyers and Colourists, 1995.
8 Shore, J. Selection dyes for cellulosic fibers. In Cellulosics Dyeing (Ed. J. Shore, Chapter 8, 352–395). Bradford, West Yorkshire, UK: Society of Dyers and Colourists, 1995.
9 Hickman, W.S. Preparation. In Cellulosics Dyeing (Ed. J. Shore, Chapter 2, 81–149). Bradford, West Yorkshire, UK: Society of Dyers and Colourists, 1995.
10 Tarakçıoğlu, I. Tekstil terbiyesi ve makinaları, Cilt 1: Tekstil terbiyesinde temel işlemler ve selüloz liflerinin terbiyesi. İzmir, Türkiye: Ege Üniversitesi Tekstil Fakültesi Yayınları, 1979.
11 Atav, R. Yün dışındaki bazı önemli protein liflerinin boyanma özelliklerinin geliştirilmesi (PhD Thesis). Tekstil Mühendisliği Anabilim Dalı, Fen Bilimleri Enstitüsü, Ege Üniversitesi, İzmir, Türkiye, 2009.
12 Onar, N. Katyonik pamuk eldesi ve uygulamaları. Tekstil ve Mühendis, 2006; 13(61), 33–40.
13 Barış, B., Atav, R. Dendrimer teknolojisi kullanılarak aromaterapi özelliğine sahip pamuklu fonksiyonel kumaş eldesi. Electronic Journal of Vocational Colleges, 2012; 2(2), 171–175.
14 Toit, L.C.D., Pillay, V., Choonara, Y.E. Nano-microbicides: challenges in drug delivery, patient ethics and intellectual property in the war against HIV/AIDS. Advanced Drug Delivery Reviews, 2010; 62(4–5), 532–546.
15 Klajnert, B., Bryszewska, M., Dendrimers: properties and applications. Acta Biochimica Polonica, 2001; 48(1), 199–208.
16 Bulut, M.O., Akar, E. Dendrimerlerin önemi ve kullanım alanları. SDU Teknik Bilimler Dergisi, 2012; 2(1), 5–11.
17 Namırtı, O., Atav, R. Tekstilde yeni bir konsept olan dendrimerlerin tarihçesi, sınıflandırılması, molekül yapısı ve özellikleri. Pamukkale Üniversitesi Mühendislik Fakültesi Mühendislik Bilimleri Dergisi, 2011; 17(2), 109–115.
18 Namırtı, O. Nanoteknoloji Ürünü Dendrimerler Kullanılarak Pamuk Liflerinin Reaktif Boyalarla Boyanabilirliğinin Geliştirilmesi (MSc Thesis). Tekirdağ Namık Kemal Üniversitesi, Fen Bilimleri Enstitüsü, Tekirdağ, Türkiye, 2013.
19 Zhang, F., Chen, Y., Lin, H., Wang, H., Zhao, B. HBP-NH2 grafted cotton fiber: preparation and salt-free dyeing properties. Carbohydrate Polymers, 2008; 74, 250–256.
20 Demir, A., Seventekin, N. Kitin, kitosan ve genel kullanım alanları. Tekstil Teknolojileri Elektronik Dergisi, 2009; 3(2), 92–103.

21 Demir, A., Öktem, T., Seventekin, N. Kitosanın tekstil sanayiinde antimikrobiyal madde olarak kullanımının araştırılması. Tekstil ve Konfeksiyon, 2008; 2, 94–102.
22 Perincek, S., Duran, K., Körlü, A.E., Bahtiyari, M.İ. Ultraviyole teknolojisi. Tekstil ve Konfeksiyon, 2007; 4, 219–223.
23 Bhatti, I.A., Adeel, S., Abbas, M. Effect of radiation on textile dyeing. In Textile Dyeing (Ed. P. Hauser, Chapter 1, 1–16). Croatia: IntechOpen Publishing, 2011.
24 Duran, K., Bahtiyari, M.İ., Ekmekçi Körli, A., Özdemir, D., Dereli, S. Doğal mucize ozon. Tekstil ve Konfeksiyon, 2006; 16(2), 75–79.
25 Prabaharan, M., Rao, J.V. Study on ozone bleaching of cotton fabric-process optimisation, dyeing and finishing properties. Coloration Technolgy, 2011; 117(2), 98–103.
26 Karahan, A., Yaman, N., Demir, A., Özdoğan, E., Öktem, T., Seventekin, S. Tekstilde plazma teknolojisinin kullanım olanakları: Plazma nedir? Tekstil ve Konfeksiyon, 2006; 16(1), 302–309.
27 Karahan, A., Ozon, UV, Atmosferik plazma kullanılarak doğal liflerin yüzeysel özelliklerinin değiştirilmesi üzerine bir araştırma (MSc Thesis). Ege Üniversitesi, Fen Bilimleri Enstitüsü Tekstil Mühendisliği Anabilim Dalı, İzmir, Türkiye, 2007.
28 New Cloth Market. Plasma treatment technology for textile industry. Link of access: https://www.fibre2fashion.com/industry-article/3884/plasma-treatment-technology-for-textile-industry, [Date of access: 19.02.2021].
29 Esteves, F., Alonso, H. Effect of CO_2 laser radiation on surface and dyeing properties of synthetic fibers. Research Journal of Textile and Apparel, 2007; 11(3), 42–47.
30 Nourbakhsh, S., Ebrahimi, I. Different surface modification of poly (ethylene terephthalate) and polyamide 66 fibers by atmospheric air plasma discharge and laser treatment: surface morphology and soil release behavior. Journal of Textile Science & Engineering, 2012; 2(2), 1–5.
31 Bahtiyari, M.I. Laser modification of polyamide fabrics. Optics & Laser Technology, 2011; 43(1), 114–118.
32 Chow, F., Chan, A., Kan, C.W. Effect of CO_2 laser irradiation on the properties of cotton fabric. Textile Research Journal, 2012; 82(12), 1220–1234.
33 Zohdy, M.H., El-Naggar, A.M., Marie, M.M., Effect of copper treatment on the dyeability of gamma irradiated acrylic fabrics with different dyestuffs. Materials Chemistry and Physics, 1999; 61, 237–243.
34 Horio, M., Ogami, K., Kondo, T. Sekimoto, K. Effect of gamma irradiation upon wool fibers. Bulletin of the Institute for Chemical Research, 1962; 41(1), 1–9.

3

Dyeing Mechanism of Cellulose Fibres

As is known, cellulose fibres have non-ionic character. However, since they are usually bleached with oxidising agents before dyeing, the alcohol groups (Cell–OH) in their structure are oxidised and turn into carboxylic acids (Cell–COOH). Therefore, they show anionic character in aqueous dyeing medium.

$$\text{Cell—COOH} \rightarrow \text{Cell—COO}^- + \text{H}^+$$

Especially in the alkali environment, the anionic character of cellulose fibres increases, as the dissociation of carboxyl groups will increase. Although cellulose fibres can be dyed with cationic (basic) dyes due to their anionic character, these dyes have not gained importance in cellulose dyeing due to the limited fastness values (especially light fastness) that can be obtained in cellulose fibres with cationic dyes.

Although cellulose fibres are anionic in dyeing in neutral and especially alkali medium, anionic dyes such as direct, vat, indigo, sulphur and reactive dyes are used for dyeing them. Therefore, electrolyte (sea salt or Glauber's salt) must be added to the dyeing liquor to eliminate the repulsion between the dye molecules and the cotton fibre, both of which carry a negative charge, as will be explained in detail in later chapters. This allows the dye to be properly absorbed by the fibres. Only dyes with a high affinity to cotton (such as direct dyes) require much less electrolyte to mask this natural repulsion between negative charges than dyes with a medium affinity to cotton (such as reactive dyes).[1]

In Figure 3.1, the dyes used in the dyeing of cellulose fibres and their percentages of use are given. As it is known, reactive dyes constitute the most important class of dyes used in dyeing cellulose-based fibres and their blends.[2]

Since cellulose fibres are hydrophilic fibres, dyeing them in anhydrous environment is not possible under normal conditions, because in these fibres, water acts not only as a solvent that allows the dye to dissolve but also as a reagent that binds to the hydroxyl groups of macromolecules in the amorphous regions of the fibres by dipole–dipole attraction forces and hydrogen bonds, increasing the cross-section of the fibres and therefore the diameter of the pores between the crystalline building blocks. In a normal water-swelled cellulose fibre:[3]

- specific pore volume: 0.53 mL/g
- specific pore surface: 390 m^2/g
- average pore diameter: 20 Å
- maximum pore diameter: 300 Å.

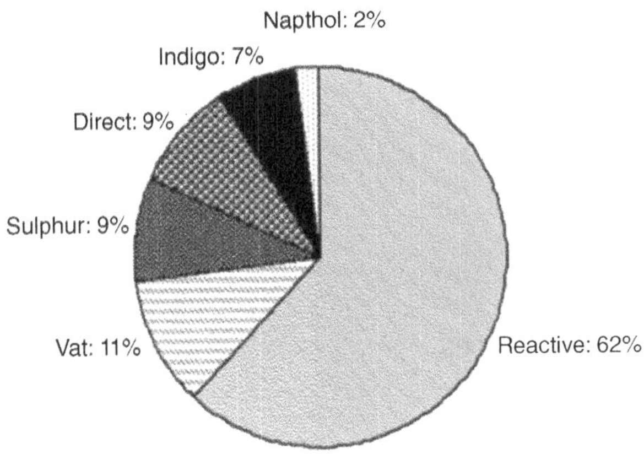

Figure 3.1 Dyes used in the dyeing of cellulose fibres and their usage amounts (%). Adapted from Ref. [2].

Since the diameters of the pores between the crystalline building blocks in dry cellulose fibres are generally smaller than the dye molecules, dye molecules or ions adsorbed on the surface of the fibres in any way cannot penetrate into the fibres. The best proof of this is that cellulose fibres cannot normally be dyed with disperse dyes using the thermosol method, because disperse dyes adsorbed on the surface of the fibres are larger than the pores in the dry fibres, they cannot diffuse into the fibres. However, if cellulose fibres are dyed with a disperse dye dispersion containing 15–20% polygleucol fibre swelling agent, the swollen pores continue to remain in their state since the polygleucol will not evaporate when the water evaporates during drying. In the next step, the fibres are dried at 210 °C for 1 minute. When heated for a period of time, dye molecules on the fibre surface can diffuse into the fibres from these swollen pores. Therefore, although cellulose fibres cannot be dyed in anhydrous environment, if the swelling function of water is provided by another substance, it is possible to dye cellulose fibres in anhydrous environment.[3]

Dyeing of cellulose fibres wetted with water and swollen pores in the dyeing liquor consists of three main steps as follows:

- Adsorption of dye particles on the surface of the fibres.
- Diffusion of dye particles into the fibres.
- Fixation of dye particles inside the fibres.[3]

3.1 Uptake of Dyes (Adsorption and Diffusion)

Although cellulose fibres have a weak (−) charge, it is not understandable at first glance that (−) charged dye anions leave the dyeing liquor and are adsorbed on the surface of the fibres. Various factors play a role in the adsorbing of the dye particles on the surfaces of the fibres, that is, in their removal from the liquor.[3]

Matter dissolved in a solution prefers to be found in regions with low energy. Since the boundary surface tension between the fibre and water is lower than the boundary surface tension of pure water, the dye particles in aqueous dye solutions generally tend to aggregate on the surface of the fibres. Beyond that, another important reason why the dye particles leave the dyeing liquor and go to the fibre surface is the attraction exerted by the fibre macromolecules on the dye molecules. Considering the structure of cellulose fibres, it is clear that there will be no ionic attraction forces (electrostatic attraction forces) between them with anionic dyes such as direct, vat, sulphur and reactive dyes used in their dyeing. There can only be attractive forces such as van der Waals bonds and hydrogen bridges between the dye particles dissolved in water and cellulose fibres. The more effective these attraction forces are, the more the dye particles are exhausted from the liquor by the fibres. For example, while the affinity of small molecule (monoazo-based) anionic acid dyes to cellulose fibres is low, direct dyes with a long continuous structure and containing conjugated double bonds have a very high affinity.[3] This shows that the affinity for cellulose fibre increases as the molecular size of the dye increases. There are two reasons for this. The first of these is that increasing the molecular size of the dye increases the dispersion attraction forces between the fibre and the dye. The second is that the dye molecule turns into a more apolar, that is, hydrophobic structure, due to the increase in the molecular size of the dye and the increase of aromatic groups such as benzene rings in its structure, because in this case, the tendency of the dye particles to escape from the water increases and the dye molecules prefer to be adsorbed on the fibre surface.[4]

To summarise, according to the classical theory explained above, the reason for the higher adsorption in large molecule dyes is the higher affinity. However, if this were the case, when the dye concentration in the liquor is increased in small-molecule dyes, the same level of adsorption can be achieved with large-molecule dyes, and even more molecules can be adsorbed on the fibre surface than large-molecule dyes, so the adsorption in small-molecule dyes can exceed large-molecule dyes. However, studies have shown that this is not the case (Figure 3.2). The reason for this is that dyes with large molecules are adsorbed in a multi-layered manner, since the association is more common.[5] For this reason, later on, the opinion that the affinity of large-molecule dyes to fibres is not due to the greater attraction forces that can occur between these dyes and the fibres but to the greater attraction forces between the dye molecules (hence the degree of association) gained weight. According to this theory, dye molecules can be adsorbed on the surface of the fibres (hence the surface of the pores), not in a single layer, but in multiple layers.[3]

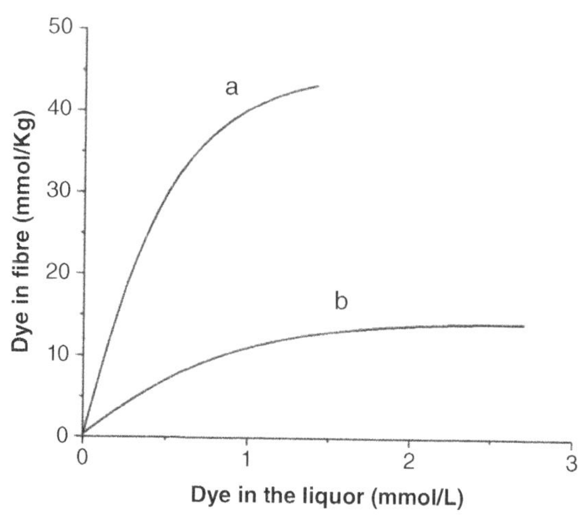

Figure 3.2 Adsorption isotherms of a direct dye (CI Direct Red 28) (a) and its half-molecular-size acid dye (b) on viscose at pH 9 at 50 °C. Ref. [5]/with permission of Society of Dyers and Colourists.

It was stated that in hydrophilic fibres, water molecules immediately enter the easily permeable regions of the fibres, causing the fibre cross-section and pores to swell. These swollen, water-filled pores (channels) that are larger in size than the dye molecules act as a pathway that allows the dye molecules to diffuse into the fibre. First, the dye molecules adsorbed on the surface (walls) of these pores move into the fibre with **Brownian** motions, at a speed that varies depending on the structure of the dye molecule and fibre and the dyeing conditions.[3]

The surface of the cellulose is always in equilibrium with the dyebath. The step that determines the dyeing rate is the diffusion of dye from the fibre surface to the interior of the cellulose. The diffusion coefficient of dye in cellulose is approximately 100 times lower than the diffusion coefficient in water. However, this ratio is not the same for all dyes. In addition, the diffusion coefficient of the dye in the fibre increases faster with temperature than the diffusion coefficient in the aqueous medium. The diffusion coefficient in fibres has a very low value in the absence of salt, increases with increasing salt concentration up to a certain maximum and then decreases.[4]

Although the penetration (i.e. diffusion) of dyes into cellulose is possible at room temperature, it occurs much faster with an increase in temperature. With a 10 °C increase in dyebath temperature, the diffusion rate more than doubles. Dye diffusion occurs only in the amorphous parts of the cotton fibre, that is, in the places (pores) where the cellulose macromolecules are loosely and randomly distributed, not tightly packed and not parallel (Figure 3.3). There is no diffusion into the crystalline regions of the fibre (tightly packed, parallel cellulose macromolecules linked by hydrogen bonds between —OH groups).[1]

Among the factors that determine dyeing equilibrium and speed are the number, distribution and accessibility of dye-binding centres in the fibres. It is assumed that in water-swellable cellulosic fibres, water-filled microscopic channels are formed between the crystallites, and dye molecules enter the fibres through these channels. Compared to

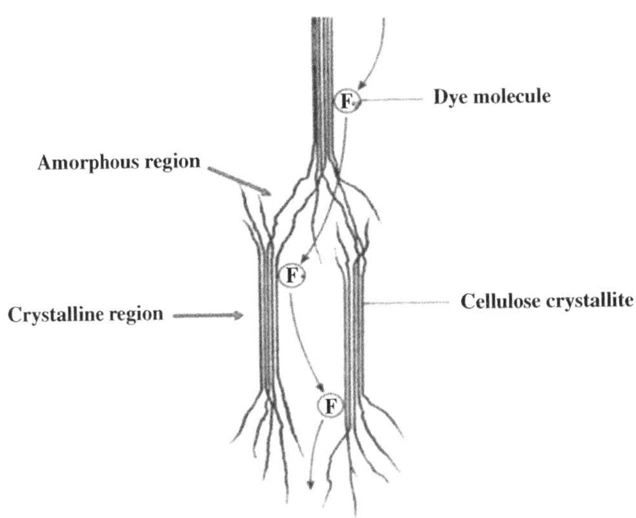

Figure 3.3 Dye diffusion into fibre. Ref. [5]/Ciba Technical Documents.

normal natural cellulose, the higher dye absorption of mercerised cellulose and regenerated cellulose is explained by their higher swelling or wider channel systems.[4]

3.2 Fixation of Dyes

Although dyes are attached to protein fibres mostly by electrostatic attraction forces, such bonds are an exception in cellulose fibres. For example, the cellulose derivative called "amine thread", which is obtained by the effect of ammonia on the immune thread or by other means, can be dyed with acid dyes because it contains basic amino groups, and electrostatic attraction forces occur between dye molecules and cellulose fibres. As a second example, if cellulose fibres are treated with substances that give them acidic properties (e.g. with tannin), they can gain affinity for basic dyes, that is, they can be dyed with them. In this case, electrostatic attraction forces occur between the dye molecules and the cellulose fibres.[3]

Apart from these exceptions, dyeing of cellulose fibres by substantive (direct), vat and sulphur dyes can be accepted as an adsorption phenomenon. The ability of dyes to dye the fibre by adsorption in this way is called substantivity. The forces that provide substantivity are hydrogen bonds and van der Waals forces. Since these forces are especially high in straight, long molecules containing a row of conjugated double bonds, such molecules are more substantive to fibres. Electrostatic attraction forces, which can form between fibre macromolecules and dyes, are not involved in the dyeing of cellulose fibres, with few exceptions. There is even some repulsion of anionic dyes due to the very weak anionic character of cellulose fibres. Reactive dyes, as the name suggests, are dyes that can bind to fibres by covalent bonds by reacting with hydroxyl groups in cellulose macromolecules. Therefore, the fixation form in the fibres is an exception in dyeing with these dyes. Thus, it turns out that only weak secondary attractive forces can be involved for the fixation of dyes on cellulose fibres. Therefore, it can be said that the attractive forces that may play a role in the binding and adhesion of dyes used in dyeing cellulose fibres, especially substantive (direct) dyes, to the fibres are H bonds and van der Waals forces. The fact that these weak attraction forces can provide the known medium wet fastness values of dyes is a situation that can hardly be explained. However, according to the theses that gained prominence, after the dye molecules or ions enter the fibres, they also form associates here. Thus, since the dye particles both grow and their solubility in water is partially reduced, they cannot easily come out (desorbed) from where they entered (sorbed) into the fibres during a subsequent wet process.[3]

The easy permeability of cotton, even at low temperature, explains why cotton dyes are not naturally resistant to washing. The dyes easily penetrate the cotton during dyeing and similarly bleed out easily during washing. This is the case even with large molecule direct dyes that have high affinity for cotton and slow diffusion into and out of the cotton.[1] Therefore, if the dyes containing ionic groups that provide solubility in water are not bonded to the fibres with strong bonds (as in reactive dyes), their fixation in the fibre is not completely satisfactory, that is, their wet fastness cannot exceed the medium level.[3] Therefore, special dyes that can either react with cellulose (forming a strong bond) or become insoluble on fibre are required to achieve high wash fastness.[1] Since cellulose fibres do not have the opportunity to bind the dye molecules to the fibres with strong bonds, except for reactive dyes, there remains an opportunity for good fixation and therefore good wet fastness values; this is to reduce or eliminate the solubility of dye particles in water. In hydrophilic fibres, it is not possible to start dyeing with hydrophobic, water-insoluble dyes from the beginning, since such dye molecules do not show any substantivity (affinity) to the fibres, so they cannot penetrate into the fibres. Pigments are hydrophobic colourants that are insoluble in water, and a superficial dyeing can only be achieved with the help of special adhesives.[3]

In the light of this information, for a cellulose dyeing with high wet fastness (well-fixed in the fibres), it is necessary to start dyeing with dyes that are water-soluble and have a certain affinity for the fibres, but after the dye particles enter the fibres, their solubility in water should be reduced or completely eliminated.[3]

One of the easiest ways to reduce the water solubility of dyes containing ionic groups (e.g. $-SO_3^-$) diffused in the fibres is to block these ionic groups. For example, if the ionic group is anionic, after dyeing, if the fibres are treated with a cationic substance containing a large alkyl group (which is hydrophobic and not soluble in water), since these cations and anions in the dye will be bound to each other by electrostatic attraction forces (ion bonds), the solubility of the dye in water will decrease, thus their wet fastness will increase.[3]

$$\text{D-SO}_3^-\,\text{Na}^+ + \text{R-NH}_3^+\text{Cl}^- \rightarrow \text{D-SO}_3^- \cdots\cdots {}^+\text{H}_3\text{N-R} + \text{NaCl}$$

This after treatment, which was carried out to improve the wet fastness of dyeings made with substantive dyes, later went through an important stage and substances that could react with cellulose fibres and even dye molecules were introduced to the market as cationic aftertreatment agents. Cationic reactant fixing agent, which is put on the market according to this principle and used in Indosol (Sandoz) dyes, act in three ways:

- by forming electrostatic attraction forces and blocking the anionic groups that provide water solubility in the dye,
- by binding to the dye molecule with covalent bonds,
- by covalent bonding to cellulose macromolecules and even by forming bridges between fibre macromolecules (thereby reducing the swelling of fibres, pores).[3]

One possibility to reduce the solubility of dyes in water is to enlarge the molecule without increasing the number of solubilising groups. This opportunity is utilised in diazonium dyes, and after dyeing, dye molecules are enlarged by diazotisation of the dye particles taken up by the fibres and coupling them with naphthols (development). After dyeing with special substantive dyes, aftertreatment with chromium or copper salts is also effective by magnifying the molecule to some extent. The attachment of metal salts, which form complexes with dye molecules, to cellulose fibres with weak adsorbtive forces may play a role in increasing the fixation. However, unlike wool fibres, metal ions are not bound to the fibres by coordinative bonds in cellulose fibres.[3]

After dyeing, the easiest way to completely eliminate the water solubility of the dye particles in the fibres is to remove the water-soluble groups. This opportunity, which is widely used in cellulose dyeing, forms the basis of dyeing with vat, indigo, sulphur and vat leuco ester dyes. Dyeing is done in the form of the dye that can dissolve in water (or in the liquor containing caustic soda), and with a process (oxidation) performed after the dye particles

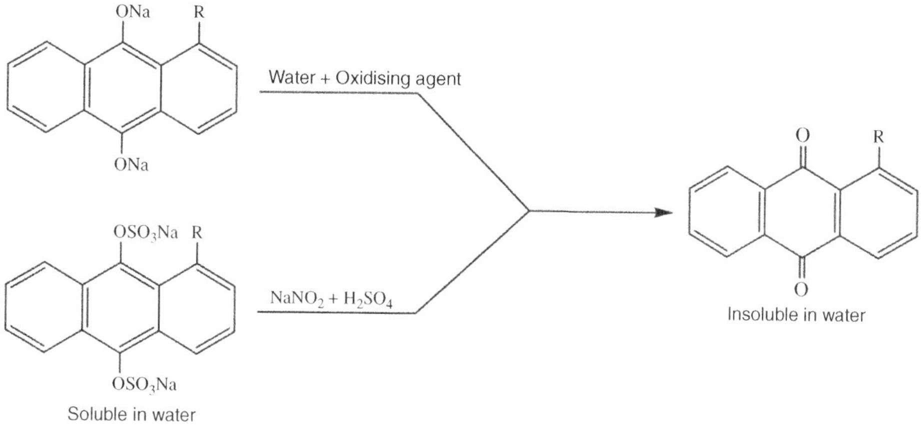

Figure 3.4 Conversion of vat (on the top) and vat leuco ester (on the bottom) dyes into water-insoluble form in fibre. Ref. [3]/Ege Üniversitesi Mühendislik Fakültesi Çoğaltma Yayınları.

are taken up by the fibres, the water-soluble properties of these dye particles in the fibres are eliminated[3] (Figure 3.4).

In the polycondensation dyes used in the past for dyeing cellulose fibres, both the removal of the water-soluble group and the growth of the dye molecule were ensured during a process performed after dyeing.[3]

$$2\text{D-SSO}_3\text{Na} + \text{Na}_2\text{S} \rightarrow \text{D-S}_3\text{-D} + 2\text{Na}_2\text{SO}_3$$

$$\text{D-S}_3\text{-D} + \text{Na}_2\text{SO}_3 \rightarrow \text{D-S-S-D} + \text{Na}_2\text{S}_2\text{O}_3$$

In fact, there are thiosulphate (Bunte salt) groups in water-soluble sulphur dyes, but since these dyes do not have affinity for fibres, the dye is first converted into thioalcoholate, which has affinity for fibres, by treating it with sodium sulfhydrate (sodium hydrogen sulphide).[3]

$$\text{D-SSO}_3\text{Na} + \text{NaHS} + \text{Na}_2\text{CO}_3 \longrightarrow \text{D-SNa} + \text{Na}_2\text{S}_2\text{O}_3 + \text{NaHCO}_3$$

After dyeing with this, as a result of an oxidation process, the dye particles in the fibres become insoluble in water.[3]

$$\underset{\text{Soluble in water}}{2\text{D-SNa}} \xrightarrow[\text{Oxidation}]{} \underset{\text{Insoluble in water}}{\text{D-S-S-D} + 2\text{Na}^+}$$

Another class of dyes developed for dyeing cellulose fibres was Alcian (ICI) dyes.[3] These dyes were water-soluble isothiourea derivatives of copper phthalocyanine containing chloromethyl group as substituent group.[6] These dye molecules contained quaternary ammonium or sulphonium groups as solubilising groups. After dyeing is done with such a water-soluble cationic dye, when the fibres are subjected to an alkali treatment or steamed (Figure 3.5), the solubilising groups in the dye particles in the fibres are broken and the water-soluble property of the dye is eliminated.[3]

The most useful possibility to obtain water-insoluble dye molecules in the fibres is based on the principle of treating the fibres sequentially with compounds that are soluble in water but form water-insoluble dye molecules when they come together. Thus, water-insoluble

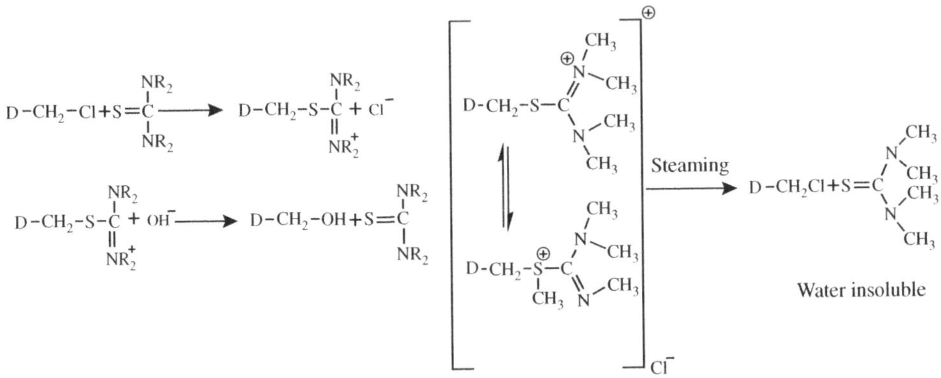

Figure 3.5 Conversion of Alcian dyes into water-insoluble form in alkaline medium (left). Ref. [6]/Society of Dyers and Colourists or when steamed (right). Ref. [3]/Ege Üniversitesi Mühendislik Fakültesi Çoğaltma Yayınları.

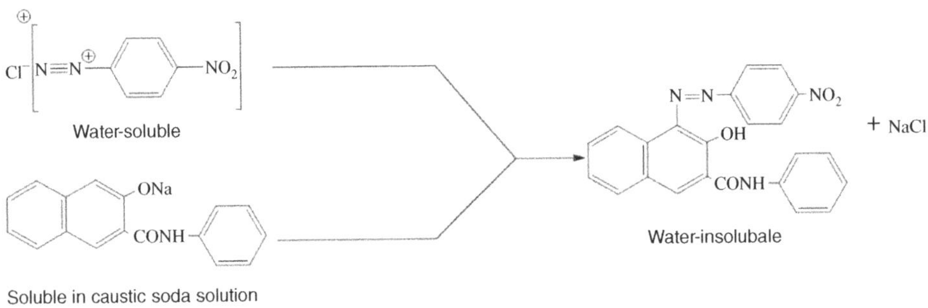

Figure 3.6 Conversion of water-soluble dye molecules into water-insoluble form within the fibre. Ref. [3]/Ege Üniversitesi Mühendislik Fakültesi Çoğaltma Yayınları.

dye molecules are formed inside the fibre, but unfortunately also on the surface of the fibre in a certain amount. This principle constitutes the basis of dyeing with naphthol (development) dyes (Figure 3.6), an example of which is given below.[3]

To summarise, unlike many other fibres in cellulose fibres, since dye molecules cannot be attached to the fibres with highly effective attraction forces (except for reactive dyes), dyeings with good fixation levels and hence having good wet fastness values can only be obtained when the water solubility feature of the water-soluble dye particles that have diffused into the fibres is decreased or eleminated.[3]

References

1 Ciba Technical Documents. Basics of fibres dyeing process, dye classes and dye bath auxiliaries, 2003.

2 Yurdakul, A., Atav, R. Boya baskı esasları. Bornova, İzmir: Ege Üniversitesi Tekstil ve Konfeksiyon Araştırma Uygulama Merkezi Yayını, 2006.

3 Tarakçıoğlu, I. Tekstil Boyacılığı I Teksiri. İzmir, Türkiye: Ege Üniversitesi Mühendislik Fakültesi Çoğaltma Yayınları, 1982.
4 Valko, E.I. The theory of dyeing cellulosic fibers. Textile Research Journal, 1957; 27, 883–898.
5 Giles, C.H. Dye-fibre bonds and their investigation. In The Theory of Coloration of Textiles (Ed. A. Johnson, Chapter 2, 97–168). Bradford, UK: Society of Dyers and Colorists, 1989.
6 Broadbent, A.D. Basic Principles of Textile Coloration. Bradford, England: Society of Dyers and Colourists, 2001.

4

Factors Affecting Uniform Dyeing in Cellulose Fibres

The most important point that distinguishes textile dyeing from other finishing processes and causes dyeing to be a process that requires more attention is that the unlevelness that will occur during dyeing can be seen with the naked eye. A normal human eye can easily see the differences in darkness more than 4%. In combination dyeings made with more than one dye, a nuance difference can be added to the colour depth differences. For this reason, the task of a good dyer, as in all finishing processes, does not end with providing dyeing with the desired properties (colour and fastness) in the most economical and ecological way, but it is also expected from him or her to ensure that this dyeing is completely uniform.[1]

Dyeing levelness refers to the even distribution of the dye throughout the substrate. An uneven distribution at the microscopic level is normal and tolerable, as visual perception still gives the impression of uniform colouration. Larger-scale (macro) unlevelness that is clearly evident on visual inspection is faulty dyeing, unless it is done intentionally to give an effect. This can take many forms, and dyeings may have differential uptake in the form of streaks, spots, crease-marks, as well as more gradual and subtle variations from side-to-side, side-to-middle, back-to-front, or end-to-end of a fabric. Assuming a correctly prepared substrate with no inherent dyeability variations present, levelness relies on the use of the correct procedure based on the substrate, dye selection (using compatible dyes), and the machinery being used (how it might provide agitation or lead to creasing of the fabric).[2]

Textile materials can be dyed at various stages during production. They can be dyed as raw stock (loose fibre), sliver, top or tow (loosely formed parallel bands of fibres or filaments), as yarn, fabric or piece, or made-up articles such as hose or garments. The stage at which dyeing takes place in the production cycle depends on cost, subsequent processings, end use of the materials and the colour fastness required. The modern trend is to delay colouration until as late as possible in the production scheme. Garment dyeing has recently become increasingly popular, as this facilitates rapid response to market trends.[3]

Moving from loose fibre dyeing to fabric dyeing, the importance of uniform dyeing increases. Because loose fibre and carding bands go through homogenisation processes while obtaining the yarn after dyeing, the irregularities that may have occurred in the dyeing become unnoticeable by being evenly distributed over the yarn section after the yarn is obtained. Dyeing irregularities in the yarns can also be partially camouflaged by using these yarns in patterned weavings or knitwear. Irregularities that do not accept any mistakes and are immediately obvious are those that occur during the dyeing of textile surfaces (fabric, knitwear, non-woven surface).[1]

Since the most widely used application methods in textile dyeing are exhaustion and impregnation, the theoretical basis of obtaining uniform dyeing in dyeing performed by these methods will be discussed separately.

4.1 Factors Affecting Uniform Dyeing in Exhaust Method

As stated in the explanation of the dyeing mechanism of the fibres, if the dyeing is continued for a sufficient time, an equilibrium occurs and from this point, no matter how long the dyeing is continued, provided that the dyeing conditions are not changed or the chemical energies are not changed, the amount of dye taken up by the fibres does not increase.[1]

$$\text{Dye remaining in the liquor} \rightleftarrows \text{Dye taken up by fibres}$$

The important point is that this equilibrium is a dynamic process. In other words, dye exchange continues from the liquor to the fibres and from the fibres to the liquor even after the equilibrium is established. However, since the amount of dye taken from the liquor and given back to the liquor by the fibres per unit time is equal to each other, the equilibrium stated above does not change as a result.[1] This is the basis of migration, which plays an important role in achieving level dyeing.[2]

As it is known, the steps of uptake of dye (adsorption and diffusion) are also bidirectional. In other words, while a certain amount of dye is continuously taken up by the fibres, a certain amount of dye is desorbed to the liquor from the fibres. However, since the chemical potential of the dye in the liquor is higher than the chemical potential of the dye in the fibres during the take-up phase, the amount of dye taken from the liquor per unit time by the fibres is higher than the amount of dye desorbed to the liquor until equilibrium occurs.[1]

It is a very important issue in terms of dyeing that some dyes leave the places where they first settled in the fibres and desorb to the liquor, both until the dyeing equilibrium occurs and after the equilibrium is established, because as it is known, chemical energy is directly proportional to concentration (or activity). As a result, the substantivity (affinity) of the dye to the fibres can be written as:

$$-\Delta\mu^{\circ} = R.T.\ln\frac{(a_L)}{(a_Ç)} = R.T.\ln K$$

In other words, the uptake of dye by the fibres occurs especially in the regions where the dye concentration is low, and the desorption of dye from the fibres to the liquor occurs especially in the regions where the dye concentration is high. As a result, over time, where more dye was initially taken than the average, dye desorbs from the darker dyed areas to the liquor, while new dye comes to the areas that are initially less dyed than the average, that is, to the lighter coloured areas. When viewed from the outside, this phenomenon is called "migration", as the dye particles seem to migrate from the darker-than-average dyed regions to the lighter-than-average dyed regions. However, the migration in question here should not be confused with the actual migration of finishing agents (including dye particles) towards the quick-drying areas or the surface of the fibres, especially during drying. Because, in the migration during dyeing made by exhaust method, the dye particles do not migrate from one place to another in the product, rather, the amount of dye taken from the liquor by the fibres in unit time and the amount of dye desorbed from the fibres to the liquor are not equal in all parts of the product. As a result, it seems as if there is a migration event.[1]

As a result of the migration event that occurs in dyeings by exhaust method, the dyeing gains a more uniform appearance as the darker-than-average dyed areas become lighter over time and the lighter-than-average dyed regions get closer to the average. For this reason, this phenomenon is also called "**post-levelling**".[1]

In the light of what has been said so far, it can be said that there are two possibilities for obtaining a uniform dyeing:[4, 5]

- Uniform uptake of the dye on the textile material from the beginning (i.e. control of the rate of dye uptake to ensure uniform dye uptake).
- Post-levelling of dye taken by the textile material (i.e. migration of the dye after it was initially absorbed non-uniformly by the fibre).

One or both of these mechanisms may play a role to a greater or lesser extent in a given dye-fibre system, but a general trend towards better fastness makes it necessary to use dyes with low or no migration ability. Therefore, the focus is on controlling the rate of dye uptake for uniform dyeing. Physical factors such as temperature and the frequency of liquor/material contact (governed by the liquor circulation rate in jet, beam or package dyeing machines) can be used to provide some control over these mechanisms. Slower heating rates generally favour more uniform dye uptake, and higher temperatures tend to increase migration or diffusion. In some cases, uniform dyeing can be affected by dyebath pH and/or the presence of electrolytes. Furthermore, the addition of levelling agents to the liquor may be used to promote uniform dye uptake.[5]

The factors affecting the two possibilities mentioned for obtaining uniform dyeing are explained separately below.

4.1.1 Uniform Uptake of Dye on Textile Material from the Beginning

A classical dyeing process by exhaust method consists of three partially overlapping steps:

- Uptake of dye by the fibres (adsorption and diffusion)
- Fixation of dye particles in the fibres
- Post-levelling of the dyeing.

The last of these steps requires that the dyeing process be continued for a long time at high temperatures, especially for dyes with good wet fastness values. For this reason, in today's conditions where energy is expensive and time is valuable, it is preferred to dye by ensuring uniform dye uptake from the beginning without the need for a levelling step afterwards. With such a conscious approach, energy savings can be achieved and the total dyeing time can be shortened. Therefore, unit machine and employee production efficiency can be increased.[1]

There are two main reasons for the non-uniform uptake of dyes by the fibres:[1]

- Affinity differences (irregularities) in the fibres
- Fast and therefore random uptake of the dye.

4.1.1.1 Affinity Differences in Fibres

If the fibres that make up the product to be dyed differ in terms of their ability to absorb and fix dyes among themselves, it becomes very difficult to obtain uniform dyeing, and sometimes even impossible.[1]

In products made from natural fibres, during the production of yarn or textile surfaces:

- the use of fibres of different quality and origin,
- the use of fibres of different diameters,
- the use of fibres with different maturity degrees,
- different external factors during the growth of the fibres.

can cause these fibres, and consequently the yarns made from them, to absorb dye unevenly during the dyeing process. For example, in cotton fibre, thin-walled raw (or dead) fibres that die prematurely show different dyeing properties compared to normal fibres. While indigo and some vat dyes (e.g. Indanthren Violet RK) do not dye these dead fibres at all, some direct (substantive) dyes dye the dead fibres darker.[1]

In addition to the differences arising from the fibres, differences can also be created in the product during textile production. This situation negatively affects the uniform dyeing. Differences in the material during textile production emerge:

- either by differences in physical effects or
- as a result of chemical damage to the fibres.

The preparation of the windings (cone, bobbin, warp beam), yarn tension differences during weaving or knitting, and fabric tension differences during various pretreatment processes, especially drying and fixation, can cause yarn or regions with different internal tensions to form in the product. Fibres with different internal tensions, on the other hand, have different dyeing and fixing properties.[1]

On the other hand, unless the fibre is uniform in whiteness, absorbency and chemical composition and has low levels of impurities (waxes, lignin, electrolytes, etc.), it is unlikely that it will take up dye in a uniform way or to the maximum extent possible. Consistent, uniform application of chemicals is a major contributor to successful preparation.[6] If an non-uniform cleaning or hydrophilisation occurs as a result of pretreatment processes, it becomes impossible to dye the product properly. During the pretreatment processes, the damage of the fibres from place to place is also an event that makes uniform dyeing significantly difficult. For example, in regenerated cellulose fibres, the outer mantle layer acts as a barrier to slow the uptake of dye. Therefore, the removal of this layer from place to place will result in the fibres being able to take up the dye faster in those regions. This will lead to non-uniform dyeing due to different dye uptake.[1]

To summarise, obtaining a proper dyeing starts with fibre selection and is only possible as a result of the conscious and careful efforts of spinners, fabric producers and pretreatmenters. Without this support and cooperation, asking a single dyer to dye properly is in many cases expecting for the impossible to be made possible.[1]

4.1.1.2 Rapid Uptake of Dye

A known basic rule in dyeing is that rapid uptake of dye will result in non-uniform dyeing. Accordingly, taking the dye properly is only possible by making it to be taken up slowly. However, on the other hand, allowing the dye to be taken more slowly than necessary is not economical as it leads to a very long dyeing time. For this reason, dyeing is typically performed at a slightly slower speed than the critical rate as much as possible. **"Critical rate"** is the limit dyeing speed at which a dyeing begins to become uneven. In this way, the dye is not only taken up properly, but also unnecessary prolongation of dyeing is prevented.[1]

The critical rate for dyeing a particular fibre with a particular dye is not a constant rate. The critical rate may be lower or higher depending on the characteristics of the dyeing device used and the way the product is found. In other words, while a cotton thread is dyed unevenly in a certain apparatus with a certain recipe in 2 hours, the same yarn can be dyed properly even when dyed with the same recipe in another apparatus in 1 hour. Because, depending on the properties of the apparatus and windings, the critical rate can be 0.5%/min in the first apparatus, while it is 2.5%/min in the second apparatus. Thus, an important possibility to ensure uniform dye uptake without prolonging the dyeing time is to increase the critical rate. For this reason, it is useful to explain the factors affecting the critical rate first. Then, the factors affecting the dye uptake rate that determine the uniform dyeing will be explained.[1]

4.1.1.2.1 Factors Affecting Critical Rate

The critical rate is determined by the operational systems. These are the features of the dyeing device and the material form.

a. The effect of the characteristics of the dyeing device

A practical dyeing process requires stirring to ensure level initial sorption and subsequent migration of the dye. In practice, either the dyebath or the substrate, sometimes both, may be moved. Depending on the machine used, process conditions may need to be adjusted. For example, a machine that agitates more gently needs a procedure designed to cause a slower initial strike of dye.[2]

The more intense and fast the contact between the textile product and the dyeing liquor, the higher the critical rate. Therefore, as the movement of the textile product accelerates in the machines, the critical rate increases as the number of revolutions in the liquor increases in the apparatuses. For example, in a jet dyeing machine where the contact with the liquor is intense and both the product and the liquor are moving, the critical rate of dyeing in the jet dyeing machine is much higher when the fabric pass speed is the same as that of a liquor, compared to the critical rate of dyeing in the winch-beck with the same recipe.[1]

If the apparatus is considered because it is easier to understand, it can be said that the critical rate changes in direct proportion to the number of circulations in the liquor per unit time, keeping all other conditions constant. The number of liquor circulations (n) is determined by the pump power (q) of the apparatus and the liquor ratio.[1]

$$q = \frac{Q}{G_T}$$

The liquor circulation through the unit material (*q*): The amount of liquor in litres passing through 1 kg of material in a minute (L/min.kg).

Liquor circulation through all the material (*Q*): The amount of liquor in litres passing through all the material in a minute.

G_T: Total material weight in the apparatus

$$n = \frac{q}{\text{L.R.}}$$

Number of liquor circulations (*n*): The number that indicates how many times per minute the entire liquor is circulated in the apparatus.

Liquor ratio (L.R.): The volume of the liquor used versus the mass of the textile material.

For example, in case of dyeing at 1:12 liquor ratio in an apparatus with a liquor circulation of 3, let's calculate the liquor transition from unit material:[1]

$$q = n \times \text{L.R.} = 3 \times 12 = 36 \frac{\text{L}}{\text{min} \cdot \text{kg}}$$

Although simple mathematical equations cannot be used in this way in machines, in many cases, the number of passes of the fabric ropes per minute (fabric passing speed: rope length) in winch-becks, twice the number of passes per minute of the fabric rope in jet dyeing machines are considered equivalent to the number of circulations in the liquor in the apparatus.[1]

The influence of the liquor circulation number or fabric pass speed on the critical speed is simply explained below.[1]

During dyeing, the dye is not always taken up by the fibres at the same speed. There is such a critical temperature (e.g. around the glassing point in man-made fibres) that dye uptake suddenly increases from this point on, and a temperature difference of a few degrees begins to cause a 100% change in dye uptake rate. If we assume that during the dyeing of a winding, the number of circulations in the liquor is 1 rpm. and that the liquor is heated with speed of 4 °C/min, in this critical region, the dye uptake rate will suddenly start to increase as the liquor passes through the winding, and for example, if the liquor is circulating from the outside to the inside at that moment and 10% of the total dye is taken up within 1 minute, since the liquor will experience a 10% loss of concentration from the outer part of the winding to the inner part, the inner part of the winding will be able to take up 10% less dye than the outer part. Therefore, the dyeing will be non-uniform (**inside–outside difference**).[1]

If the number of liquor circulation in the apparatus is 8 rpm instead of 1 rpm and if 10% of the total dye was taken in 1 minute, since the amount of dye taken during one cycle of the liquor would be 10% : 8 = 1.25%, only this much difference would occur between the outside and the inside of the winding. Dyeing will appear uniform as the human eye cannot detect it. As a result, although the dye is taken up with a rate of 10%/min in both cases, in the second case, as the critical rate increases due to the higher liquor circulation rate, the dye uptake rate remains lower than the critical rate, and thus a uniform dye uptake can be achieved.[1]

b. The effect of the material form

In general, it can be said that it is difficult to obtain uniform dyeing from loose fibres to finished products in textile dyeing. There are two possibilities for dyeing wound yarns: hank and package dyeing. It has long been known that hanks can be dyed more easily than threads wound in package form.[1]

In yarn dyeing, production of standard packages in terms of weight, diameter and density, maintaining constant cycle times and press packing to even out any slight variability of winding are crucial. Uniform density is essential to prevent unlevelness caused by channelling.[6] The density of the windings affects the critical rate of the dyeing and thus the uniform dyeing. In principle, as the density of the windings (weight of textile material per unit volume) increases, the critical rate also increases, provided that it remains within certain limits. In Figure 4.1, the dye uptake curves and critical rates of dyeings of the same type of material with the same recipe and under the same liquor circulation number but with packages prepared in different densities are shown.[1]

In dyeings made with packages with a winding density of 0.17 kg/dm^3 (i.e. loose) dyeing starts to become unlevel when the dye uptake rate exceeds 1.2%/min, in windings with a winding density of 0.30 kg/dm^3 (i.e. tighter), the dyeing is still uniform even if the dye uptake rate is increased to 2.1%/min. The reason why the critical rate increases as the winding density increases is explained as follows. As the density increases, more of the liquors passing through the windings is able to contact the textile material. Therefore, the relative velocity between the surface of the fibres and the liquor increases, and thus uniform winding temperature and dye concentration can be better achieved. However, the validity of this positive correlation, in which the critical rate increases with the increase in winding density,

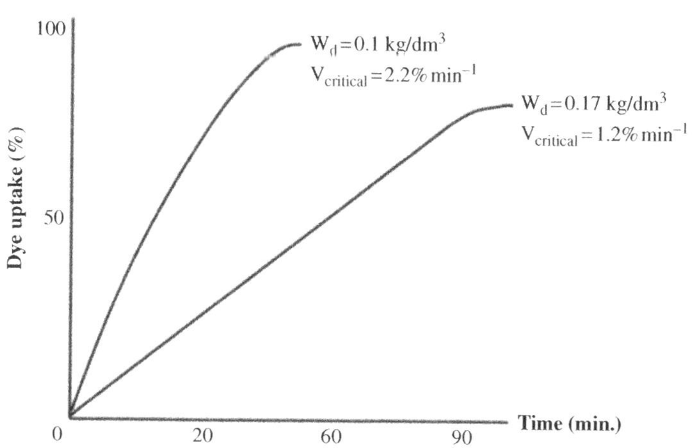

Figure 4.1 Effect of winding density on dye uptake rate and critical rate. Ref. [1].

is limited. Because keeping the winding density excessively high increases the resistance encountered by the liquor (the pressure difference between the inside and the outside of the windings increases depending on the circulation direction) and thus reduces the number of circulations in the liquor. Increasing the pump power excessively in order to keep the speed of the liquor high despite the high resistance of the windings causes deformation in the windings, formation of channels and even yarn breakages.[1]

Accordingly, it is beneficial to prepare the windings as dense as possible for bulky materials (such as textured polyester, high-bulk [HB] polyacrylonitrile, wool yarn) due to their structure, in terms of both increasing the critical rate and increasing the amount of goods that can be dyed in a batch. On the other hand, care should be taken not to prepare the windings of materials that are tight due to their structure or made of fibres that swell or shrink excessively. Otherwise, due to swelling or shrinkage that may occur during dyeing, the windings may compress excessively, making the liquor circulation difficult.[1] Cotton yarn can be wound more densely than staple viscose, but uniformity of the winding is the most important factor. Air must be removed from the packages with steam or hot water, using a non-foaming wetting agent, before dyeing is begun.[7]

4.1.1.2.2 Factors Affecting Dye Uptake Rate

For a dyeing, if the operational parameters such as the type of dyeing device, liquor ratio, liquor and/or material movement and speed, and the material form are constant, the critical rate of dyeing is also constant. In this case, the task of the dyer for a uniform dyeing is to ensure that the dyeing rate is slightly lower than the critical rate by choosing the recipe-related parameters such as pH, electrolyte (salt), levelling agent, temperature/time adjustment correctly.[1]

During normal dyeing, dye adsorption usually occurs at a sufficient rate on the surface of the fibres (hence the swollen pores), but these particles cannot diffuse into the fibres at the same speed. In other words, diffusion is the slowest step during the dye uptake and therefore the step that determines the dye uptake rate. Therefore, the most effective adjustment of the dye uptake rate can be achieved by adjusting the diffusion rate. However, since diffusion is a step following the adsorption step of dye particles on the surface of the fibres, the rate and amount of diffusion can be affected by the rate and amount of adsorption. Therefore, the rate and amount of dye uptake are ultimately affected by the rates and amounts of adsorption and diffusion.[1] For this reason, in this section, the effects of factors such as substantivity of the dye, dye concentration, temperature, pH of the liquor, electrolyte and auxiliary on the rate and amount of adsorption and diffusion will be explained separately in general.

a. Factors affecting adsorption

- ***Substantivity of the dye:*** According to the exhaust method, in order for the dye particles to be moved from the liquor and adsorbed on the surface of the fibres, these particles should prefer the fibres to the liquor. Although the tendency of dye particles that lower the surface tension of the water, to accumulate at the fibre–liquor interface where the lowest surface tension is present, plays an important role in this regard, the attraction forces between the dye-dye and dye-fibre molecules and macromolecules also play an important role. The more effective the attraction forces (such as electrostatic, dipole-dipole, induction, hydrogen bond) and the less effective the repulsive forces (the repulsion of ions with the same charge), the faster and more dye adsorption becomes. However, the same conditions also increase the degree of association of dye molecules or ions. On the other hand, since very large aggregates cause difficulties in penetrating the fibres, in this case, although adsorption increases, dye uptake slows down and decreases, especially at low temperatures. However, in dyes having small molecules and hence having less effective intermolecular attraction forces, when the conditions for optimally sized aggregates (associates) to be diffused to the fibres are provided and therefore the substantivity of the dye particles to the fibres is increased, both adsorption and diffusion will accelerate and increase, so dye uptake is accelerated and increased.[1]

The high adsorption rates of dyes with large molecules, which have high intermolecular attraction forces and thus can form large aggregates, increase the risk of inside–outside differences, especially in the dyeing of windings (coil, warp beam, fabric beam). In such cases, instead of starting the dyeing with the form with high substantivity to the fibres, it is of great benefit to start the dyeing with the form that has no or very little affinity for the fibres, if any, and to ensure that the dye is dispersed properly on the material first. This possibility can be exploited, for example, in dyeing cellulose fibres with vat dyes. While the water-soluble sodium leuco form of the vat dyes has a high affinity for cellulose fibres, the water-insoluble vat acid (anthrahydroquinone) and normal pigment (anthraquinone) forms have no affinity for the fibres. Therefore, if the dyeing is started with the vat acid or pigment form of the dye, initially, the dye particles will only adhere to the fibres mechanically. That is, if the liquor circulation is adequate the dye distribution on the material will also be uniform. In the next step, when the dye molecules are converted into the form with their affinity for the fibres, they will first be adsorbed and then diffuse into the fibres.[1] For example, in semi-pigmentation process, the dyebath is set with dye and chemicals at 15–20 °C and dyeing is initiated immediately. The temperature is then gradually raised, during which period dye that is not yet vatted becomes distributed in insoluble form in the substrate. Vatting begins simultaneously and the leuco compound formed is taken up by the fibre. With this procedure a more uniform dyeing is obtained from the outset compared with the leuco process.[7]

To summarise, when dyeing with low-substantivity dyes with small molecules, it will be possible to obtain uniform dyeing even if dyeing is done in conditions where adsorption will be fast. However, it is more difficult to obtain uniform dyeing in dyeings made with large molecule dyes with high substantivity. While dyeing with this type of dyes, dyeing conditions should be adjusted so that the rate of adsorption and therefore dye uptake is not too high.

- ***Dye concentration:*** As the dye concentration in the liquor increases, the number of dye molecules and ions in the unit volume will also increase, so dye molecules and ions come closer to each other, and thus the effectiveness of secondary attractive forces, which can only be effective at short distances, increases, and the degree of association increases. Therefore, as the liquor ratio decreases (concentration per unit volume increases), adsorption and the amount of dye taken up by the fibres increases even when dynamic equilibrium is achieved.[1] Higher (or longer) liquor ratios provide more uniform dyeing, but are less efficient in terms of energy, water and chemical consumption.[2]

In pale and light shade dyeing, since the amount of dye to be adsorbed to the fibres is low, if the adsorption and diffusion of dye is not adequately retarded, especially if there are differences in affinity in the fibres or in temperature and concentration in the liquor, the risk of unlevel dyeing is high.[1] The heavier the depth of shade and closer to the saturation value for the fibre under the dyebath conditions (pH, electrolyte, etc), the more likely is a level dyeing. Thus, for heavy shades, relatively little control is needed in the dyeing process. However, pale shades are the most challenging to produce uniform dyeing.[2]

- ***Temperature:*** Since adsorption is an exothermic event, the adsorption decreases as the temperature is increased. For this reason, equilibrium given below shifts to the liquor side as the temperature rises and therefore the dyeing efficiency decreases.[1]

$$\text{Dye remaining in the liquor} \rightleftarrows \text{Dye taken up by fibres}$$

At higher dyeing temperatures, the dyeing efficiency is lower because the amount of dye remaining in the bath is greater, but since there is more energy in the system, there is better levelling.[2] For this reason, dyeing is finished at high temperatures, even if dyeing is started at low temperatures, in order not to prolong the time for the dyeing equilibrium to be reached and to ensure that post-levelling (migration) is high. In this case, since the equilibrium will be slightly shifted to the liquor side, a precaution that can be taken to reduce the amount of dye remaining in the liquor is to lower the temperature a little before finishing dyeing. Figure 4.2 shows the effect of decreasing the dyeing temperature at the end of dyeing on increasing dye uptake.[1]

Although keeping the temperature low increases the adsorption, it can decrease the dye uptake rate more or less. It is not possible to obtain sufficient dyeing after a normal dyeing period of a few hours at low temperatures, especially in dyes with large molecules and large aggregates (with a high degree of association). However, it is also possible to dye in the cold with some dyes that have small molecules and therefore do not have a high degree of association.[1] Dyeing operations are usually

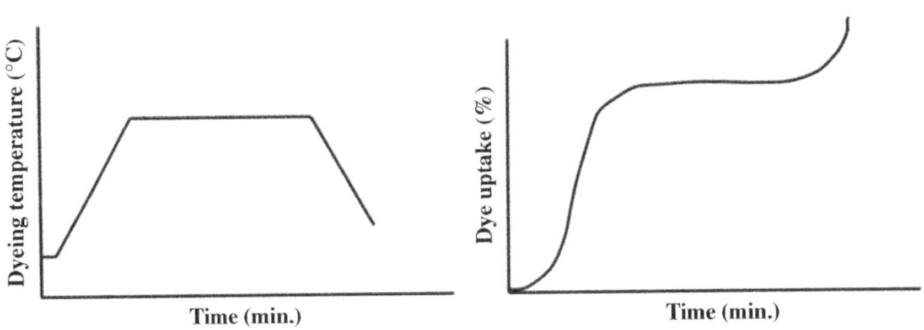

Figure 4.2 The effect of lowering the temperature at the end of dyeing on increasing dye uptake. Ref. [1].

carried out at a temperature that allows rapid and uniform dyeing for a period of time that will ensure adequate penetration into the fibres.[2]

- ***pH value:*** The pH value of the dyeing liquor decreases or increases the adsorption by affecting, on the one hand, the ionity of the fibres, on the other hand, the ionity and degree of association of the dye.[1]

  It has been previously stated that there are carboxyl groups formed by the effect of various oxidising factors in cellulosic fibres. When an alkali substance (e.g. soda ash) is added to the dyeing medium, the (–) charge of the fibres increases, as the dissociation of carboxylic acid groups will increase, and thus the adsorption of anionic dyes slows down.[1]

$$\text{Cell-COOH} + \text{OH}^- \rightarrow \text{Cell-COO}^- + H_2O$$

  The change in the pH value of the dyeing liquor also has a significant effect on the degree of association of the dye molecules. For example, association in vat dyes occurs as a result of the formation of H bonds between phenolic —OH groups in the form of vat acid of the dye. Therefore, as the strength and concentration of the alkali used are increased, since the —OH groups in the dye molecules will transform into —ONa, the possibility of the dye molecules forming H bonds with each other is eliminated and the degree of association and therefore adsorption decrease.[1]

- ***Electrolyte:*** As mentioned before, in ionic dyes, the addition of electrolyte to the liquor reduces the dissociation of the dye molecule, and therefore aggregation increases.[1]

$$\text{D-SO}_3\text{Na} \rightleftarrows \text{D-SO}_3^- + \text{Na}^+$$
$$\text{NaCl} \rightarrow \text{Na}^+ + \text{Cl}^-$$

  The decrease in the dissociation of the dye molecules, on the other hand, causes an increase in the degree of association since it will reduce the electrostatic repulsive force between the dye molecules (ions). The reduction of the dissociation of the dye molecules (i.e. the presence of dye molecules in the form of molecules instead of ions) also increases the adsorption by reducing the effect of electrostatic repulsive forces in the form of (–) charge repelling (–) charge, which may be effective between fibre and dye in cellulose fibres.[1]

- ***Auxiliary:*** Textile auxiliaries that promote uniform dyeing are called "levelling agents". According to the mechanism of action, levelling agents can be divided into two:

  a. levelling agents that promote uniform uptake
  b. levelling agents that promote post-levelling.

  On the other hand, levelling agents can be divided into two according to their affinities:

  a. levelling agents with affinity for fibres
  b. levelling agents with affinity for dyes.

  In terms of their chemical structures, the levelling agents can be classified into five groups as in normal surfactants:

  a. Anionic levelling agents
  b. Cationic levelling agents
  c. Non-ionic levelling agents

d. Non-ionic/cationic levelling agents
e. Non-ionic/anionic levelling agents.
Depending on the hydrophilicity and ionicity of the fibres and dyes, levelling agents show different mechanisms and degrees of action in different fibre-dye combinations.[1]

Non-ionic levelling agents are mostly used in cellulose fibres and these levelling agents have affinity for anionic dyes used in dyeing cellulose fibres.[1] Dye-substantive levelling agents prevent dye absorption by the fibre by complex formation with the dye in the dyebath. This complex is of such a molecular size that it cannot penetrate into the fibre. As the dyeing temperature is gradually increased, the complex becomes less and less stable and the dye is gradually released to enter the fibres.[3] However, these levelling agents can increase the amount of dye remaining in the liquor by slowing down and reducing the adsorption of the dye on the fibre surface. For this reason, if levelling agents that do not have a high affinity for the dye and, as a result, release the dye anions when the temperature is increased, this disadvantage is partially eliminated. Otherwise, in the later stages of dyeing, it may be necessary to add a special anionic levelling agent to the liquor in portions so that the non-ionic levelling agent prefers to bind with the anionic auxiliary and releases the part of the dye it was holding in the liquor.[1]

b. Factors affecting diffusion
If the diffusion phenomenon is simplified as "the movement of dye particles from the swollen fibre pores towards the inside of the fibres (but also outwards from the fibres in a certain amount) with Brownian motions", it becomes evident that there are three main elements that determine the diffusion rate:

- the size of the dye particles
- the states of the pores in the fibres
- the amount of energy hiding the motion of the dye particles.

Instead of examining these items one by one, the effect of factors such as substantivity of the dye, temperature, pH of the liquor, electrolyte and auxiliaries on these steps will be explained below.[1]

- ***Substantivity of the dye:*** As the substantivity (affinity) of a dye to fibres increases, it is normal for the amount of dye taken out of the liquor by the fibres to increase in equilibrium. However, this situation should not be misinterpreted as the dyes with high substantivity are taken up more easily and quickly by the fibres.[1]

Dyes with high substantivity to fibres are generally large-molecule dyes that can create effective intermolecular attraction forces. Therefore, their degree of association is also high. Their adsorption on the surface of the fibres is fast and high (there is a risk of uneven adsorption), but their diffusion into the fibres is slow and less since they cannot fit through the pores. However, if conditions that reduce the degree of association are chosen, dye diffusion also increases.[1]

Since the increase in substantivity to the fibres changes in many cases in parallel with the increase in the mutual attractions between the dye molecules and the fibre macromolecules, that is, the mutual attraction forces, dyes with high substantivity begin to bind to the macromolecules during the dye uptake. This is a matter that makes it difficult to be taken properly.[1]

- ***Temperature:*** The most significant factor affecting the diffusion of dyes into fibres is temperature, because temperature change affects the condition of both fibres and dyes.[1]

  In hydrophilic fibres, in which the swelling of the fibres and pores play an important role, the swelling of the fibres and therefore the pores also decrease slightly as the increase in temperature reduces the association of water molecules. This situation can be particularly advantageous during the dyeing of tightly woven fabrics, densely wound fabrics and products made of regenerated cellulose fibres with high swelling ability, allowing for uniform dye penetration into the yarns and fabrics. For this purpose, contrary to the custom, dyeing is started at high temperature and excessive swelling of the fibres is sought to be prevented. On the other hand, in hydrophilic fibres, as the temperature is increased, the attraction forces between the fibre macromolecules in the amorphous regions will decrease, and it will be easier for the dye particles to enter between the macromolecules. Therefore, despite the shrinkage of the pores, it cannot be said that the increase in temperature makes the fibres more difficult for dyes to diffuse.[1]

  The effect of temperature increase on the dye is much more pronounced. In dyes with large molecules and high degree of association, it is not possible for large dye aggregates (associates) adsorbed on the fibre surface at low temperatures to fit into the pathways (pores) into the fibres, and the number of individual dyes or small aggregates in the dynamic dye association equilibrium is very low. Its diffusion into fibres is slow and less. However, as the degree of association decreases as the temperature is increased (dyes are generally monomolecular or mostly in dimer form at boiling temperature), the obstacle that the dye particles cannot fit into the pores disappears, and therefore dye diffusion accelerates and increases.[1]

  On the other hand, in dyes with small molecules and weak associate formation ability, the dye particles adsorbed on the fibre surface are smaller than the swollen pores in the fibre, even at cold, so the aforementioned incompatibility problem does not exist. Increasing the temperature decreases the adsorption on the one hand and completely eliminates the association on the other hand. Thus, although the diffusion of small dye molecules (or ions) into the fibres becomes easier, their desorption from the fibres is also facilitated. Therefore, when the temperature increases in small molecule dyes, the amount of dye taken up by the fibres for equilibrium can decrease. On the other hand, even if there is a rapid uptake of dye in small molecule dyes, the risk of unlevel dyeing when working with these dyes is low, since levelling (migration) begins effectively after the dye is taken.[1]
- ***pH value:*** The change in the pH value of the dyeing medium affects the adsorption as well as the diffusion to some extent. This is because the change in pH value changes the state of the functional groups in the fibres, on the other hand, it also affects the dissociation and association of the dye. With the effect of pH, the rate and amount of not only adsorption but also diffusion can increase as the number of ions that attract the dye ions in the fibres increases or the number of repulsive groups decreases.[1]

  On the other hand, the effect of pH value on the dissociation of the dye molecule and thus the degree of association decreases or increases the diffusion depending on the molecular size of the dye. Decreased dissociation of dye molecules (increase in the forms of sulphonic acid, carboxylic acid, phenolic vat acid) slows down and reduces diffusion in dyes with large molecules and high associate forming ability. Because in this case, the aggregation of dyes with large molecules and high associations increases, making it difficult for them to enter. However, in case of a slight increase in aggregation in small molecule dyes, since it will be difficult to get out of the fibre (desorption), diffusion may accelerate to some extent and increase as a result.[1]

- ***Electrolyte:*** Since the addition of electrolyte reduces the dissociation of ionic dyes and therefore increases the degree of association, its effect on diffusion is as stated above. In other words, while it slows and decreases diffusion in dyes with large molecules and high association ability, it accelerates and increases diffusion in small molecule dyes.[1]
- ***Auxiliary:*** If levelling agents with affinity for the dye are used in dyeing, the diffusion rate into the fibres becomes slower, since the dye-levelling agent complex formed is larger than the dye molecules alone and the affinity of this complex to the fibres is lower. On the other hand, if auxiliaries with affinity for fibres are used, they slow down the diffusion as well as the adsorption of the dye to the fibres, since they will bind to the functional groups in the fibres prior to the dye molecules and/or ions.[1]

4.1.2 Post-levelling of Dyeing

It was previously stated that dye exchange continues from liquor to fibres and from fibres to liquor while dynamic dyeing equilibrium is being established and after it is achieved. After the equilibrium is established this exchange continues in such a way that the amount desorbed and absorbed per unit time is equal to each other. The important point here is that the amount of dye desorbed to the liquor per unit time by the fibres and thus absorbed from the liquor varies greatly depending on the dye-fibre system and dyeing conditions. In some dyes, the amount of dye absorbed and desorbed per unit time is so high that no matter how unevenly the dye is taken at the beginning, it becomes uniform in a short time afterwards. Conversely, in some dyes, the amount of dye absorbed and desorbed per unit time is so low that even if dyeing is continued for hours, the unlevelness that occurs during dyeing cannot be levelled out afterwards. Another unfortunate situation for dyers is that dyes that have good levelling and therefore are easy to achieve uniform dyeing generally have low wet fastness values and dyes with good wet fastness values generally have insufficient levelling abilities.[1]

The situation mentioned above, of course, does not arise as a result of coincidence. Both the wet fastness values of the dyes and their levelling abilities are closely related to the fixation degree of the dye with the fibres. As the degree of fixation increases, wet fastness values also increase, but the subsequent levelling (migrating) ability of the dye decreases.[1] The following describes the ability of dyes to form bonds with fibres, their solubility in water and the effects of fibre condition on post-levelling.

4.1.2.1 Effect of Dye's Ability to Bond with Fibre Macromolecules

The greater and more effective the affinity of a dye to the fibres and the related bonds it can form with the fibre, the lower the levelling ability of that dye. The ability of the dye to form bonds with fibre macromolecules may vary depending on the dyeing pH. In this case, it is possible to reduce the bonding of the dye to the fibres by adjusting the ambient pH value, and thus to increase the post-levelling.[1]

In addition to the pH value, a very effective factor that increases the post-levelling by reducing the bonds between the dye and the fibre is temperature. As the temperature is increased, since the bonds between the dye and the fibre on the one hand and the bonds between the dye-dye on the other hand will decrease (as the degree of association of the dye in the fibre will decrease), the post-levelling also increases. Since diffusion, and thus both sorption and desorption, will accelerate with an increase in temperature, post-levelling increases. For this reason, the post-levelling step in dyeing is usually done while the dyeing continues at high temperature.[1]

Some levelling agents also have an effect on wool that will increase the levelling ability of the dyes. These types of levelling agents generally have affinity for dyes, and the resulting dye-levelling complex has both a lower ability to bind to fibres and a higher water-solubility compared to the dye particles alone. Thus, since the mobility (desorption and sorption) of the formed particles is higher, the post-levelling is also better. Another group of levelling agents, on the other hand, has a preventive effect (reducing the degree of association) of dye molecules from forming large aggregates in the liquor and within the fibres. As mentioned before, the reduction in size of the dye particles facilitates both the sorption and desorption of the dye into the fibres, that is, it increases the post-levelling ability.[1]

4.1.2.2 Effect of Dye Solubility in Water

In the dyeing of cellulose fibres, dyes can only be attached to the fibres by weak secondary attractive forces. However, good fixation, that is, good fastness values, can be achieved by reducing or completely eliminating the solubility of the dye particles in water after they have penetrated into the fibres. As it can be understood from this, during dyeing of hydrophilic fibres in aqueous medium, as the solubility of the dye in water decreases, the fixation and thus the fastness increases, but the post-levelling ability decreases. Therefore, in the case of dye types that are initially applied in a water-soluble form and later rendered insoluble within the fibres (e.g., vat dyes), if one intends to take advantage of the post-leveling ability, this step should definitely be carried out while the dye is still in a water-soluble state. Even after dyeing is complete (the dye is converted into water-insoluble form in the fibres), for any reason, if the dyeing needs to be levelled or the colour tone is to be lightened afterwards, the dyes must be returned to the water soluble form (e.g. vatting) beforehand. If this cannot be done (such as with naphthol dyes), once the dyeing is complete levelling cannot be achieved afterwards. However, it may be possible to remove the dyes by degrading the dye molecules.[1]

4.1.2.3 Effect of Fibre Condition

In fibres containing ionic groups, the levelling ability of the dye can be affected by changing the ionic groups.[1] For example, in the dyeing of cellulose fibres with direct dyes, the addition of alkali to the medium increases the anionic character of the fibres, so the displacement and levelling ability of anionic dye molecules is higher in dyeing processes performed in alkali medium.

In natural fibres where the dye diffuses into the fibres through swollen pores, substances that increase the swelling of the pores enhance the post-levelling, while substances that reduce the swelling of the pores (e.g. forming cross-linking bonds between macromolecules) diminish it.[1]

4.2 Factors Affecting Uniform Dyeing in Impregnation Method

Unlike dyeing by exhaust method, it is not possible for the dyeing to be levelled after dyeing by impregnation method. This is because migration does not occur as the movement of dye within the fibres from one place to another, but rather as a result of the continuous transfer of some dye from the fibres to the liquor and from the liquor back to the fibres (desorption-sorption). Therefore, in methods such as pad-batch (impregnation-cold batching), pad-roll (impregnation-warm batching), pad-steam (impregnation-steaming), even though the product is wet, i.e., the fibres are swollen during holding or steaming, no leveling occurs because there is no liquor available for dye to transfer to and from the fibres. However, in these methods, if an intermediate drying is done before the dye is fixed to the fibres after impregnation, some dye can be transported to the quick-drying regions at the beginning of drying (until about 25–30% water remains), as water flows from other areas to the quick-drying regions (**true migration**). This is a phenomenon that does not increase levelness, on the contrary, adversely affects it.[1] For this reason, dye should be applied to the fibres uniformly from the beginning in dyeing by impregnation method. As it is known, the moisture in a wet textile material can be divided into five groups as follows:

1. Dripping water (removed during squeezing between rollers)
2. Surface water
3. Capillary water
4. Swelling water
5. Crystal water (hygroscopic moisture).

Since capillary + swelling + crystal water constitute 45% of the weight of the fibres, the remaining amount is the surface water amount after this amount is subtracted from the taken liquor (e.g. for pick-up value of 70%, the surface water content is 25–30%). The rate of removal of surface water depends on the dryer used (conditions such as the amount and speed of the impinging air). In the first phase of drying, the surface water evaporates, and in the second phase, the capillary water on the surface of the fibres evaporates. In the third phase, the swelling water in the fibres diffuses to the surface and evaporates. Two forms of migration are known:

a. ***Migration on the fabric surface:*** As a result of this migration, which occurs as the migration of dyes from the slow-drying (more humid) parts to the quick-drying (drier) parts of the fabric surface, there is a difference in darkness along the width of the fabric (between the two edges or the edge and the middle). For this type of migration to occur, there must be a continuous liquid layer (film) on the surface of the fabric through which the dye can move. Since such a layer exists only in the first phase of drying (as long as surface water is present), this type of migration can only occur in the first phase.
b. ***Fibre or yarn migration:*** It is the migration of dye from the inside of the fibres, yarns or fabric towards the outside. If this type of migration occurs uniformly throughout, it does not pose a major problem. As a whole, it causes a dyeing that has low penetration into the fibres and therefore appears darker. However, non-uniform fibre migration

leads to dyeings with varying depths of shade at different points on the fabric. This type of migration can occur in the first phase of drying and at the beginning of the second phase.[1]

During impregnation with dye, the lower the pick-up, the less the amount of surface water to be removed in the first phase, thus reducing the risk of migration. The correct operation of the dryer used, the drying amount to be equal at every point of the fabric, and the addition of an appropriate antimigration agent to the liquor are measures to reduce the risk of migration. Another measure taken to prevent migration is to place a 2 meter pre-drying channel in front of the dryer. In this channel, usually infrared radiation (IR) are used, sometimes with hot air. The aim is to ensure that surface water is removed as quickly as possible, thus reducing the time during which migration can occur.[1] In impregnation dyeing, the number of parameters affecting the dyeing result is quite high. Among these, the first ones that come to mind are as follows:[8]

- The properties of the fabric (density, weight, etc.)
- The pretreatment processes of the fabric
- The temperature of the fabric
- Moisture content of the fabric
- The pH of the fabric
- Tension of the fabric
- Transition speed of the fabric
- Fixation time
- Fixation temperature
- Squeezing pressure of the rollers
- Elasticity of the rollers
- Diameter of the rollers
- The width of the rollers
- The temperature of the rollers
- The temperature of the liquor
- The temperature of the supplemental liquor
- Water quality
- Type of alkali used
- Type of dye used
- Auxiliaries used
- Temperature of the processing water
- The width and volume of the vessel
- Season of dyeing
- Washing operations

The differences that may occur in the above-mentioned parameters adversely affect the levelness and reproducibility of the dyeing.

Physical properties of the fabric such as ***density, weight, weave structure*** (plain, twill, satin, etc.) also affect its dye-uptake ability, and therefore the colour that can be obtained under the same conditions. This is especially important in industry when it is desired to obtain the right dyeing in one go. While all the properties of the fabric are constant, it should be taken into account that a change in only one of the physical properties such as density, weight and weave structure may change the dyeing result, and accordingly, the dyeing recipe should be revised when even one of these properties changes.[8, 9]

Appropriate and effective ***pretreatment*** is very important especially in pad-batch dyeing. The products should not contain any oil residues, sizing and bleaching materials, and the fabric should have a neutral pH. Another important point is that the wetting and absorption properties should be optimal and uniform in all parts of the fabric.[8, 9]

Temperature is also a very important parameter in dyeing by impregnation method. The temperature of the process water used, ambient air, impregnation liquor, foulard rollers, etc., in addition to the temperature of the fabric entering the dyeing process, can also affect the dyeing result. Possible temperature differences should be eliminated by blowing cooled air on the fabric entering to the foulard or passing it over the fabric cooling rollers. Similarly, if the dye solution is prepared at a temperature of 60 °C, it must be cooled before being combined with alkali in the dosing pump (e.g. in reactive dyes). Otherwise, the risk of hydrolysis will increase. In addition to the impregnation step, temperature is also very important in the fixation step. For example, it is crucial for the levelness and reproducibility of the dyeing that the temperature of the environment where the batches are kept is constant and homogeneous in dyeing by pad-batch method. Similar to the temperature, ***moisture*** differences on the dyeing fabric also have an effect on the dyeing result. The fact that the moisture content of the edges and the centre of the fabric is different can show itself as a wing (right–left) difference in dyeing.[1, 8, 9]

Pick-up values should be around 60–80% for cotton and 80–100% for viscose. It is recommended that the ***impregnation time*** should not exceed 1 second for cottons (by adding a wetting agent to the liquor if necessary), and 1.5–4 seconds for regenerated cellulose.[1]

The ***water quality*** used in dyeing is another factor affecting the results. As a general rule, soft or demineralised water with a neutral pH should be used. Especially when working with sodium silicate in reactive dyeing, it is necessary to use soft water to prevent harsh handle or to prevent silicate precipitation in the machine. These precipitates may form stains on the product. In order to prevent water-insoluble formations in alkaline dyeing medium, hard water must be softened, demineralised or a sequestering agent should be added. The point to be considered when choosing a ***sequestering agent*** is that it does not chelate the metal in metal complex dyes. In this case, colour as well as fastness values may be adversely affected. For this reason, it is not appropriate to use ethylenediamine tetra acetic acid (EDTA)-based products. In addition, the dispersing capacity of the sequestering agent must be known. It is stated that polycarboxylate/phosphonate-based materials are particularly effective in silicate-containing processes and aftertreatments.[8, 9]

There are two main problems encountered in dyeing made by impregnation method. The first of these is the problem of dyeing the beginning parts (e.g. the first 50–60 m) of the fabric darker and the latter parts lighter in case the dye used is substantive to the fibres, which is called "**tailing**". In order to reduce the head-tail difference in the fabric during impregnation, the volume of the padding vessel should be kept small and the liquor replenishment time should be less than 7 minutes. In regenerated cellulose, on the other hand, "**negative tailing**" problem may be encountered (Figure 4.3). The reason for this is if the fibres are given less than 1.5 seconds in the impregnation step, the swelling of the fibres will not be complete in such a period of time (time should be allowed to overcome the resistance created by the outer layer in viscose fibres), unlevel dyeings may occur and dyeings

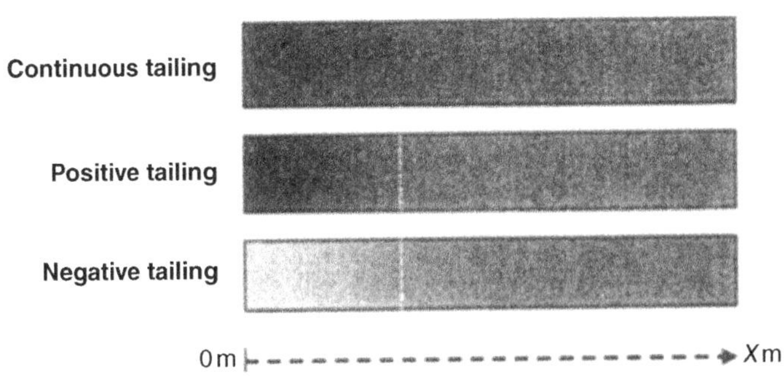

Figure 4.3 Tailing problem in pad-batch dyeing. Ref. [8]/Dystar.

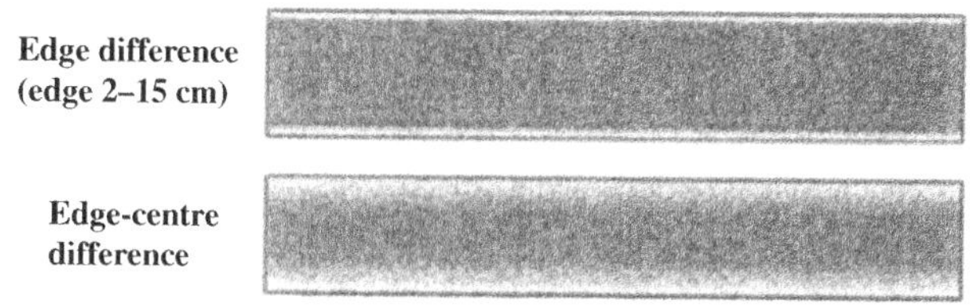

Figure 4.4 Edge difference problem in pad-batch dyeing. Ref. [8]/Dystar.

that have not fully penetrated into the fibres may occur, and since the dye concentration in the liquor will increase over time, light and dark dyeing may occur. This is called "negative tailing".[1,8]

The second error that can occur in dyeing by impregnation method is the problem that arises from the application of different pressure to the edges and centre parts as a result of the curving of the padding roller, which is called the **edge-centre-edge difference (wing difference or listing)** (Figure 4.4). In this case, the more squeezed edges are dyed light, and the less squeezed centre parts are dyed darker. The degree of curvature occurring in the rollers under a certain pressure can be calculated by the formula:[10]

$$F_m = \frac{q.l^4}{E.J.K}$$

F_m = central curvature of the roller (cm)
q = Linear pressure (kg/cm)
l = Distance between support points (cm)
E = Modulus of elasticity (kg/cm^2)
J = Moment of inertia (cm^4)
K = A constant corresponding to the specific loading (stress) condition (384).

Since the q, E and J values in this formula are material or operating conditions dependent, it is not possible to change them drastically. K is already a constant number. So, the way to reduce curvature is to increase the number of support points. Based on this logic, new padder roller constructions (floating rollers, bendable rollers, etc.) have been developed that can squeeze equally at the edges and in the centre.[10] A uniform liquor pick-up is the most important prerequisite for faultless dyeings.[7]

The edge–centre–edge variation may arise not only from differences in the squeezing pressure applied by the rollers, but also from irregularities resulting from the sanding process used to remove silicate residues, or from temperature variations present in the material to be dyed.[8] Furthermore, a uniform liquor pick-up can only be ensured if there is no variation in absorbency across the width and length of the fabric. For this reason, the pretreatment processes should also be carried out continuously in open-width.[7]

References

1 Tarakçıoğlu, I. Tekstil Boyacılığı I Teksiri. İzmir, Türkiye: Ege Üniversitesi Mühendislik Fakültesi Çoğaltma Yayınları, 1982.

2 Bide, M. Dyeing. In Kirk-Othmer Encyclopedia of Chemical Technology (Chapter 5, 1–115). Hoboken, NJ, USA: Wiley, 2016.
3 Broadbent, A.D. Basic Principles of Textile Coloration. Bradford, England: Society of Dyers and Colourists, 2001.
4 Yurdakul, A., Atav, R. Boya baskı esasları. Bornova, İzmir: Ege Üniversitesi Tekstil ve Konfeksiyon Araştırma Uygulama Merkezi Yayını, 2006.
5 Baldwinson, T.M. Classification of dyeing and printing auxiliaries by function, In Colorants and Auxiliaries, Organic Chemistry and Application Properties, Second Edition, Volume 2 – Auxiliaries (Ed. J. Shore). Manchester, UK: Society of Dyers and Colourists, 2002.
6 Hickman, W.S. Preparation. In Cellulosics Dyeing (Ed. J. Shore, Chapter 2, 81–149). Bradford, West Yorkshire, UK: Society of Dyers and Colourists, 1995.
7 Latham, F.R. Dyeing with vat dyes. In Cellulosics Dyeing (Ed. J. Shore, Chapter 5, 246–278). Bradford, West Yorkshire, Uk: Society of Dyers and Colourists, 1995.
8 Dystar. Technical documents (Reactive continuous dyeing).
9 Dystar. Levafix®, Remazol® Reactive dyes catalogue.
10 Tarakçıoğlu, I. Tekstil terbiyesinde aplikasyon yöntem ve makinaları, Ege Üniversitesi Emel Akın Meslek Yüksek Okulu, 1998.

5

Dyeing with Direct (Substantive) Dyes

The characteristic of direct dyes is that they dissolve in water and can be taken up directly by cellulose fibres from the aqueous solution (liquor) without the need for any special treatment. For this reason, these dyes are called **"direct dyes"**.[1] On the other hand, dyeing of cellulose with direct dyes is based on adsorption, and the ability to dye fibre by adsorption is called substantivity. For this reason, these dyes are also called **"substantive dyes"**. The interactions that provide the substantivity (affinity) of these dyes towards cellulose fibres are van der Waals forces and hydrogen bonds.[2]

5.1 Structure and Classification of Direct Dyes

In terms of their molecular structures, most of the substantive dyes are diazo and polyazo dyes. In other words, they are dyes obtained by coupling an aromatic diazonium salt containing at least two diazonium groups with aromatic amines or phenols. In Figure 5.1 the reactions related to the production of a direct dye are given.[1]

The solubilising groups in these dyes are usually sulpho and rarely carboxyl groups. Accordingly, substantive dyes are anionic dyes.[1] These dyes can be represented by the general formula $D(SO_3Na)_n$ ($n = 1 - 5$). Other substituents in their structure are hydroxyl, amino, amido, etc., groups.[3]

More than 75% of all direct dyes are non-metallised azo structures, and every shade is available in this group. The majority of these are diazo in colours ranging from brighter yellow to blue, or polyazo species predominating in darker greens, browns, greys and blacks. About 50% of all direct dyes are in the diazo structure, a typical example is Colour Index (CI) Direct Blue 1 dye. Copper complex direct dyes are mostly duller violets, navy blues and blacks derived from the diazo and polyazo subclasses. These are mostly used in burgundy, brown, grey, navy blue and black recipes. Stilbene and thiazole dyes, which are the only non-azo chromogens in colours ranging from yellow to red, have some similarities to their diazo and polyazo structures. Stilbene direct dyes are mostly yellow, orange or brown. A typical example is CI Direct Yellow 12. Most thiazole dyes are yellow dyes, such as CI Direct Yellow 59 dye. Dioxazine and phthalocyanine chromogens are used in bright blue colours with high light fastness. The triphenodioxazine ring system forms the basis of some important blue direct dyes such as CI Direct Blue 106. Phthalocyanine direct dyes

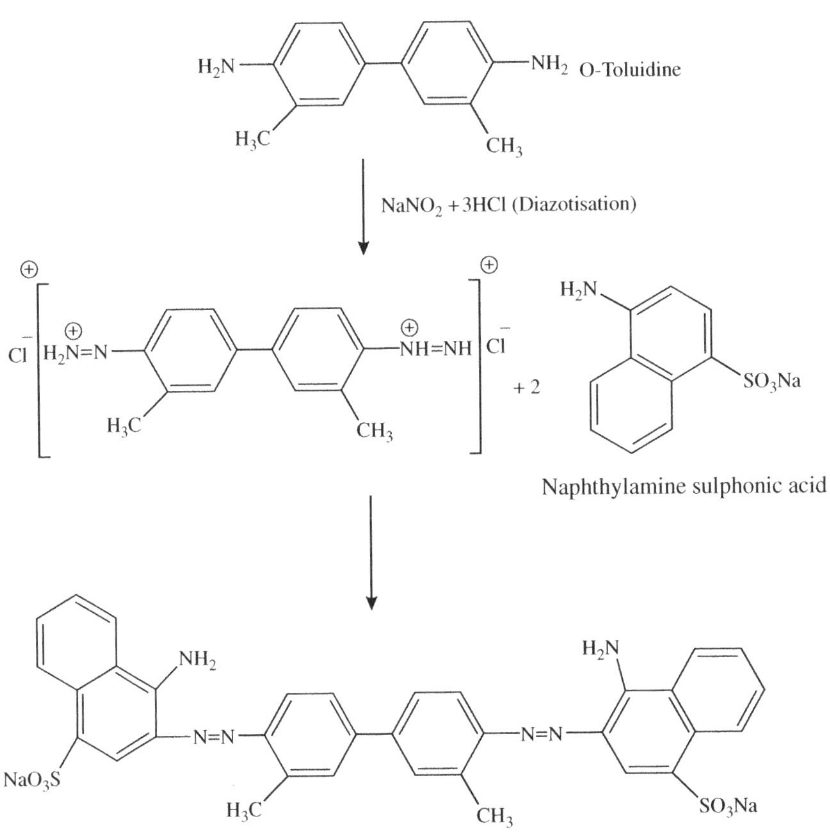

Figure 5.1 Synthesis of benzopurpurine 4B dye. Ref. [1]/Ege Üniversitesi Mühendislik Fakültesi Çoğaltma Yayınları.

are water-soluble sodium salts of sulphonated copper phthalocyanine. They give bright turquoise blue colours with high light fastness, but their wet fastness is low unless after-treatment with resin is applied.[4] Direct dyes are divided into three main classes according to Society of Dyers and Colourists (SDC):

- ***Class A (dyes with high levelling ability):*** These dyes have small molecular size, and even the addition of salt from the beginning of dyeing does not prevent proper dyeing. In order to ensure sufficient dye uptake by the fibres, it is necessary to use more salt than in other classes.
- ***Class B (salt-controlled dyes):*** These dyes have poor levelling and migration ability, and the addition of salt should be done after the dyeing temperature (usually between 90 °C and boiling temperature) is reached.
- ***Class C (salt- and temperature-controlled dyes):*** These dyes also have poor migration and levelling ability, but proper dyeing is achieved with salt and temperature control. Salt should be added in portions after reaching the boiling temperature.[2,4]

The migration and salt sensitivity properties of direct dye classes are given in Figure 5.2 comparatively. As can be seen from Figure 5.2, the migration capabilities of these dyes can

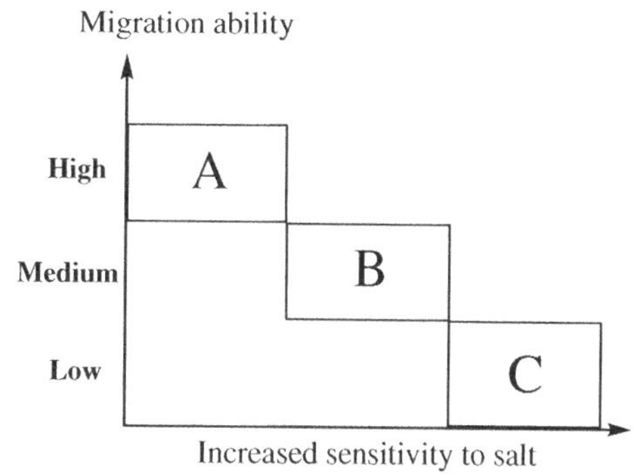

Figure 5.2 Migration and salt sensitivity properties of direct dye classes. Ref. [2].

be listed as A > B > C. So, as a result of the migration test, their staining of the adjacent fabric will decrease in the order of A > B > C. In the migration test, an adjacent fabric of equal weight is attached to the dyed fabrics, and the process is carried out with liquors containing 10 g/L Na_2SO_4 at 98 °C for various periods of time (e.g. 20–40–60–80 minutes) and as a result of the test, the migration abilities of the dyes are determined by evaluating the dye migration to adjacent fabrics. The dyes containing 10 g/L Na_2SO_4 are treated at 98 °C for various times (e.g. 20–40–60–80 minutes). Higher dye transfer to the adjacent fabric indicates that the migration ability of the dye is higher.[5]

Since the migration and levelling properties of class A, B and C dyes are different from each other, dyes in the same class should be combined. Class A and B can be combined; class B and C can be combined. However, A and C should never be combined. Beyond that, the temperature-dependent uptake curves of the selected dyes should be parallel to each other when making combinations. Otherwise, the desired colour and nuance cannot be obtained, since one of the dyes will be more dominant than the other.[5]

5.2 Fastness Properties of Direct Dyes

Since the forces that play a role in the binding of direct dyes to cellulose fibres are weak interactions such as van der Waals forces, dipole–dipole forces and hydrogen bonds, their wet fastness is low, especially in dark colours. However, it is possible to improve wet fastness to a certain extent with aftertreatments.[2] The light fastness of dyeings with direct dyes on cellulosic fibres ranges from poor to fairly good, but some copper complex direct dyes have very good light fastness.[6]

5.3 End-uses of Direct Dyes

Since dyeing with substantive dyes is cost-effective and simple, this dye class was used extensively in the dyeing of cotton and regenerated cellulose fibres.[1] Dyes in this class are generally used for dyeing materials which are not washed frequently such as decoration fabrics, velvet and lining.[2] In addition, they were used in dark colours such as black which require high dye concentrations, due to their low cost. However, today, this dye class is not preferred in dark colours due to its low wet fastness values. The main advantages and disadvantages of direct dyes can be listed as follows:

Advantages

- ✓ Dyeing process is easy.
- ✓ It is possible to correct non-uniform dyeing.
- ✓ There is no need for long washing processes after dyeing.
- ✓ Light fastness is generally good.
- ✓ Wastewater load is low.
- ✓ The dye liquor can be reused (standing bath).[2]
- ✓ By selecting alkali-resistant dyes, scouring, bleaching and dyeing can be done in a single bath.[7]

Disadvantages

- ✓ Wet fastness is limited (for materials that have not been aftertreated).[2]

5.4 Dyeing with Direct Dyes

Direct dyes can be applied using both exhaust and impregnation (pad-batch, pad-jig, pad-steam) methods.[2]

5.4.1 Dyeing by Exhaust Method

It is necessary to use soft water when working with these dyes. Precipitation may form when hard water is used. Calculated amount of dye is taken and made into paste with soft hot water and then completely dissolved with boiling water by mixing. It is also beneficial to boil it for a while.[2] Soft water should be used in the dyeing process as well as in the dissolving process. When hard water is used, precipitates may form. Presence of ions such as Ca^{++}, Mg^{++}, Cu^{++} and Fe^{++} in water can cause colour shifts, harsh fabric handle and decrease in hydrophilicity. For the removal of these metal ions, the use of an appropriate sequestering agent is required. Today, acrylate- and phosphonate-based sequestering agents are used in dyeing. The use of the correct sequestering agent in dyeing processes is of great importance.[5] The use of ethylene diamine tetra acetic acid (EDTA) and nitrilo triacetic acid (NTA) should be strictly avoided, especially with metallised direct dyes containing copper, because EDTA causes colour deviations by demetallising the metal of the dye in metal-containing dyes. In addition, even very little residue of chemicals containing chlorine used to purify water can adversely affect many dyes. Therefore, if there are chlorine residues in the water, they should be removed with antichlorinating agents (sodium bisulphite [$NaHSO_3$], sodium thiosulphate [$Na_2S_2O_3$]).[8]

At the beginning of the dyeing process, the dye molecules migrate to the fibre and are adsorbed on the fibre surface. In the second step, the dye begins to diffuse into the easily permeable amorphous regions of the fibre. Generally, the diffusion step is the slowest step in dyeing, and the bath temperature is increased to accelerate diffusion. However, in this case, the adsorption decreases. Thus, while the portion of dye that binds to the fibre decreases, the portion of dye remaining in the liquor increases. As a result, the equilibrium shifts to the fibre side in cold and to the liquor in hot. After a certain period of time, an equilibrium is established between the dye adsorbed on the fibre and that desorbed from the fibre, after which there is no benefit in prolonging the dyeing.[2] Although this equilibrium can sometimes be shifted completely to the fibre side depending on the type of dye used and dyeing conditions (i.e. all the dye in the liquor can be adsorbed by the fibres), in many cases 1/4 or even 1/3 of the dye remains in the liquor. In such cases, especially baths used for dark shades can be reused after adding dye and other consumed substances. This is called **"dyeing in a standing bath"**. Since it is often difficult to determine the amount of decrease in the liquor, the amount of additions to be made in many enterprises is based on experience. Replenishment of 70–75% of the liquor gives good results in many cases.[1]

Dyeing can be done in an alkali or neutral environment, by exhaust method with direct dyes. Accordingly, the following chemicals are added into liquor:

a. Dyeing in alkali medium

In light shades	In dark shades
5%	1–2% Calcined soda ash
2–5%	10–20% Calcined sodium sulphate or common salt
4–10%	20–40% Crystalline sodium sulphate

b. Dyeing in neutral medium

In light shades	In dark shades
2–5%	10–20% Calcined sodium sulphate or common salt
4–10%	20–40% Crystalline sodium sulphate

It is also beneficial to add 0.5–1% wetting agent to the liquor. Dyeing begins by inserting the material to be dyed into the liquor at 40–50 °C. Depending on the depth of shade and the dyeability of the material, either no salt is added to the liquor initially or a small amount is added. The liquor is slowly heated to boiling temperature, and dyeing is continued at the boil for 45–60 minutes. The required amount of salt is added to the liquor in small portions. Adding the majority of the salt before the liquor starts to boil has a negative effect on the penetration of the dye into the fibres and the levelness of the dyeing.[1]

For dyeings performed in low liquor ratios and with materials such as tightly woven fabric or hard-twisted yarn, which are difficult to dye properly and penetrate, some dye manufacturers recommend starting the dyeing by entering the material into a salt-free boiling liquor. Then, salt is added gradually in small portions and dyeing is continued in the boiling liquor. In order to increase the amount of dye taken up by the material, some catalogues recommend cooling the bath before terminating the process.[1]

The factors affecting the dyeing equilibrium in dyeing processes with direct dyes are temperature, liquor ratio, salt and alkali. Each of these is described separately below.[2]

a. Temperature: In the dye catalogues, the percentage of dye taken up by the fibre at various temperatures for each dye is given in graphs. When these graphs are examined, it is seen that the uptake of various dyes by the fibre varies significantly.[1] For example, for CI Direct Blue 218 dye, the maximum exhaustion temperature is around 100 °C, while the maximum exhaustion temperature of CI Direct Yellow 12 dye is around 20 °C. In some dyes, the exhaustion reaches its maximum at a certain temperature and decreases again if the temperature is increased further. For example, for CI Direct Orange 72 and CI In Direct Red 81 dyes, the maximum exhaustion is reached at 80 °C and 60 °C, respectively, and when the temperature rises above this, the dye exhaustion decreases.[8] The variation of the uptake of various direct dyes with temperature is shown in Figure 5.3.

As can be seen from Figure 5.3, for some dyes, it is seen that the amount of dye taken up by the fibre increases with increasing temperature and the highest amount of dye is taken up at 100 °C. In others, this increase is rapid at first but reaches a maximum at 40–60 °C, and a further increase in temperature does not lead to a change. In a third group, the amount of dye taken up by the fibre increases up to 60–70 °C and starts to decrease again as the temperature rises further. In this case, the critical point is the selection of dyes with similar uptake curves depending on temperature in the combination, as stated before. Otherwise, a few degrees difference in the liquor temperature will cause a different dyeing nuance.[1]

Although dyeing with substantive dyes is usually performed around the boiling temperature, dyeing can sometimes be done at temperatures above 100 °C (HT conditions)

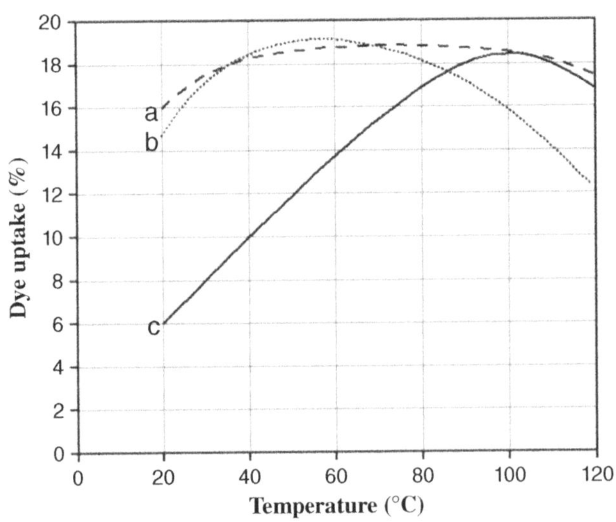

Figure 5.3 Variation of uptake of various direct dyes with temperature. Ref. [1].

when dyeing cotton or viscose yarn bobbins, warp beams or rayon fibres.[1] When PES/CO blends are to be dyed with a Disperse/Direct dye combination, the PES component is dyed first, followed by the cotton component as the direct dyes will be damaged during dyeing of PES under HT conditions. However, if the cotton part is to be dyed first, then the stability of the direct dye to HT conditions must be tested.[5]

The advantage of dyeing under HT conditions is that the diffusion ability of the dye and the levelling of the dyeing increase at temperatures above 100 °C. Thus, it is possible to obtain a uniform, well-penetrated dyeing in a short time. This means saving time and steam consumption. Not all substantive dyes are resistant to dyeing under HT conditions (around 120 °C). In many dye catalogues, the durability of dyes under HT conditions is stated.[1] For example, dyes such as CI Direct Blue 80, CI Direct Yellow 105 and CI Direct Orange 39 are resistant to HT conditions, while CI Direct Yellow 44, CI Direct Red 80 and CI Direct Red 83 dyes do not have HT stability.[8] The stability of the dye under HT conditions also depends on the pH of the liquor. Generally, operating in a neutral environment is preferred. In cases that require the addition of soda ash, it is useful also to add 1.5 g/L ammonium sulphate to the liquor to buffer it.[1]

It should be noted that the amount of dye taken up by the material to be dyed begins to decrease at temperatures above 100 °C. Therefore, after the dyeing at 120 °C is finished, the liquor must be cooled to temperatures below 100 °C without removing the material. In the meantime, since the fibres will absorb some more dye, it is important that the cooling rate and final temperature are consistent in each batch to ensure reproducibility.[1]

b. **Liquor Ratio:** As the liquor ratio increases, the equilibrium shifts to the bath, thus the dyeing efficiency decreases as the amount of dye remaining in the liquor increases.[2]
c. **Salt (Electrolyte):** During dyeing, dye molecules are adsorbed by cellulose fibres, not dye anions. Therefore, when salt is added to the liquor, the amount of dye that binds to the fibre increases as the dissociation of the dye will decrease.[2] Effect of salt on dye uptake can be seen from Figure 5.4.

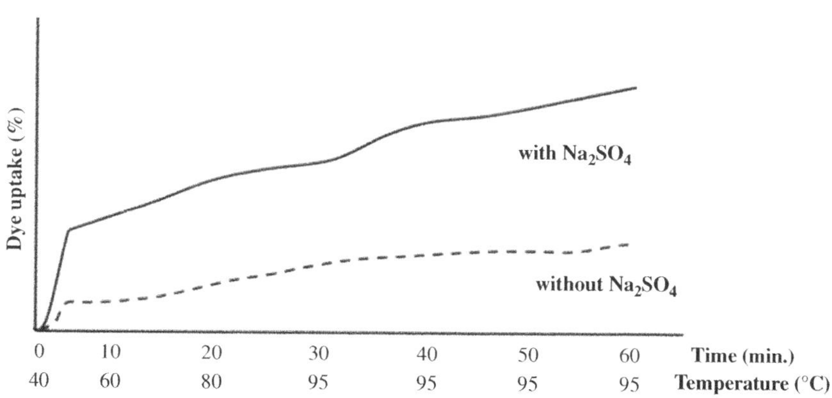

Figure 5.4 Effect of salt on dye uptake. Ref. [2].

The enhancement of dye uptake by salt can be explained by two mechanisms:

- As it is known, salt molecules are present in a dissociated state in aqueous solutions. For example, sodium chloride dissociates into ions in water to give Na^+ and Cl^- ions. The positively charged sodium ion moves to the fibre/water interface, effectively lowering the negative zeta potential, thus removing the barrier that prevents the dye/fibre interaction. This is due to the neutralisation of the negative charge of the fibre by the cations (Na^+) of the salt. When the negative charge on the fibre surface decreases, the electrostatic repulsive forces between the anionic dye and cellulose decreases, consequently the dye approaches the fibre surface more easily and the dye uptake increases.[2]
- The second effect of salt, which increases the dye uptake, is due to the increase in the aggregation of the dye. This phenomenon can be explained according to Le Chatelier's principle as follows:

$$\text{D-SO}_3\text{Na} \rightleftarrows \text{D-SO}_3^- + \text{Na}^+ \quad [1]$$
$$\text{NaCl} \rightarrow \text{Na}^+ + \text{Cl}^- \quad [2]$$

As can be seen in equation [1], the dye dissociates into its ions in the aqueous medium, when salt is added to the medium, the sodium cation concentration in the liquor increases, and the equilibrium in equation [1] shifts to the left, in accordance with the action-reverse reaction principle (Le Chatelier) and hence aggregation increases, dissociation decreases. As aggregation increases, dye uptake increases, since the aggregated form of the dye can approach the fibre more easily than the dissociated form.[2]

Three types of salts are commonly used in dyeing processes with direct dyes. They are:

- Sodium chloride (common salt) (NaCl),
- Anhydrous (calcined) sodium sulphate (Na_2SO_4) and
- Sodium sulphate decahydrate (crystalline sodium sulphate) ($Na_2SO_4.10H_2O$).

The first two salts are used in equal amounts, whereas the third requires approximately double the dosage of the others. This is because crystalline sodium sulphate contains 10 moles of water of crystallization, which constitutes about 56% of the molar mass of the salt. Although the usage amounts of anhydrous sodium sulphate and sodium chloride salts are the same, their effects on the dyeing result may be different.[9]

Since salt increases the dye uptake, it should be used more in dark shades and less in light shades. As the amount of salt used increases, the dye uptake increases up to a certain point. In addition, as the dye concentration in the liquor increases, the amount of dye taken up by the fibre also increases. Therefore, the need for salt decreases as the dye uptake increases in dyeings made using less water or in other words at lower liquor ratios.[5] It is recommended that the salt concentration in the liquor should not be less than 4 g/L and not more than 40 g/L. Salt is added at the beginning of dyeing in dyeing devices with high circulation speed or in case of using class A dyes, and 10 minutes after reaching the dyeing temperature for class B and C dyes. Adding salt should be done in 3–5 portions according to the depth of shade with 10-minute intervals between each portion. The portions of salt should be unequal, but gradually increasing (e.g., 1/7, 2/7, 4/7).[8]

d. **Alkali:** Soda ash is usually added to the liquor as an alkali. The addition of soda ash to the liquor generally facilitates dye dissolution, maintains solubility, and promotes uniform dyeing. For this reason, the use of soda ash is preferred in cases where there is a solubility problem and for dyeing in light shades. As it is known, soda ash exhibit alkaline properties due to hydrolysis and its solution in water contains hydroxyl (OH^-) ions. With the effect of these hydroxyl ions, the dissociation of carboxyl groups in cellulose increases.[1]

$$\text{Cell—COOH} + OH^- \rightarrow \text{Cell—COO}^- + H_2O$$

Thus, the (−) charged (D-SO_3^- dye anion) dye is repelled by the (−) charged fibre. Since the increase in the (−) charge of the cellulose causes it to repel the anionic dye more, the dye uptake is retarded (by decreasing the uptake rate) and levelness is achieved. For this reason, more soda ash should be used in light shades and less (or no) soda ash in dark shades.[2] Cotton fabrics can be dyed in an alkaline or neutral medium depending on the situation, whereas dyeing in a neutral medium is preferred for regenerated cellulose.[1]

5.4.2 Dyeing by Impregnation Method

Direct dyes are generally less suitable for continuous dyeing than for exhaust dyeing. The head-tail difference (tailing) is a serious problem in continuous dyeing, and achieving adequate diffusion requires extended time in pad-batch or pad-steam methods. Whenever possible, great care should be taken to select dyes with similar uptake properties. The wet fastness of direct dyes may be lower when applied by continuous methods compared to conventional exhaust dyeing, but this property can be improved by adding salt or increasing the impregnation temperature.[4]

Most of the substantive dyes can be used in semi-continuous and continuous dyeing methods. However, the process of impregnating the dye into the fabric in the padding mangle requires attention. Since substantive dyes have affinity for cellulose fibres, the risk of tailing is high.[1] Class A direct dyes tend to migrate during drying and steaming. Class C dyes show rapid uptake. For this reason, class B dyes are often the best choice for continuous dyeing.[7] It is beneficial that the vessel of the padder is as small as possible, in terms of accelerating the consumption and regeneration of the liquor. Even a horizontal padder may be preferred. The pick-up value should be 60–80%. Since the immersion time is short, the wettability of the fabric should be good and the drying should be uniform. It is also beneficial to add a suitable wetting agent in the liquor. The temperature of the impregnation liquor should be 30–40 °C in light shade dyeing. However, since the solubility of some substantive dyes is low in cold and increases when heated, the temperature of the impregnation liquor should be kept between 60 and 80 °C for medium and especially dark shades.[1]

Information on the fundamentals of dyeing by various semi-continuous and continuous dyeing methods with direct dyes is given below.

5.4.2.1 Pad-batch Method

Although the cold pad-batch method is not recommended for regenerated cellulose, it gives satisfactory results on cotton. After the dye is dissolved in hot water, urea is dissolved in water at 30–35 °C in a separate container, and after it is combined with the dye solution, auxiliaries (such as a highly effective cold wetting agent) are added. The urea concentration in the liquor is 100–150 g/L. Impregnation is done at 20–30 °C in the padder and the material is wrapped in rolls after being impregnated with dye liquor, covered with polyethylene foil so that no air enters, and then left to dwell while being rotated. Batching (fixing) time varies between 8 and 24 hours depending on the dyeing depth.[1]

5.4.2.2 Pad-roll Method

The pad-roll method is a preferred dyeing method, especially for linings made of rayon. In addition to the dye and wetting agent, 80–120 g/L urea and 1 g/L sequestering agent are added to the impregnation liquor. After impregnation, the material, which is heated by passing through the IR channel, is fixed in the thermo-batching room for 2–8 hours. Generally, 2 hours for light shades, 3–5 hours for dark shades, 8 hours for dark blue and black are sufficient. Meanwhile, the dry temperature should be 80–90 °C and the wet temperature should be 5 °C lower than the dry. Although it is sufficient to rinse the fabric coming out of the thermo-batching room after an air passage in light shade dyeing, a good soaping is essential for dark shade dyeing.[1]

5.4.2.3 Pad-jig Method

This semi-continuous method, which is based on the fixation of the dye applied to the fabric impregnated with the dye solution in the padder by treating it with boiling salt solution in the jigger, can also be economical in dyeing small lots. Compared to conventional jigger dyeing, the penetration of the dye into the fibres is superior. Generally, there is no need for intermediate drying after impregnation, but if the fabric coming out of the padder will not be transferred to the jigger immediately, intermediate drying may be recommended. Intermediate drying must be uniform, otherwise there is a risk of dye migration. There is excess (5–30 g/L) sodium sulphate or common salt in the fixing liquor in the jigger. In very dark shades and when the liquor ratio is low, not all of the salt is added at the beginning; instead, some of it can be added later. In dyeing medium and dark shades, some dye is added to the fixation liquor from the impregnation liquor. Usually, 10–40 mL/L addition is sufficient. Fixation is carried out at the boil, and the fixation time depends on the dyeing depth and the size of the batch. Generally, 4–8 passages are sufficient, and the fixation time should not be less than 20 minutes. After fixation, the material is thoroughly rinsed and dried.[1]

5.4.2.4 Pad-steam Method

In this fully continuous dyeing method, which is mostly used for dyeing large batches, the fabric impregnated with the dye solution in the padder is steamed for 1–5 minutes. Steaming is performed with saturated steam at 100–105 °C in continuous steamers. The steaming time depends on the material to be dyed and the dyeing depth, as well as the dye used. Many dye manufacturers classify their dyes into groups as fast (1–2 minutes), medium (2–3 minutes) and slow (3–5 minutes)-fixing dyes in their catalogues. It is beneficial to add urea and a sequestering agent in the impregnation liquor. Especially in cases where the steaming temperature drops below 100 °C, it is beneficial to increase the amount of urea, and even to add a thickener such as sodium alginate to the liquor.[1]

5.4.2.5 Pad-dry-thermofixing Method

In this method, the fixation of substantive dyes can be achieved directly with the application of dry heat, as in the thermosol method. With the liquor containing urea, antimigration agent and fixation accelerator, after the cotton is padded at ambient temperature, whereas regenerated cellulose fibres are padded at 50–60 °C with 80–90% pick-up, subsequently, drying and fixation are carried out at 120–130 °C for 60–120 seconds. The fixation accelerator increases the solubility of the dyes, the swelling of the fibres and the energy provided during fixation.[1]

5.5 Post-dyeing Aftertreatments with Direct Dyes

Direct dyes are inexpensive, but their wet fastness is low, especially in dark colours, as their bonding to the fibres occurs only through weak secondary attraction forces (van der Waals attraction forces and hydrogen bridges). Therefore, after dyeing their fastness needs to be improved.[2] For this purpose, the following agents are used in aftertreatment processes:

- metal salts (copper sulphate, potassium dichromate),
- formaldehyde condensation products,
- cationic synthetic resin precondensates,
- cationic post-treatment agents (fixing agents) and
- reactive cationic aftertreatment agents.

Apart from these, simply rinsing the fabric with water containing salt (10 g/L sodium sulphate or sodium chloride) after dyeing can be beneficial in terms of preventing migration that may occur while the fabric is held in a wet state between dyeing and drying, thereby ensuring better fastness.[1] Nowadays, cationic fixing agents are generally preferred because metal salts, formaldehyde condensation products and cationic synthetic resin precondensates cause environmental problems.[2]

a. Aftertreatment with metal salts: This process results in metal complexation, which causes blocking of the solubilising groups of the dye, increased aggregation, decreased solubility and improved washing fastness. Light fastness also generally improves. The catalytic effect of traces of heavy metal ions present can cause fibre damage, especially in the presence of detergents containing oxidising agents. This application also makes the colour of the dyeing duller.[10] Molecular structures of some special direct dyes are suitable for forming complexes with copper ions. Therefore, both light and wet fastness values increase when the material dyed with this type of dye is treated with copper sulphate. However, the colour shade of the dyeing changes.[1]
b. Aftertreatment with formaldehyde: The wet fastness of dyeings made with some special substantive dyes also increases after a post-treatment with formaldehyde. These are dyes containing free amino or hydroxyl groups in their molecules. There is not much change in the colour shade of the dyeing at the end of the treatment carried out

$$D\text{-}SO_3^- Na^+ + F^+X^- \rightarrow D\text{-}SO_3^- \text{----}\ {}^+F + NaX$$

(FX: Cationic fixing agent with general formula $R\text{-}NH_3^+Cl^-$)

Figure 5.5 Mechanism of dye fixation with cationic fixing agents. Ref. [2].

at 50–60 °C for half an hour with a liquor containing 1–2% formaldehyde and 1–2% acetic acid (30%).[1]

c. Aftertreatment with cationic synthetic resin precondensates: In this rare post-treatment method, the dyed material is treated with cationic precondensates of aminoplasts (a kind of synthetic resin). During this process, electrostatic attraction forces occur between anionic dyes and cationic synthetic resin precondensates. Urea/formaldehyde, melamine/formaldehyde and dicyanamide/formaldehyde preparations are used as precondensates. As a result of these procedures performed with these, wet fastness values are improved.[1]

d. Cationic aftertreatment agents (fixing agents): These products increase wet fastness values by blocking the solubilising sulpho groups of the anionic dye molecule. By blocking the solubilising groups in the dye structure, the solubility of the dye in water is reduced and an increase in molecular size is achieved.[1,2]

While cationic fixing agents increase the wet fastness values (washing and water fastness) of the dyed product, they generally decrease the light fastness values. In addition, chemicals used for fastness improvement can cause colour change in the dyed fabric. Therefore, this issue should also be considered in the selection of aftertreatment chemicals.[2]

e. Reactive cationic aftertreatment agents: Cationic post-treatment agents containing a normal hydrophobic group effectively increase some wet fastness values such as water, sea water and perspiration fastness, but they can only increase the washing fastness by 1–2 grades. This is because especially during washing with plenty of soap and detergent, some cationic aftertreatment agents prefer to bond with soap or detergent anions by leaving the dye anions. As a result, since the blocking effect on the dye molecules in the groups that provide water solubility diminishes, some dye may bleed into the washing water. Reactive cationic post-treatment agents can make covalent bonds with the fibre, and since the bond they form with the dye is strong, during home laundering, the problem of dye loss from the fabric is eliminated as the cationic product does not release the dye anion. Accordingly, in order for the effect of blocking the water-soluble anions in the dye molecules to be permanent, the cationic aftertreatment material used must be well fixed to the fibres. Since there is no possibility other than covalent bond formation for a good fixation in cellulose fibres, these post-treatment agent molecules must be attached to the cellulose fibres by covalent bonds. The first products introduced in this area were water-soluble reactive quaternary ammonium compounds. In the past, these materials were added to the dyeing liquor from the beginning, that is, a single-bath process was carried out. This cationic auxiliary, which reacts with the hydroxyl groups in the cellulose fibres during dyeing, enables the fibres to acquire a slightly cationic feature, and therefore, as dye anions are attached to the fibres by electrostatic attraction forces in addition to the secondary attraction forces, as in the case of wool dyeing in an acidic environment. Since the water-soluble groups in the dyes are also blocked, the fixation of the dye increases significantly. As a result of this, even the washing fastness of the dyeing at 90 °C can be improved to grades of 4–5.[1]

References

1 Tarakçıoğlu, I. Tekstil Boyacılığı I Teksiri. İzmir, Türkiye: Ege Üniversitesi Mühendislik Fakültesi Çoğaltma Yayınları, 1982.
2 Yurdakul, A., Atav, R. Boya baskı esasları. Bornova, İzmir: Ege Üniversitesi Tekstil ve Konfeksiyon Araştırma Uygulama Merkezi Yayını, 2006.
3 Aspland, J.R. A Series on dyeing, Chapter 2/Part 2: Direct dye structure. 1991; 23(12), 30–32.
4 Shore, J. Dyeing with direct dyes. In Cellulosics Dyeing (Ed. J. Shore, Chapter 3, 152–186). Bradford, West Yorkshire, UK: Society of Dyers and Colourists, 1995.
5 Giles, C.H., Duff, D.G., Sinclair, R.S. Giles's Laboratory Course in Dyeing (bs. 4). Bradford, UK: Society of Dyers and Colourists, 1989.
6 Koh, J. Dyeing of cellulosic fibers. In Handbook of Textile and Industrial Dyeing (Ed. M. Clark). Cambridge, Birleşik Krallık: Woodhead Publishing Limited, 2011.
7 Broadbent, A.D. Basic Principles of Textile Coloration. Bradford, England: Society of Dyers and Colourists, 2001.
8 Aspland, J.R. A Series on Dyeing, Chapter 1/2: Direct dyes and their application. 1991; 23(11), 41–45.
9 Yurdakul, A., Atav, R., Arabacı, A. Tekstil boyacılığında kullanılan sodyumklorür ve sodyumsülfat tuzlarının reaktif boyamadaki performanslarının karşılaştırılması. TÜBİTAK projesi, 2004.
10 Sekar, N. Direct dyes. In Handbook of Textile and Industrial Dyeing (Ed. M. Clark). Cambridge, Birleşik Krallık: Woodhead Publishing Limited, 2011.

6

Dyeing with Vat (Indanthren) Dyes

Vat dyes are basically indigo or anthraquinone derivatives that are insoluble in water and are rendered soluble during the dyeing process (Figure 6.1).[1] Vat dyes in the form of water-insoluble pigment (vat pigment) are converted into their water-soluble leuco form by reduction in an alkali medium (vatting process) during the dyeing process. The colour of the leuco form, which has an affinity for the fibre, is different from the original pigment, and the leuco form regains its original colour through the oxidation process after dyeing. Thus, the dye is converted back to its water-insoluble initial form; therefore, the wet fastness of vat dyes is high.[2]

Before the discovery of synthetic dyes, natural dyes obtained from various vegetable, animal and mineral sources were used. At that time, the blue colour was only provided by indigo obtained from *Indigofera tinctoria* or dyer's woad obtained from *Isatis tinctoria*. This dye was of great value because of its high light fastness. Blue-coloured indigo, which is not water-soluble, was fermented in wooden vats and thereby made water-soluble. This process, which is called vatting, is the origin of the name of the vat dyes. Animal and vegetable fibres were dyed with greenish-yellow indigo dissolved in the vat, and then the dyed fibres were exposed to air oxygen to obtain the colour of the original indigo pigment.[3]

Another natural vat dye is purpura, a brominated derivative of indigo. Purpur is a purple dye obtained by extraction from the secretion of Murex Snail *(Murrex brandaris)*, a sea gastropod living on the eastern Mediterranean coast. This colour, called Tyrian purple, was very valuable in those years due to its high light fastness and its difficult production, and as a very expensive dye, it was used in the robes of Persian, Babylonian and Roman emperors and royalty.[3–5]

The discovery of synthetic vat dyes has a history of more than 100 years. In 1901, while Rene Bohn was trying to obtain the anthraquinone derivative of indigo, he discovered this dye group by accident. Bohn obtained a water-insoluble black dye by dissolving β-amino anthraquinone in potassium hydroxide (Figure 6.2), but this dye was not found suitable for use in dyeing. Then, this mixture was treated with hydrosulphite in an alkaline medium, and it was observed that it assumed a dark blue, bright colour. With the resulting mixture, cotton was dyed and a brighter blue colour was obtained. In addition, it has been observed that the fastness values obtained are much better than those of natural indigo. Bohn named this dye "Indanthren", inspired by the words "indigo" and "anthracene".[6]

Later, the synthesis conditions were changed, the colour depth was increased and other colours were obtained. Thus, the discovery of other vat dyes was accelerated.[6] The dye range was completed in 1912. Vat dyes, the number of which reached 135 in 1980, were

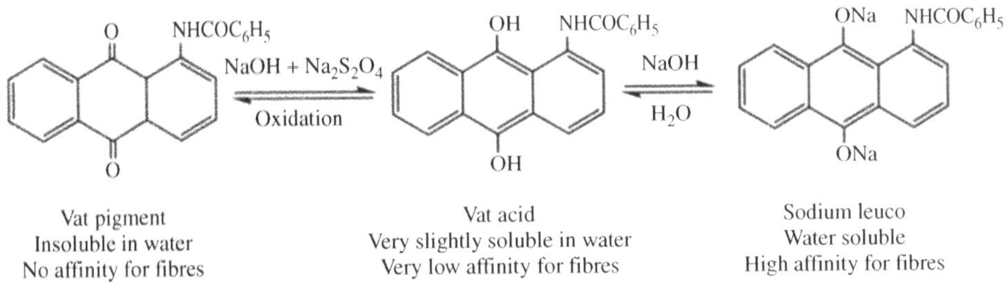

Figure 6.1 The chemical structure of vat dyes and the species formed by the reactions occurring in dyeing with them. Ref. [1]/Ege Üniversitesi Mühendislik Fakültesi Çoğaltma Yayınları.

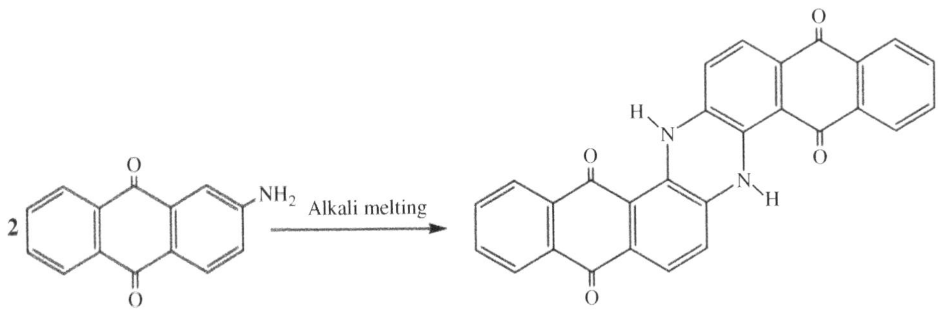

Figure 6.2 Synthesis of vat dye (Indanthren Blue RS). Ref. [6]/Dystar.

selected in terms of performance properties over time. Their number decreased to 40 in 2002.[7] Studies in recent years have focused on developing processes instead of producing new dyes. In particular, studies are underway to develop alternatives that will reduce the use of hydrosulphite (pre-reduced form dye production, reduction in nitrogen atmosphere, reduction in an ultrasonic reactor, etc.) or eliminate it (electrochemical reduction, electro-catalytic hydrogenation [ECH], etc.) in order to eliminate the disadvantages caused by the hydrosulphite used for the reduction of the dye.

6.1 Structure and Classification of Vat Dyes

Except for indigo, the vat dyes used in dyeing are mainly anthraquinone derivatives and condensed aromatic ring systems with a conjugated double bond system. They are single or mixed dyes, each containing usually two, four or six reducible keto groups. The chemical structures of some typical vat dye classes are shown in Figure 6.3.[8]

Indanthrones are blue dyes (e.g. CI Vat Blue 4, 6 and 14). They can cause problems in dyeing, as they are prone to excessive reduction and excessive oxidation. However, they still constitute one of the most important classes of vat dyes due to their impressive colour, excellent fastness and reasonable cost. Flavantrone (CI Vat Yellow 1) and its derivatives are among the oldest synthetic vat dyes like indanthrones. Despite some disadvantages (e.g. slow oxidation, poor wash fastness and tendency to photochromism), these dyes are still considered

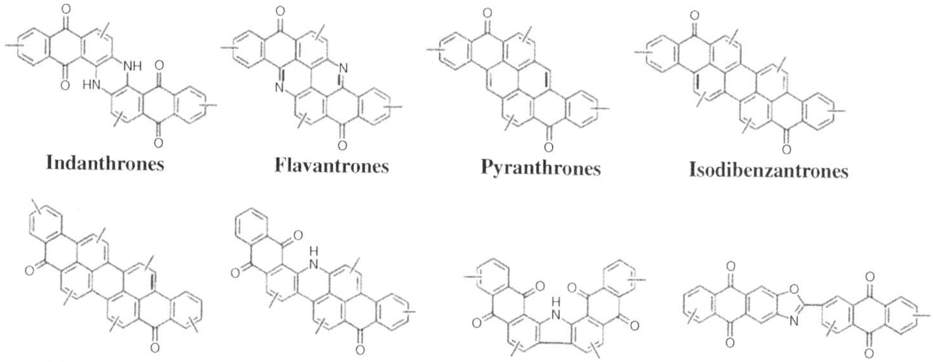

Figure 6.3 Chemical structures of some typical vat dye classes. Ref. [8] Society of Dyers and Colourists.

important because of their good light fastness. The same applies to the pyranthrone species (e.g. CI Vat Orange 2 and 9). Although these dyes have some disadvantages (low light fastness of the non-brominated product and fibre damage during dyeing), they are still widely used due to their good levelling properties, high colour strength and moderate cost.[8]

Isodibenzantrone (isoviolantrone) dyes include some deep purple colours (e.g. CI Vat Violet 1 and 9), which have high colour strength and good resistance to bleaching, but lower fastness to rubbing, hot pressing, and water staining. Dibenzanthrones (violantrones) form an important large group of dyes such as dark blues, navy blues, greens and blacks (e.g. CI Vat Blue 16, 19, 20 and 22, Green 1, 2 and 9 and Black 9). These dyes have many of the disadvantages of isodibenzantrones but have generally retained their importance because of good levelling properties, excellent fastness to light and wet processing, and because some colours cannot be produced by other means with these properties.[8]

The Benzantron acridone species represent another major class, including CI Vat Green 3 and 13 and Black 25. These are olive greens, greens, browns and greys. Although they are not easy to dye, they have excellent fastness and provide a certain degree of protection of the fibre against light-induced degradation.[8]

The class of anthraquinone carbazole dyes is of great importance for colours ranging from yellow and orange to brown, khaki and olive green (e.g. CI Vat Orange 11 and 15, Brown 1, 3 and 44, Green 8 and Black 27). These dyes are extremely important because of their good levelling properties and excellent fastness to light, wet processing and chlorine. Finally, the main representative of the anthraquinone oxazole class is CI Vat Red 10, a strong bright red with good levelling ability and excellent fastness.[8]

Other vat dyes with important structural elements include:

1. perylene tetracarboxylic diimide derivatives, e.g. CI Watt Red 23 and 32;
2. imidazole derivatives, e.g. CI Vat Yellow 46 (a bright and lightfast yellow);
3. triazinylaminoanthraquinone derivatives, i.e. reaction products of aminoanthraquinones and cyanuric chloride. CI Vat Blue 66 is a prime example with its excellent levelling properties and good chlorine fastness. Unlike blue dyes, indanthrone is not susceptible to over reduction.[8]

The chemical structure of the dye affects the properties of the leuco form in the dyeing process (e.g. heat stability in vatting, substantivity, absorption rate, diffusion into the fibre and levelling properties). It is also an important factor in determining the resulting dyeing properties, including colour and fastness properties.[8] Due to the differences between the vat dyes in terms of their molecular structures and their affinities for the fibre, there are differences between the temperatures they need for the vatting process in order to make them soluble in water. Therefore, vat dyes are divided into three classes as follows:[1,2]

- cold **(IK)**,
- warm **(IW)** and
- hot **(IN and IN Special)**.

Among these, the importance of IK dyes with low affinity has decreased today.[8] This classification is a classification method used by a German company, which is the largest manufacturer of vat dyes in the world, and various dye manufacturers can also use other letters (such as Hot: A, Warm: B and Cold: C) and numbers (Hot: I, Warm: II and Cold: III) in classification according to dyeing temperature in their catalogues.[1]

6.2 Fastness Properties of Vat Dyes

Since the vat dyes, which are reduced to their water-soluble (sodium leuco) form during the dyeing process, are oxidised again after dyeing and converted to their water-insoluble initial form (vat pigment) in the fibres, the wet fastness values of dyeing made with them (like washing, boiling with soda ash, mercerisation, hypochlorite, peroxide and perspiration) are very good. Light fastness is also outstanding in most of them.[1] The light fastness of vat dyes in light shades is around 5–6, and in dark shades 7–8.[3] Only easily reduced dyes may experience fastness problems during boiling in alkaline environment. This is true for flavanthrones (e.g. CI Vat Yellow 1) and pyranthrones (e.g. CI Vat Orange 9). In these cases, the fibre itself has a reducing effect. Vianantrone dyes (e.g. CI Vat Blue 20) are also susceptible in this respect. Increases in molecular size and the number of reducible keto groups lead to increased resistance to this effect. The problem can be avoided by using a mild oxidising agent (e.g. Ludigol). Indanthrones are sensitive to hypochlorite. In this case, excessive oxidation leads to the formation of azine groups and the dyes become greener and duller. CI Vat Blue 4 is more prone to this problem than CI Vat Blue 6.[8]

In products dyed with vat dyes, besides fading due to light, sometimes fibre damage can also be encountered. Some anthraquinone-based vat dyes in colours such as yellow, orange and red (e.g. CI Vat Orange 2) can reduce the strength of the fibres they dye under the effect of light. This is because these dyes act as an oxidation catalyst. A part of the vat dye is first reduced to the vat acid form with the effect of rays, and then this reduced part is oxidised by the effect of air oxygen, and in the meantime compounds such as hydrogen peroxide are released as shown in Figure 6.4. These released peroxides also cause oxidative damage to cellulose. Blue and violet dyes, on the other hand, do not harm the fibres as they absorb in the longer wavelength spectrum region (lower energy).[1] In fact, dyes CI Vat Green 3 and Blue 4, 6 and 14 have the opposite effect of preventing fibre degradation when exposed to light.[8]

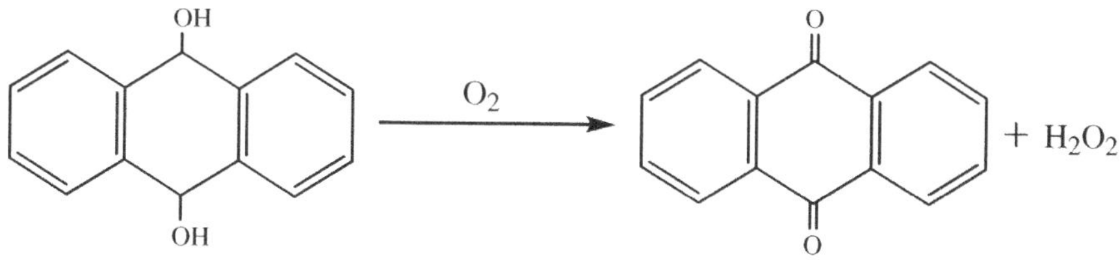

Figure 6.4 Reduced vat dye interacts with air oxygen to release hydrogen peroxide. Ref. [1]/Ege Üniversitesi Mühendislik Fakültesi Çoğaltma Yayınları.

For the reasons explained above, when a dye that damages fibres is combined with a dye that does not damage fibres and has good light fastness, the light fastness of the latter decreases. Because in this case, the peroxides released damage the dye molecules oxidatively instead of the cellulose. This issue requires attention, especially when producing green with yellow–blue combination.[1] The behaviour of combination dyeings in this respect depends on the choice of dye, the mixing ratio and colour depth, the moisture content of the material, the UV content of the exposed light, and the substrate.[8] However, unless the wrong dye combination is made, very high light fastness values can be obtained even in light colours.[1]

6.3 End-uses of Vat Dyes

Vat dyes are used for dyeing and printing of all cellulosic fibres as well as polyester/cellulose blends.[6] Although the reduced leuco form has affinity for wool and polyamide fibres, its use in these fibres has not gained importance due to the damage of wool fibres under dyeing conditions (strong alkali environment) and fastness problems in dyeing polyamide fibres.[8]

As mentioned before, reactive dyes constitute the largest class among dyes used in cellulose dyeing. This is followed by vat dyes with the second largest share. If a comparison is made in terms of value, the vat dyes again rank second after the reactive dyes.[6]

Roughly 11% of vat dyes are used in printing and 89% in dyeing. Continuous processes are the most important application area, with a share of approximately 54%. The continuing popularity of vat dyes is due to the fact that their properties cannot be obtained with any other dye group.[7]

Vat dyes generally exhibit high fastness properties. Due to its high wet fastness, it is used in products resistant to heavy washing conditions, in dyeing industrial, military and medical textiles, sewing thread and similar materials.[2] In addition, this dye group is used in dyeing textile materials to be used in construction of awnings, tarpaulins, umbrellas, tents, military and civil uniforms, etc., since the products coloured with vat dyes have high resistance to external weather conditions (such as light and humidity).[6] Also, due to its high fastness properties, it is used in high-quality outerwear, sports and daily wear, shirting and upholstery fabrics, as well as yarn dyeing for woven and knitted fabrics.[7]

In light of this information, the advantages and disadvantages of vat dyes can be summarised as follows.

Advantages

- ✓ Very high fastness values.
- ✓ Absence of dye hydrolysis.
- ✓ Not sensitive to pretreatment defects (such as H_2O_2 residue on the fabric).
- ✓ Less water consumption compared to reactive dyeing.
- ✓ Shorter dyeing time as less washing is done compared to reactive dyeing.[2]

Disadvantages

- ✓ Dye is expensive.
- ✓ Not suitable for bright nuances (red, turquoise).
- ✓ Low wet rubbing fastness in dark colours.
- ✓ Higher cost of chemicals and auxiliaries than reactive dyeing.
- ✓ Wastewater load (dye, hydrosulphite, caustic soda, auxiliary and adsorbable organo halogens [AOX]).[2]

In Table 6.1, usage areas of vat dyes are given in comparison with reactive dyes.

6.4 Forms of Vat Dyes

Vat dyes are available in the market in powder, paste, granule or liquid forms. They also have different particle sizes and concentrations within themselves. In powder brands, dye pigments are coarser than in other forms and therefore require longer vatting times.

Table 6.1 Comparison of usage areas of vat and reactive dyes (++: High, +: Medium, –: Low).

	End-use	Vat dye	Reactive dye
Yarn	Coloured weaving	++	+
	Coloured knitting	+	++
	Sewing-embroidery thread	++	+
Weaving	Outerwear/t-shirt	++	++
	Casual wear	+	++
	Sportswear	+	++
	Upholstery	++	+
	Workwear	++	–
	Civil uniform	++	–
	Military uniform	++	–
	Medical textile	++	–
Knitting	Casual wear	–	++
	Sportswear	+	++
	Outerwear/evening dress	–	++
	Military underwear	++	+

Ref. [6]/Dystar.

This is because the contact surface on which the reducing agent acts is smaller compared to dye forms consisting of finer particles. Powder brands cannot be used in methods in which the dye is transferred to the product in the form of pigments without vatting.[1]

Fine powder, micro powder and paste brands have smaller pigments than normal powder brands. However, they are not always suitable for pigmentation methods.[1]

In brands such as colloisol (powder, paste and liquid colloisols), microdisperse powder, ultradisperse powder, dye particles are present in the finest form. The use of these brands in the dyeing of the wound yarns according to the pigmentation method in the apparatus and in all continuous methods gives the best results.[1]

6.5 Dyeing with Vat Dyes

With vat dyes, dyeing can be done by both exhaustion and impregnation methods. As the size of dyeing lots decreases, dyeing by exhaust method becomes more important. Woven fabrics made of cellulosic fibre or polyester/cellulose mixtures can be dyed with dyes in jet or overflow or yarns in bobbin dyeing apparatus. Pre-pigmentation method is the most commonly used method in bobbin dyeing for concentrations up to 3%. The conventional leuco process is more suitable for depths above 3%.[3]

In general, continuous methods are preferred for dyeing large batches with vat dyes. All fabrics made of cellulosic fibres and polyester/cellulose blends can be dyed using the pad-steam process. Pretreatments (e.g. desizing, scouring and bleaching) are essential for a uniform dyeing. Good reproducibility also depends entirely on these processes.[3]

6.5.1 Dyeing by Exhaust Method

Dyeing made by exhaust method with vat dyes consists of making the dye soluble in water (vatting), dyeing and post-treatment (rinsing, oxidation, soaping and rinsing). Below is information about all the steps of dyeing separately.

6.5.1.1 Vatting and Dyeing

The process of converting vat dyes into water-soluble form is called "vatting". Vatting is done by using reducing agent (hydrosulphite) and alkali (sodium lye) (Figure 6.5). Depending on the type of dye, the amount of reducing agent and alkali that should be used and the vatting temperature and duration vary.[2]

As can be seen from the reaction given above, vat dyes in the form of water-insoluble pigments are converted to water-soluble leuco form by reduction with hydrosulphite in alkali medium (in the presence of caustic soda).[1] The rate of the reaction is determined by

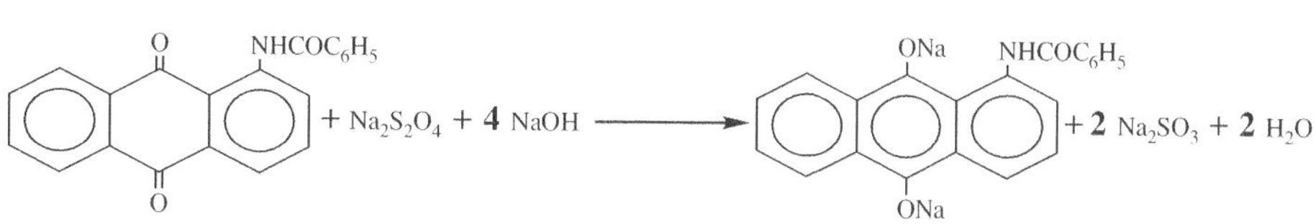

Figure 6.5 Reduction reaction of vat dye with hydrosulphite in alkaline medium. Ref. [1]/Ege Üniversitesi Mühendislik Fakültesi Çoğaltma Yayınları.

the relationship of the oxygen atom of the keto (C=O) groups in the structure of the dye with the reducing agent. In a heterogeneous system, the dissolved reducing agent attacks the dispersed dye particles. This group disintegrates when the double bond electrons in the keto oxygen leave the bond. Finally, slippage occurs at the double bond to stabilise the dye in its enolate form. Trials showed that vatting rate:

- doubles with an increase of 10 °C in temperature,
- does not depend on hydroxyl ion concentration above pH 12,
- increases as the reducing agent and dye concentration increases and
- depends on the crystal form and particle size of the dye.[8]

Vatting of dyes can be done in two different ways:

- either in the dyeing liquor **(in a long liquor)**
- or separately with a smaller volume of concentrated solution **(stock vat)**.[2]

Some vat dyes are suitable for vatting only in a long liquor, while good results can be obtained with some vats only in a stock vat. In both cases, the dyes must first be mixed with water; this process is done in different ways depending on the form of the dye. In normal powder brands, the dye is made into a paste by mixing with wetting agent and hot water (50 °C). In liquid colloisol brands, the dye can be easily diluted with water at the desired rate.[1]

a. ***Vatting in a long liquor:*** The dyeing bath is heated up to 25–60 °C depending on the type of dye, and the dye dispersion mixed with water is added to the liquor by filtering and stirring. Then the required amount of caustic soda and hydrosulphite are added to the liquor. If the liquor is mixed well, vatting takes 15–20 minutes for powder brands, 10–15 minutes for fine powder and paste brands, 5–10 minutes for colloisol, etc., brands.[1]

   In the vat dye catalogues, on the page of each dye, the class of the dye (IK, IW, IN or IN Special), which vatting method or methods are suitable for (vatting in a long liquor, vatting in a stock vat), the method that can be used for its oxidation (water rinsing, air oxidation, peroxide oxidation, potassiumbichromate/acetic acid oxidation and hypochlorite oxidation) are indicated. A few dyes specified in the catalogues must definitely be vatted in a long liquor.[1]

   In the dye catalogues, the concentrations of caustic soda, hydrosulphite and salt (only for IK and IW classes) to be used in dyeings to be made according to method of vatting in a long liquor for IK, IW and IN class dyes are given in separate graphs. In these graphs, the liquor ratio is 1:2 to 1:40 for dyeing, the abscissa dyeing depth (%) and the ordinate is the caustic soda (38 °Bé) (mL/L), hydrosulphite (g/L) and salt concentration (g/L) shows.[1]

   For a better understanding of the subject, it will be more useful to explain the calculations for dyeing according to vatting in a long liquor method with an example:

   Example: Use the Indanthren catalogue to make the necessary calculations for dyeing the 100 kg cotton material in 1:20 liquor ratio with an IK class dye at a concentration of 2% according to **vatting in a long liquor method**.

   First, the caustic soda concentration graph in the catalogue for the IK class is opened, and the point with 2% dyeing depth in the abscess is marked on the graph. This point is then aligned with the straight line corresponding to the liquor ratio of 1:20. When a perpendicular is drawn from this point to the ordinate where it intersects the line, the figure (5.6 for these values) gives the caustic soda concentration (38 °Bé solution) in mL/L that should be used for this dyeing.

Similarly, for the IK class, the hydrosulphite concentration graph in the catalogue is opened, and the point with 2% dyeing depth in the abscissa is marked on the graph. This point is aligned with the line corresponding to the liquor ratio of 1:20. The figure obtained by drawing a perpendicular to the ordinate at the intersection (2 for these values) gives the hydrosulphite concentration in g/L that should be used for this dyeing.

Finally, the salt concentration graph in the catalogue for the IK class is opened, and the point with 2% dyeing depth on the abscissa is marked on the graph. This point is aligned with the line corresponding to the liquor ratio of 1:20. The figure obtained by drawing a perpendicular to the ordinate at the intersection (22 for these values) gives the salt (sodium sulphate) concentration in g/L that should be used for this dyeing.

A prescription is issued by performing the procedures described above. To dye 100 kg material in a 1:20 liquor ratio, the liquor volume will be 2000 L. When the calculation is made considering the liquor volume and chemical concentrations, the required amounts of hydrosulphite, caustic soda and salt are 4 kg., 11.2 L and 44 kg, respectively. Since the dyeing depth is 2%, The amount of dye to be used for 100 kg. material is 2 kg. Thus dyeing is initiated by taking 1988.8 L of water (= 2000 L–11.2 L), and dissolving 4 kg hydrosulphite, 11.2 L caustic soda (from a 38 °Bé solution) and 44 kg. sodium sulphate in it. After a while, dye is added and the material is subjected to the dyeing process at 25–30 °C (IK class) for 60 minutes. In Figure 6.6, a sample dyeing graph and dyeing recipe for a dyeing process to be carried out with vat dyes in jets or overflows are given.

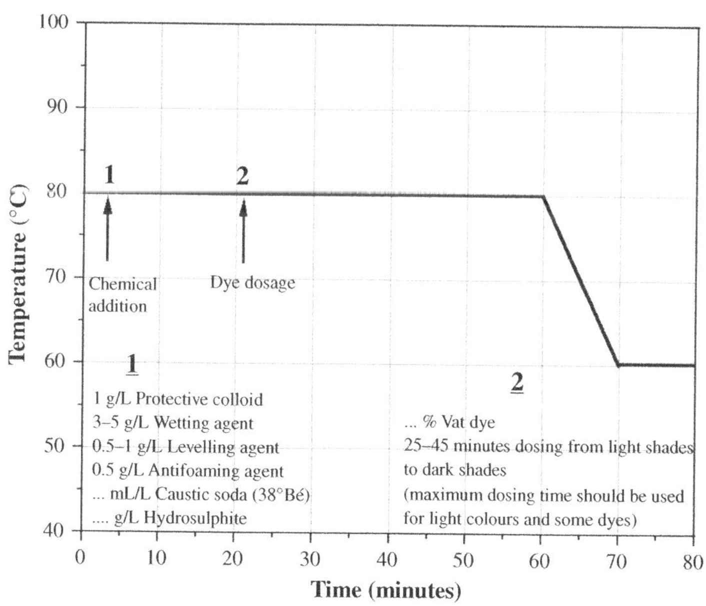

Figure 6.6 Dyeing graph and dyeing recipe for a dyeing process to be carried out with vat dyes in jets or overflows. Ref. [9]/Dystar.

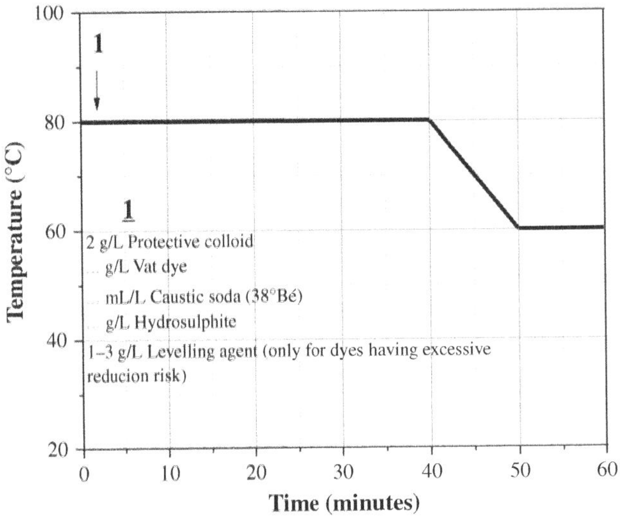

Figure 6.7 A sample dyeing graph and dyeing recipe used in bobbin dyeing at depths greater than 3%. Ref. [9]/Dystar.

In Figure 6.7, a sample dyeing graph and dyeing recipe are given for a dyeing process to be carried out with vat dyes in bobbin dyeing apparatus at depths greater than 3%.

b. ***Vatting in a concentrated stock vat:*** In this method, after the dye is concentrated in a separate container, the solution containing the vatted dye is poured into the main dyeing liquor containing some caustic soda and hydrosulphite by filtration. It is stated in the dyeing catalogues that some dyes are suitable only for dyeing according to the stock vatting method. Apart from this, it is a safer method compared to the first one, especially when powder brands are used. In addition, when working according to special dyeing methods such as temperature steps and vat acid, which will be explained later, method of vatting in a concentrated stock vat is preferred.

When nuance adjustment is to be made, the nuancing dye to be added should be added to the liquor after being vatted in the stock vat.[1]

Due to the differences in the structures of the dyes, the vatting recipes in the stock vat vary. For this reason, when dyeing with any dye is intended according to method of vatting in a stock vat, first, the page belonging to that dye should be found from the dyeing catalogue, and it should be checked whether the said dye is suitable for vatting in a stock vat, and if appropriate, according to which recipe (no. 1, 2 or 3) it is suitable for vatting. In Table 6.2, the amounts of caustic soda, hydrosulphite, water, and vatting temperatures and times in the three main vatting recipes are given.[1]

Depending on the dyeing depth, the required amount of dye is vatted in water containing hydrosulphite and caustic soda in the amounts specified in the table, at the temperature and time specified in the table. In the meantime, it is beneficial to add auxiliaries (e.g. protective colloid substance) to the solution in order to increase the durability of the stock vats.[1]

After the dye is vatted in a stock vat, the dyeing process is started. For this purpose, a dyeing recipe is prepared as described above in the dyeing section according to the method of vatting in a long liquor. From all the amount of hydrosulphite and caustic soda that is specified to be put into the liquor in the recipe, the amounts consumed for vatting in a stock vat are subtracted, and the remaining amounts of hydrosulphite and caustic soda are added to the liquor heated up to the starting temperature of dyeing. After filtration, the dye solution prepared in the main vat is added. Then, the material to be dyed is loaded into the device, and it is subjected to the dyeing process for 60 minutes at a temperature suitable for the dye (25–60 °C).

For a better understanding of the subject, it will be more useful to explain the calculations for dyeing according to method of vatting in a concentrated stock vat solution with an example. For this, for the example given above, let's calculate the recipe for dyeing according to method of vatting in a stock vat.

Example: Make the necessary calculations using the Indanthren catalogue for dyeing 100 kg cotton material in a liquor ratio of 1:20 with an IK class dye at a concentration of 2% **according to method of vatting in a stock vat**. (The dye is suitable for vatting with the recipe no. 2 given in Table 6.2.)

First, the amount of dye needed for 2% dyeing of 100 kg material is calculated, and the result is 2 kg. The amounts of hydrosulphite conc. (kg), caustic soda (38 °Bé) (L) and water (L) given in Table 6.2 are the amounts required for vatting 1 kg of dye. Therefore, these values are multiplied by 2 to get results for 2 kg of dye. According to this, for 2 kg dye, 1.5 kg hydrosulphite conc., 6 L caustic soda (38 °Bé), 100 L water are required.

For the dyeing process, first, 2 kg of dye is vatted in 100 L water containing 1.5 kg hydrosulphite conc. and 6 L of caustic soda (38 °Bé) at 60 °C for 10 minutes. Then a recipe is issued for vatting in a long liquor. Dyeing recipe for vatting in a long liquor was already prepared above for this dye and the amounts of hydrosulphite, caustic soda and salt required to dye 100 kg of material at a liquor ratio of 1:20 were found as 4 kg, 11.2 L and 44 kg, respectively. Since there are 2 kg of dye, 1.5 kg hydrosulphite conc. and 6 L of caustic soda (38 °Bé) in the stock vat, these values should be subtracted from the amounts that need to be added to the main dyeing liquor. According to this; hydrosulphite conc. amount is: 4 – 1.5 = 2.5 kg, caustic soda (38 °Bé) amount is: 11.2 – 6 = 5.2 L, water amount is: 2000 – 100 – 11.2 = 1888.8 L. Therefore, 1888.8 L of water is taken and 2.5 kg of hydrosulphite, 5.2 L of caustic soda (38 °Bé) and 44 kg of sodium sulphate are dissolved in it, then the temperature is adjusted to 25–30 °C

Table 6.2 The amount of caustic soda (38 °Bé), hydrosulphite (concentrated), water and vatting temperatures and times in the three stock vat recipes.

Amount of dye (kg)	Hydrosulphite amount (kg)	Amount of caustic soda (L)	Amount of water (L)	Temperature (°C)	Time (min.)
1	0.5	1.5	50	50	10
1	0.75	3	50	60	10
1	1.5	6	100	60	10

Adapted from Ref. [9].

(IK class) and the stock vat solution is added in it and the material is subjected to dyeing for about 60 minutes.

6.5.1.1.1 Chemicals Used in Vatting and Dyeing Processes and Their Functions

Since there are differences between the molecular sizes and affinities of the vat dyes toward the fibres, some of them are dyed at low temperatures and some at high temperatures. Likewise, some require less reducing agent or alkali, while others require more reducing agent or alkali. Again, some dyes require salt, while others do not need salt in order for the dye in the liquor to be sufficiently absorbed by the fibres.[1]

The amount of reducing agent (hydrosulphite), alkali and salt used according to the dye type and dyeing temperatures are given in Table 6.3.

As a result of X-ray analysis, the average dimensions of the vat dyes are as follows:

- Cold dyes (IK): 8–10 Å
- Warm dyes (IW): 10–20 Å
- Hot dyes (IN): 20–40 Å.

Since the dimensions of the spaces (areas that can be easily penetrated) between the micelles in cotton fibres are up to 20–40 Å when the fibres are swollen, the most suitable sizes for the aggregates that can enter these areas are the particles formed by the combination (association) of the following numbers of molecules:[1]

- Cold dyes (IK): 4–5 molecules
- Warm dyes (IW): 2–4 molecules
- Hot dyes (IN): 1–2 molecules.

Therefore, in order to make a level but efficient dyeing, since the degree of association is already low in small molecule (8–10 Å), IK dyes, it should be ensured that the degree of association is not further decreased by keeping the temperature low. On the other hand, since the degree of association is high in IN dyes with large molecules (20–40 Å), the aggregates should be broken down as much as possible by raising the temperature. IW class dyes are between IK and IN class dyes in terms of molecular size and dyeing temperature.[1,2]

As it is known, in order for the dye molecules and aggregates adsorbed on the surface of the fibres to diffuse into the easily permeable regions of the fibres, their sizes must be slightly smaller compared to the sizes of these easily permeable (amorphous) regions.

Table 6.3 Dyeing temperatures and hydrosulphite, alkali and salt usage amounts according to dye type in vat dyes.

Dye class	Dyeing temperature	Hydrosulphite amount	Alkali amount	Salt amount
IK	20–30 °C	Low	Low	High
IW	40–50 °C	Medium	Medium	Low
IN	50–60 °C	High	High	—
IN special	60–80 °C	High	High	—

Adapted from Ref. [2].

Substantivity is optimal when the size of the aggregates matches the size of the easily permeable regions, because although diffusion of small molecules and aggregates to these regions is easiest, it is also easier for them to come out without fixation. Large aggregates, on the other hand, cannot fit inside without breaking down. Large aggregates must be broken down (deaggregated) in order to penetrate into the fibre. This is because the diameter of the aggregate must be smaller than the diameter of the fibre pores in order for the dye molecules to diffuse into the fibre, that is, large aggregates cannot diffuse into the fibre. Since the dissociation of large aggregates will require time, the high degree of aggregation causes the diffusion rate to be slow. Therefore, diffusion is the main step that determines the rate of dyeing. The most effective method to reduce the degree of association is to raise the temperature. When the temperature is increased, the hydrogen bridges that hold the dye molecules forming the aggregate together are partially broken down. On the other hand, easily permeable regions become larger due to the mobility of cellulose macromolecules with the effect of temperature. For this reason, hot dyeing is done with large-molecule IN-class dyes with high aggregation. In IK class dyes with a low degree of association, it is not necessary to raise the temperature, since there is no problem of dissociation of the aggregates. Dyeing with these gives better results in the cold.[1]

Therefore, dyeing conditions should be chosen to increase the degree of association slightly for dyes with small molecules and low association degree. This is why dyeing process in small molecule vat dyes is done at cold (20–30 °C). When dyeing rayon and viscose (in some special cases, cotton) only with IK class dyes, it is recommended to start the dyeing process at 40–50 °C, to take advantage of the better levelling ability at this temperature, and then to continue the dyeing process in the cold in order to maximise the amount of dye taken by the fibres.[1]

For dyes with large molecules and high degree of association, the dyeing conditions should be chosen to reduce the degree of association. This requires dyeing process at hot (50–60 °C or 60–80 °C for IN special dyes) for large molecule IN class vat dyes. The tone, nuance and even some fastness values of the dyeing change in dyeing made at high temperatures with some vat dyes. In order to prevent such situations, the highest temperatures that can be applied with each dye are also specified in the catalogues. Some blue and green dyes (Indanthren Blau RS and the like) are sensitive to over reduction. This sensitivity of them increases at temperatures above 60 °C. However, it is possible to work with these dyes up to 80 °C by slightly increasing the amount of hydrosulphite and adding sodium nitrite equivalent to 15–20% of the hydrosulphite amount in the liquor, along with the same amount of triethanolamine.[1]

In the explanations above, the effect of temperature on diffusion has been taken into account in determining the dyeing temperature for sufficient dye uptake and uniform dyeing. An issue that may come to mind here is that as the adsorption will decrease in case the temperature is increased, the dynamic equilibrium of "dye remaining in the dye liquor ↔ dye taken up by the fibres" may shift to the liquor side, resulting in a decrease in the total amount of dye taken. This will mean a decrease in dyeing efficiency. However, this is valid if it is waited until the equilibrium is established. If the dyeing is finished before the equilibrium is established, the total amount of dye taken is still high even though the temperature is increased. This can be clearly seen from Figure 6.8.[1]

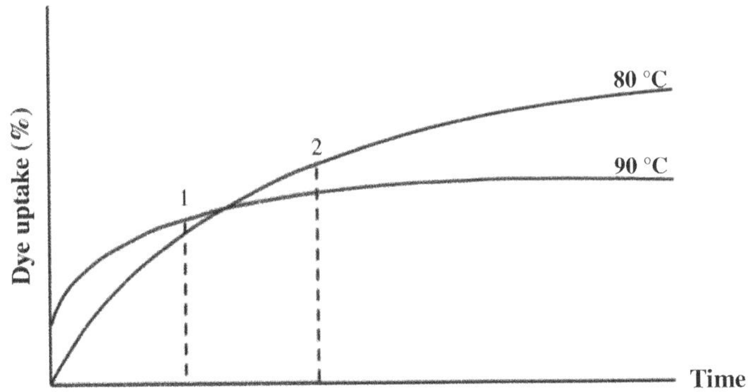

Figure 6.8 The uptake curves of an IN vat dye by the fibres at two different temperatures. Ref. [1].

When the dyeing is terminated at the point marked with (1) in Figure 6.8, the amount of dye taken up at 90 °C is higher than at 80 °C, due to the high diffusion and therefore the overall dyeing rate. However, if the dyeing is extended well and terminated after equilibrium to occurs (at the point marked with 2), the amount of dye taken in dyeing at 90 °C will be less than that taken at the lower temperature.[1]

There is a close connection between the molecular structures of vat dyes and the chemical amounts (hydrosulphite, caustic soda, salt, etc.) needed during dyeing. It is useful to explain these relationships here in order to understand the reason for the operations in the application.[1] For this reason, the chemicals used in dyeing processes with vat dyes, their functions and appropriate usage amounts are explained separately below.

a. Reducing agent

The leuco potential of all vat dyes under standard conditions is between −650 and −1000 mV. Therefore, satisfactory vatting (reduction to the enolate leuco form) can only be achieved with a reducing agent that has a more negative reduction potential than that. The reducing agent commonly used in dyeing with vat dyes is sodium hydrosulphite, called "hydro". The chemical formula of hydrosulphite, also known as sodium dithionite, is $Na_2S_2O_4$. Other reducing agents are used for special purposes in vat dyeing, or they are not widely accepted in practice. These include hydroxyalkylsulphinates, which find use in vat printing and dyeing at temperatures above 100 °C. Thiourea dioxide is a strong reducing agent but is as sensitive to atmospheric oxygen as hydrosulphite in alkaline solutions. Although freshly prepared alkaline solutions of thiourea dioxide appear relatively stable, there is rapid formation of formamido sulphinate. In this form it is more easily oxidised than hydrosulphite. Sodium borohydride, on the other hand, reacts too slowly for vat dyeing. There are also vatting processes using the easily biodegradable chemical dihydroxyacetone.[8]

The stabilisation of hydrosulphite added to the dyeing liquor as a reducing agent in aqueous medium is high if there is no air in the medium. However, with the increase in temperature, if atmospheric oxygen is present, the decomposition of hydrosulphite increases rapidly. In this case, sodium dithionite is oxidised by consuming sodium hydroxide and decomposed to form sodium sulphite and sodium sulphate. Therefore, in practice, more sodium ditionite is used than is required for dye reduction, with the aim of stabilising the dyeing system.[10] Below is the reaction for the decomposition of sodium dithionite in alkaline medium (in the presence of sodium hydroxide) by reaction with atmospheric oxygen.

Alkaline environment:
$Na_2S_2O_4 + 2NaOH + O_2 \rightarrow Na_2SO_3 + Na_2SO_4 + H_2O$

In practice, the reducing agent content in the dyebath should be such that it can rapidly and economically reduce all vat dyes to a water-soluble form. While the alkaline hydro solution is mostly decomposed for this purpose, a certain amount is decomposed by reacting with the air oxygen as shown above.[3] This decomposition is especially greater at high temperatures. Therefore, the amount of reducing agent (hydrosulphite) required by the vat dyes mainly depends on the effect of the dyeing temperature they require (which depends on the molecular structures of the dyes) on the oxidative decomposition of hydrosulphite with air oxygen.[1] In addition, the amount of reducing agent needed also depends on factors such as:

- the number of reducible groups in the structure of the dye,
- the molecular weight of the dye and
- pure dye content of the dye.[8]

Since the dyeing process with IN class vat dyes is carried out at high temperatures, the hydrosulphite decomposition with atmospheric oxygen is more and therefore the amount of hydrosulphite required to be added to the liquor is higher than the IK class vat dyes. This difference is due to the fact that the oxidative decomposition of hydrosulphite increases rapidly with increasing temperature.[1] This is clearly seen in Figure 6.9.

In a liquor containing 10 g/L hydrosulphite and 20 mL/L NaOH (38 °Bé), 10% of the hydrosulphite is decomposed at 20 °C after 60 minutes, while 75% is decomposed at 98 °C. Of course, these numbers may be slightly more or less depending on factors such as the movement of the liquor and the size of the surface in contact with the air. Therefore, taking

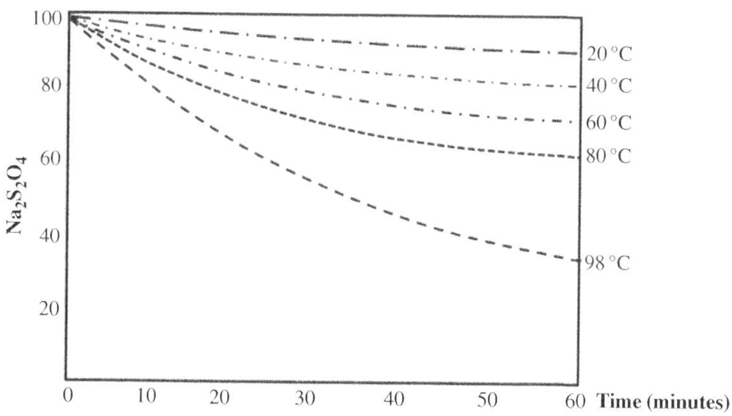

Figure 6.9 Decomposition curves of hydrosulphite over time at different temperatures. Ref. [1].

into account that some of the hydrosulphite will be consumed by atmospheric oxygen when dyeing with vat dyes, extra hydrosulphite should be added to the liquor.[1] The amount of this excess depends on the application conditions, especially the effect of some factors that determine the oxidation rate:

- **Temperature:** With all other conditions constant, oxidation accelerates as the dyeing temperature increases.[8]
- **Dithionite concentration:** The more concentrated the initial dithionite solution, the longer it takes for the dithionite to become inactive, given other conditions being constant.[8]
- **Movement of liquor and air:** The more the liquor moves in the presence of atmospheric oxygen, the faster the degradation takes place. This is especially valid for hank dyeing machines, winch-becks and apparatus.[8] As the movement of the liquor and the material, and therefore the contact with the air, increases, especially when working at high temperatures, the oxidative decomposition of the hydrosulphite, hence the consumption amount of the hydrosulphite, increases. This is why machines with extensive air contact are not very suitable for dyeing with vat dyes.[1]
- **Specific surface area:** The larger the specific surface area (i.e. the ratio of surface area to volume) of the liquor, the faster the sodium dithionite is oxidised. Since the time required for half oxidation is inversely proportional to the specific surface area, oxidation can take hours in a large vessel with a relatively small surface area. On the other hand, due to its high specific surface area, dithionite liquor impregnated on a fabric may undergo oxidation within 30–60 seconds depending on its concentration and prevailing conditions.[8]
- **Presence of dyes:** All other conditions being constant, the reducing agent oxidises more rapidly in the presence of leuco dyes.[8]

On the other hand, under the conditions normally applied in practice (pH 12–14), the oxidation of sodium dithionite does not depend much on the alkali content of the liquor.[8]

Sodium thiosulphate and sodium bisulphite in neutral medium when in contact with water while the anhydrous hydrosulphite is completely stable; in alkaline environments, it hydrolytically decomposes to form sodium thiosulphate and sodium sulphite. The reactions related to the hydrolytic decomposition of sodium dithionite are given below.[8]

Neutral environment:

$$2Na_2S_2O_4 + H_2O \rightarrow Na_2S_2O_3 + 2NaHSO_3$$

Alkaline environment:

$$2Na_2S_2O_4 + 2NaOH \rightarrow Na_2S_2O_3 + 2Na_2SO_3 + H_2O$$

As can be understood from the above reactions, especially at high temperatures, not only oxidative degradation occurs, but also degradation by hydrolysis. For example, when hydrosulphite is heated at 100 °C for 1 hour in a nitrogen atmosphere, 17–18% decomposition occurs, and when heated in contact with air, 77–78% decomposition occurs. That is, 17–18% of the degradation is due to hydrolytic and 60% to oxidative degradation. Although it is possible to reduce the degradation of hydrosulphite up to 10% with additions such as triethanolamine, it would be more appropriate to use other reducing agents instead of hydrosulphite in dyeing at high temperatures.[1]

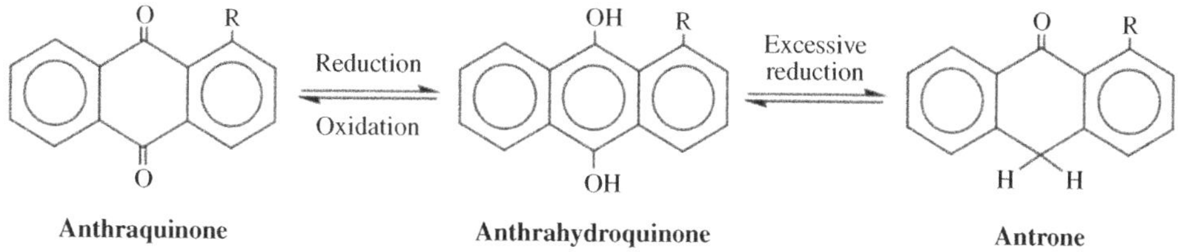

Figure 6.10 Reactions related to excessive reduction of vat dyes. Ref. [1]/Ege Üniversitesi Mühendislik Fakültesi Çoğaltma Yayınları.

The most important point during dyeing with vat dyes is the use of reducing agent as necessary. If less reducing agent is used than needed for the reduction of the dye, vatting cannot be performed efficiently, and as a result, it becomes difficult to obtain an efficient and uniform dyeing. On the other hand, using more reducing agent than necessary causes "over reduction", which creates a problem. Over reduction, as shown in Figure 6.10, is the further reduction of the anthraquinone dye, which has been converted to the anthrahydroquinone (vat acid) form during the reduction process, resulting in the formation of antrone derivatives instead of remaining in the vat acid form. Since it is difficult for these antrone derivatives to be oxidised again to form the original dye the dye that has been over-reduced deteriorates.[1]

An indicator paper called "vat yellow" or "Indanthren yellow" is used to see if there is enough hydrosulphite in the liquor. The colour of the vat yellow paper should turn blue within three seconds at the latest after being dipped in the liquor. If it does not change colour, it is understood that there is not enough hydrosulphite in the liquor, and hydrosulphite is added to the liquor. Since there must be sufficient hydrosulphite in the liquor during the entire dyeing period, it is necessary to check the liquor frequently with vat yellow paper. The addition of 1 L of caustic soda (38 °Bé) is also essential with the addition of every 1 kg hydrosulphite conc.[1]

Since there is a risk of over reduction in some dyes (especially Indanthren Blau RS, etc.), it is necessary to pay special attention not to exceed the amount of hydrosulphite required. The presence of excess hydrosulphite than desired in the vatting liquor can only be determined by measuring the redox potential of the liquor. This measurement is made with the help of a potentiometer and the values are in mV. The potential required for the optimal vatting of a dye is called the "leuco potential". The potential in the vatting liquor is due to the electrons released in the oxidation of hydrosulphite to sodium sulphite. These electrons provide the reduction of the vat dye.[1] For this reason, in the classical dyeing system of vat dyes, the wastewater contains decomposition products of dithionite as well as alkali. In particular, sulphite and sulphate, which are dithionite decomposition products, are formed in high concentrations. This situation causes the following problems:

- Bacterial poisoning in wastewater due to sulphites.
- Abrasion of concrete in anaerobic processes in wastewater due to sulphites and sulphates.

Hydrosulphite and its oxidation products not only cause increased pollution in wastewater but also cause indirect damage by accumulating on the device and pumps.[10]

Hydrosulphite is still a very important and inexpensive reducing agent used in the traditional vat dyeing process and is always overused for process safety. For this reason, it is tried to optimise the amount of hydrosulphite with ready-made recipe calculation programs.[3]

The introduction of dyebaths and rinsing water into the canals after vat dyeing causes various environmental problems because the reducing agents required for dyeing are oxidised to species that are very difficult to regenerate. Therefore, many entrepreneurs are trying to replace environmentally unsuitable sodium dithionite with ecologically more suitable alternatives. Among these alternatives, the first thing that comes to mind is the replacement of sodium dithionite with an organic reducing agent (glucose, α-hydroxy ketones, etc.) whose oxidation products can be biodegraded.[11]

In order to reduce the amount of hydrosulphite needed by increasing the vatting efficiency the vatting process can be carried out in an ultrasonic reactor. Ultrasound reduces the average molecular size of vat dyes by breaking down large molecules. In the study conducted with a total of seven vat dyes in three different chemical structures, benzantrone, vialontrone and anthraquinone-based, it was determined that ultrasound reduced the molecular size of each dye. Before ultrasound application, vat dyes may contain particles larger than 14 microns, whereas even the largest particles are smaller than 2 microns when ultrasound is applied.[12] As a result, the reaction rate increases in vatting, because the particle diameter decreases, the active surface area where the reaction takes place increases.

To reduce or eliminate the use of hydrosulphite beyond the methods mentioned above, there are two more possibilities:

- Electrochemical reduction and
- Electrocatalytic hydrogenation

Information about these methods is provided below..

I. Electrochemical reduction

Electrochemistry is an advantageous method for the reduction of dyes because it minimises the consumption of chemicals and reduces the waste load. Efforts are still underway to get the reduction rate and efficiency high enough to make electrochemical reduction industrially viable. The step limiting the rate of electrochemical reduction is electron transfer from the cathode surface to the surface of the dispersed dye pigment by solid–solid contact.[13]

Electrochemical reduction can be done in two different ways: direct and indirect.

- ***Direct electrochemical reduction:*** This method is based on the mechanism of the formation of a radical anion as a result of a reaction between the dye and the leuco dye, followed by the electrochemical reduction of this radical. The amount of leuco dye required to initiate the reaction is formed by a conventional reaction (e.g. by adding some water-soluble reducing agent). However, once the reaction has started, no further reducing agent is required, and the subsequent steps of the reaction proceed autonomously.[11, 14]

For the success of the process in industrial application, the choice of electrode materials plays an important role. It is stated that copper is the most suitable material tested so far in terms of reaction rate, whereas titanium is not a suitable material for this process due to its very low current density. Lead, on the other hand, shows a good catalytic property for radical reduction but produces a higher current than other materials due to its dissolution during electrolysis.[11]

- ***Indirect electrochemical reduction:*** Indirect electrolysis is based on the principle that the reduction process is carried out using a regenerative mediator (electron transfer provider) (iron II complexes) at the cathode. Complexes formed by iron with triethanolamine or gluconic acid are used as mediators.[13]

  In electrochemical reduction, the reducing agent is electrons from the cathode instead of the electrons of the dithionite. The iron complex acts as a mediator in this system and reduces the dye in its divalent form. Meanwhile, Fe (II) is oxidised to Fe (III), and then Fe (III) is reduced back to Fe (II) in the cathode section of the electrolysis cell. Fe (II) then reduces the vat dye again.[15]

  In the literature, it is stated that the dyeing results obtained as a result of the reduction of vat dyes with the indirect electrochemical reduction technique are comparable to the classical method using sodium dithionite.[16]

II. Reduction by ECH

Reactor performance is still low in both direct and indirect electrochemical reduction methods, and there is a need for the use of some chemicals. ECH method has been developed recently in order to increase the ecological compatibility of the dyeing process. While working with this method, there is no need to use any reducing agents, and therefore, the waste load due to chemical consumption is reduced. ECH is based on the electrochemical reduction of water to generate adsorbed hydrogen (H_{ad}).[17]

$$H_2O + e^- \rightarrow H_{ad} + OH^-$$

This adsorbed hydrogen reacts with the organic substrate (dye) on the surface of an electrically conductive powder metal catalyst (e.g. Raney nickel or black platinum).[17] Studies show that the reduction processes of vat dyes can be done by using hydrogen activated by palladium (Pd) catalyst in alkaline liquor at room temperature.[18]

b. Alkali

Another substance required for making the dye soluble in water (vatting) in dyeing with vat dye is alkali. Caustic soda (sodium hydroxide) is used as an alkali. The pH of the liquor should be around 12–13. Otherwise, a water-insoluble enolic acid leuco compound may form, which has no affinity for fibres.[8]

In vat dyeing, just as the molecular sizes of the dyes and their associated degrees of aggregation are taken into account when determining the dyeing temperature, these factors are also considered when selecting the type and amount of alkali. Because, in addition to temperature, one of the most important factors affecting the degree of association is the type and amount of alkali. Association in vat dyes occurs as a result of the formation of H bridges between phenolic hydroxyl groups in the vat acid form of the dye.[1]

As the strength and concentration of the alkali used in dyeing with vat dyes increase, the presence of dye molecules in the form of leuco form instead of the vat acid form increases. This is clearly seen in Figure 6.11.[1]

Since the association occurs as a result of the formation of H bridges between the phenolic hydroxyl groups in the vat acid form of the dye, the degree of association and,

Figure 6.11 The transformation of the dye in the form of vat acid to leuco form with the effect of alkali. Ref. [1]/Ege Üniversitesi Mühendislik Fakültesi Çoğaltma Yayınları.

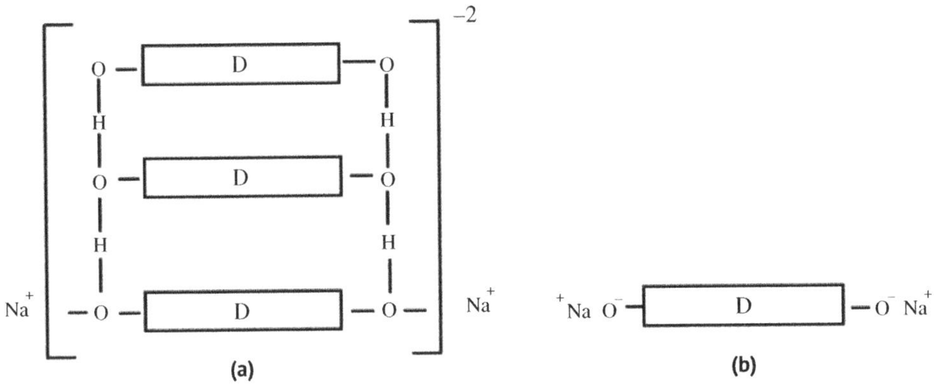

Figure 6.12 (a) An aggregate formed by the combination of three dye molecules and (b) a single dye molecule. Ref. [1]/Ege Üniversitesi Mühendislik Fakültesi Çoğaltma Yayınları.

accordingly, the adsorption decreases as the number of free phenolic hydroxyl groups decreases in the presence of alkali. Depending on the strength and amount of alkali, as the number of —OH groups in the structure of dye molecules decreases, and the number of —ONa groups increases, the degree of molecular association and aggregate formation decreases.[1]

In the formula given in Figure 6.12a, an aggregate containing three molecules is depicted. Here, the whole aggregate has a "−2" valence. However, as can be seen in the formula in Figure 6.12b, a single molecule also has a "−2" valence. This shows that as the degree of aggregation increases, the negative valence per unit molecule decreases. Since cellulose fibres are (−) loaded, the higher the negative valence of the dye, the more difficult the adsorption becomes. As a result, as the degree of aggregation increases, the negative valence decreases and the adsorption rate increases.[1]

Since alkali reduces association in vat dyeing, using weaker or smaller amounts of alkali in IK class (cold) dyes with small molecules and a low degree of association prevents excessive dissociation of aggregates. Conversely, strong or concentrated alkali will completely dissociate small aggregates into individual molecules. On the other hand, it is tried to reduce the degree of association by using stronger or more alkali in the IN class dyes with large molecules and high degree of association. Because in these dyes, in order to convert the free phenolic hydroxyl (—OH) groups into sodium salt (—ONa), the H bridges must first be broken. This can be achieved by using high concentrations of strong alkalis. IW class dyes fall between the IK and IN dyes in terms of alkali amount used.[1,2]

Table 6.4 summarise the percentages of dye taken up by the fibres during 2% dyeing at 45 °C for 45 minutes with dyes of low and high degree of association by using a strong alkali (NaOH) and a weak alkai (Na_3PO_4).[1]

Table 6.4 Percentage of dye taken up by the fibres in dyeings with low and high degree of association dyes by using strong and weak alkalis.

Dye	Degree of association	10 mL/L NaOH (38 °Bé)	20 g/l Na_3PO_4
Indanthren Gelb 5K	Low	33%	73%
Indanthren Oliv T	High	86%	55%

Adapted from Ref. [1].

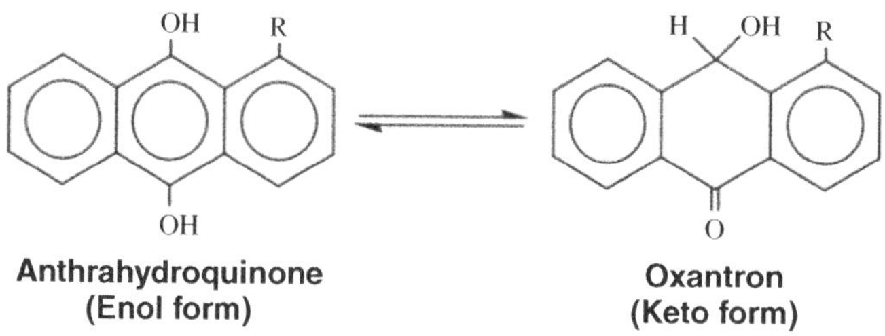

Figure 6.13 Reaction of conversion of enol form of vat dye to keto form. Ref. [1]/Ege Üniversitesi Mühendislik Fakültesi Çoğaltma Yayınları.

When Table 6.4 is examined, it is seen that the dye with a low degree of association is more readily absorbed by the fibres when dyed in the presence of weak alkali, whereas the dye with a high degree of association is better absorbed by the fibres when dyed in the presence of strong alkali. It is generally not objectionable for the amount of caustic soda in the liquor to be slightly less or more than the amount calculated in the prescription. A small deviation from the calculated caustic soda amount is generally acceptable; however, some dyes are sensitive to over- or under-alkalisat (Indanthren Blau RS and IN Special group dyes). Such dyes are indicated in the catalogues.[1]

Addition of excess alkali and high vatting temperature cause undesirable saponification and therefore defective dyeing. For example, the acyl group of acylamino anthraquinone dyes can be removed as a result of saponification. Generally, dyes with a tight molecular structure (condensed ring structure) require more alkali than those with an open structure. As the number of carbonyl groups to be reduced in dye molecules increases, the alkali requirement also increases.[1] In addition, atmospheric oxygen should be taken into account in determining the amount of caustic soda. This is because caustic soda is consumed during the reaction of hydrosulphite with atmospheric oxygen, as well as during the vatting process.[8] Using more or less alkali than necessary can have negative effects. The use of less alkali causes the enol form formed under normal conditions to turn into keto form, as shown in Figure 6.13, the keto form is difficult to revert back to the enol form or to be oxidised.[1]

c. Salt

In vat dyeing, neutral salts such as sodium sulphate or sodium chloride are used to increase the affinity of the leuco dye to the fibre.[8] The use of salt increases the degree of association, increasing the rate of adsorption and uptake of dye. Because, as can be seen in Figure 6.14, the addition of salt to the liquor reduces the dissociation of the dye in the form of sodium

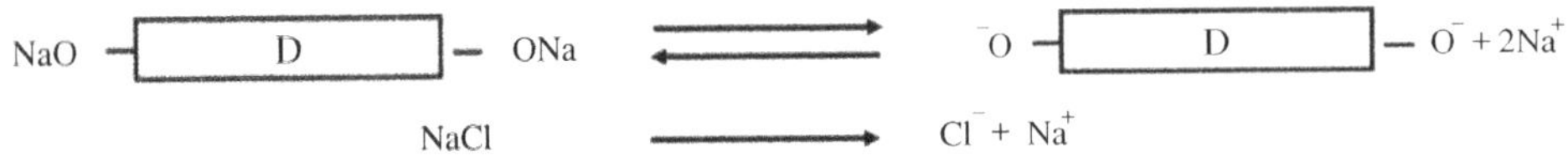

Figure 6.14 The effect of salt on the dissociation of the dye in the leuco form.

leuco according to the action-opposite reaction (Le Chatelier) principle, in which case the equilibrium of "dye remaining in the liquor ↔ dye taken up by the fibres" shifts to the fibre side, as the total negative charge will decrease. However, in case of using more salt than necessary, too much association will slow down the diffusion, and dyeing difficulties may arise.[1]

As can be understood from the explanations above, for IK class dyes with low association, salt should be added to increase the rate of adsorption and therefore the amount of dye taken from the liquor. On the other hand, there is no need to use salt for IN class dyes with already high association. In fact, it is harmful because the aggregates are too large, making diffusion difficult. IW class dyes are among the IK and IN dyes in terms of the amount of salt usage.[1,2]

The salt amounts stated in the catalogues are generally for calcined sodium sulphate and table salt. If crystal sodium sulphate is used, the amount should be double that as stated in the catalogue. For dyeing mercerised cotton, half the amount of salt specified in the catalogues should be used. No salt is required for dyeing rayon and viscose. It is recommended to add 5–10 g/L salt towards the end of dyeing only for very dark shades. In addition, it is beneficial to add salt half an hour after the start of dyeing in order to ensure that the dyeing is uniform and to prevent the risk of precipitation of the sodium leuco compound in light-coloured dyeings, dyeing mercerised cottons and dyeings where IN and IW dyes are combined.[1]

d. Auxiliaries

In vat dyeing, depending on the situation, it is beneficial to add in the liquor auxiliary chemicals such as:

- if the material to be dyed has not undergone adequate pretreatment, wetting agent to emulsify the waxes in the raw cotton and to ensure adequate penetration of the dye liquor into the substrate,
- dispersing agent to prevent agglomeration of undissolved particles in the liquor,
- a sequestering agent (complexing agent) to retain alkaline earth ions from the process water and cotton and
- levelling agent (non-ionic products that form complexes with leuco dyes) to ensure uniform dyeing.[8]

6.5.1.1.2 *Obtaining Uniform Dyeing with Vat Dyes*

For uniform dyeing with vat dyes, first, considerations should be given to such matters:

- the material to be dyed should be thoroughly pretreated,
- dyes having high post-levelling ability should be used and
- dyeing conditions should be adjusted optimally.[9]

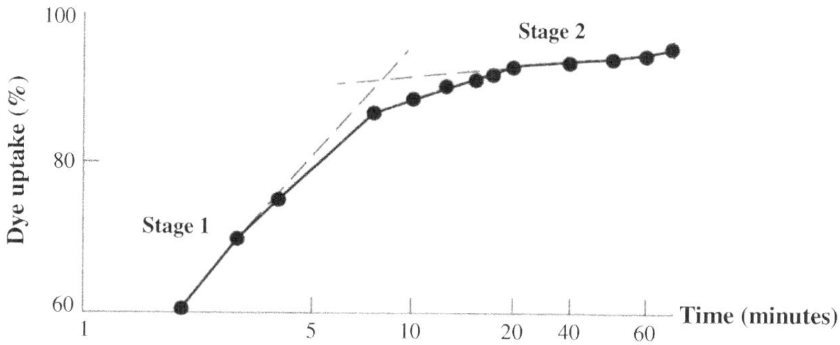

Figure 6.15 Exhaustion behaviour of a leuco vat dye. Ref. [8]/with permission of Society of Dyers and Colourists.

When a cellulosic textile material is introduced into the alkaline dyeing liquor containing leuco vat dye, the dye is absorbed from the liquor into the fibre until the equilibrium state is reached after a certain time. The higher the substantivity of the dye towards the fibre, the higher its concentration in the fibre at equilibrium and the lower it is in the dyebath. Rapid exhaustion even at low temperatures, is characteristic of vat dyes. The dyeing kinetics study shows that the dye is absorbed in two stages (Figure 6.15).[8]

Most of the dye (usually 80–90%) is exhausted in about 10 minutes (stage 1), after which the rest is slowly taken up (stage 2). As a result, there is a danger of uneven dyeing of the substrate after the first stage, where the dye is initially absorbed by the more accessible outer layer of the fibre. This causes ring dyeing. The second stage is the diffusion of the dye into the inner core of the fibre. At this stage, the dye concentration in the outer layer of the fibre drops again, and only a small amount of dye can be taken up from the bath. Therefore, there are two stages in vat dyeing kinetics: the first of these is characterised by the rapid exhaustion of the liquor, largely due to the substantivity of the leuco dye for the fibre, and the latter by the diffusion of the dye into the interior of the fibre.[8]

In the first stage, the dye exhaustion rate is almost independent of the liquor circulation rate. On the other hand, it is significantly affected by the textile material. For example, the larger the surface area of the substrate in contact with the dyebath, the faster the first step will be completed. In addition, experimental studies have shown that the initial dye uptake speed is increased by low dye concentration, short liquor ratio and higher dyeing temperature. In practice, the faster the dye is absorbed onto the fibre at this initial stage, the greater the risk of obtaining an uneven dyeing.[8]

The penetration of dyes into the fibres takes place in the second stage. The rate of this is determined by the diffusion of the leuco dye from the outer layer of the fibre to the inner part. Dyes with good levelling properties have higher penetration values than those with limited levelling ability. The diffusion rate is highly dependent on the dyeing temperature and increases exponentially with increasing temperature. Studies have shown that dyes with similar structure and therefore similar substantivity for cellulosic fibres diffuse more slowly (e.g. CI Vat Blue 20 compared to CI Vat Green 1) when dimerised in the liquor. Such dyes also do not level easily, so the levelling property is dependent not only on the substantivity but also on the diffusion rate.[8]

If a certain amount of cellulosic fibre is put into a leuco dye solution of a known concentration and a constant temperature, dyeing equilibrium is reached after a while, that is, the dye is distributed at a certain ratio between the solution and the fibre. Dye uptake usually ranges from 85 to 95%. The substantivity of the dye for the fibre depends on the dye structure and does not appear to be associated with the tendency to form leuco dye aggregates in the liquor, especially since it has been found that dyes which tend to form dimers in solution (e.g. CI Vat Blue 20) are mostly present in the monomolecular form once they have exhausted onto the fibre.[8]

Since the primary reason for the uneven uptake of the dye is the rapid uptake of the dye by the fibres, the rate of uptake of the dye should be slowed down in order to increase the uniform uptake. Three important factors affecting the rate of dye uptake can be listed as:

- temperature,
- levelling agent and
- the affinity of the dye for the fibres.[1]

Therefore, the precautions to ensure uniform vat dyeing can be examined in three groups. Details regarding these are given below.

a. Setting the temperature

To ensure slow uptake, the temperature must be low, and to ensure post-levelling, the temperature must be high.[1] In the absence of non-ionic levelling agents, leuco dyes present a risk of excessive dye uptake at the beginning and consequently uneven dyeing, as mentioned earlier. This risk is especially great when dyeing mercerised cotton and especially regenerated cellulosic fibres, as well as dyeing in pastel colours and short liquor ratios. If the liquor is taken up too quickly, even at room temperature, while other parameters (i.e. substrate, dyeing equipment, recipe) are constant, the only thing that can usually be done to achieve uniform dyeing is to provide conditions that promote levelling. As noted earlier, the rate of diffusion of dye into fibre increases significantly with increasing temperature. At the same time, substantivity decreases, resulting in an increase in dye concentration in the liquor. This facilitates the migration of the leuco dye as shown in Figure 6.16. Figure 6.16 illustrates the dye transfer to the liquor and the undyed adjacent fabric at certain time intervals when fabric, when the fabric dyed with 4% CI Vat Green 1 dye is processed with a blank vat dyeing liquor at different temperatures.[8]

Although levelling accelerates with increasing temperature, it is quite clear that much more dye remains in the dyebath at 95 °C compared to 60 °C. Therefore, the temperature needs to be lowered in order for the colour to reach its maximum depth after the levelling phase.[8]

Even dyes which are rapidly taken up by the fibres at normal temperatures, slow down considerably at 12–18 °C. If the temperature is slowly increased to the normal dyeing temperature after a while, it can be ensured that the dyeing does not take too long and the post-levelling ability can be increased. The name of the dyeing method based on this principle is "Temperature Steps".[1]

In the temperature steps method, vat dye is used, and the initial temperature is kept low to ensure uniform dye uptake. The dye vatted in the main vat is filtered and placed in the cold liquor (containing calculated amounts of caustic soda and hydrosulphite). After the

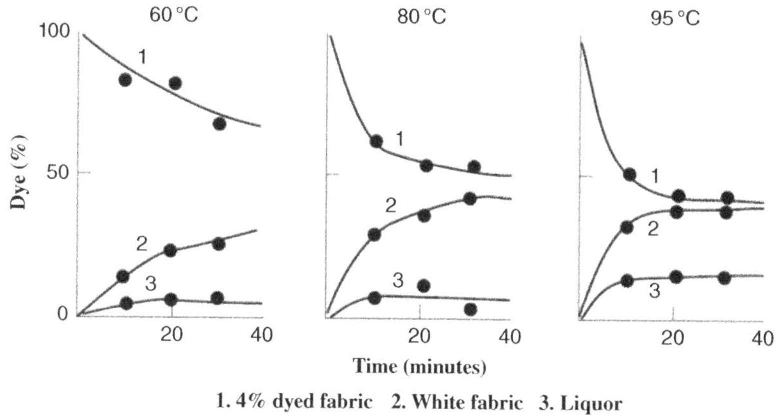

Figure 6.16 Levelling of CI Vat Green 1 dye at various temperatures. Ref. [8]/with permission of Society of Dyers and Colourists.

dyeing process is done at 18 °C for 15–20 minutes, the temperature is raised at a rate of 1 °C/min up to 35 °C and 2 °C/min from 35 °C to 65 °C (for IN type vat dye). The dyeing process is carried out for 30 minutes at this temperature. After dyeing, the fabric sample is subjected to rinsing–oxidation–soaping–rinsing processes, respectively.[1]

Sometimes, despite the use of the above-mentioned special dyeing methods and the addition of auxiliaries to the liquor, it may not be possible to obtain optimal dyeings in terms of levelness and penetration into the fibres with vat dyes at normal (up to 80 °C) temperatures. Especially in mercerised cotton, regenerated cellulose and linen products, the difficulties encountered due to the high affinity of the dye for the fibres can be eliminated in most cases only by dyeing at HT conditions (90–115 °C). By raising the temperature above 90 °C, the diffusion rate of the vat dye molecules and aggregates increases, and thus the levelling properties are also improved. Thus, even with dyes with poor levelling ability, it is possible to obtain uniform dyeings that have penetrated into the fibres under HT conditions. In addition, the affinity of most of the vat dyes for cellulose fibres decreases at high temperatures. For this reason, it is possible to correct uneven dyeing by working under HT conditions. In dyeing performed under HT conditions, the colour of cotton or linen is partially whitened by the effect of the alkaline reducing vatting liquor. Especially if a complexing agent is added to the liquor, a very good bleaching effect is obtained. Thus, it is possible to dye the raw product, which is not very dirty, directly in light shades under HT conditions. The challenge of working with vat dyes in HT conditions is finding a suitable reducing agent that works well and is stable at these temperatures. Since some vat dyes are sensitive to high temperatures, it is not recommended to dye them by HT method. Dye manufacturers specify such dyes in their catalogues.[1]

HT leuco dyeing method is recommended for dyeing medium and dark shades. First, the dyes are vatted with caustic soda and suitable reducing agent at 60–90 °C (60 °C for dyes that are at risk of over reduction) for 5–10 minutes. Dyeing begins after adding glucose (in dyes that are at risk of over reduction) and other auxiliaries. The liquor temperature is increased as quickly as possible to the dyeing temperature (95–115 °C) and dyeing is continued at this temperature for 30–60 minutes. In dyeing by HT method, after the process carried out at the dyeing temperature, the temperature must be reduced before the dyeing is completed. As it is known, the temperatures at which various dyes are absorbed most by the fibres are different. Cooling to this temperature (50–80 °C) is beneficial in ensuring that as much dye as possible is absorbed from the liquor by the fibres. It may be recommended to cool at a rate of 2 °C/min. After the liquor is cooled to 80–90 °C, it is beneficial to add 1–2 g/L hydrosulphite to the liquor, especially to improve rubbing fastness. When working with IW dyes by HT method, salt can be added to the liquor. It is recommended to add salt towards the end of dyeing as a measure to increase levelness.[1]

b. Using levelling agent

Unless long dyeing times are not acceptable, the problem of uneven dyeing cannot always be solved by adjusting the temperature alone. In this case, it is inevitable to use levelling agents that can form complexes with dyes. The purpose of adding such auxiliaries is to shift the dye uptake equilibrium in favour of the dyebath to delay the rate of dye uptake by the fibre or to promote the migration of dye that has already been taken up.[8]

These auxiliaries include alkylene oxide or imine polymers that can complex with leuco dyes. Although this interaction normally occurs with monomolecular leuco dyes, it can also occur with dyes that form dimer in aqueous media.[8] Levelling agents, which are used in dyeing with vat dyes and have affinity for dyes, combine with dye molecules and particles, reducing their mobility on the one hand, and blocking the forces that provide their affinity for the fibres, on the other. As a result, they reduce the rate of dye exhaustion.[1] In addition, they help uneven dyeing to be levelled at high temperatures.[8]

The formation of a dye-auxiliary complex most likely begins with an ion-dipole attraction (similar to the interaction between leuco dye and cellulose) between the hydrated dye anion and the strongly hydrated polar polymeric chain. These relatively strong forces bring the dye and polymer chain molecules together. Thus, the π-electron system of the dye can interact more with the dipoles of the polymeric substance. These bonding forces depend on the size of the dye molecules and their planar structure. The strength of the interaction between the dye and the auxiliary determines whether the auxiliary acts as a stripping agent (e.g. polyvinylpyrrolidone) or as a levelling agent (e.g. fatty alcohol polyoxyethylene). Because of the similarity in the forces exerted on the dyes by the substrate and the auxiliaries, the auxiliaries generally have poor levelling properties and a pronounced effect on dyes with high substantivity. However, such products show only limited retarding effect when used dyes with lower substantivity and good levelling properties. These auxiliaries can be used in higher concentrations when particularly necessary (e.g. for dyeing light colours and dyeing at low liquor ratios). If necessary, the retarding effect of the auxiliaries can be reduced by raising the temperature.[8]

c. Starting dyeing with the form of the dye with low affinity towards the fibre

Some of the various forms of some dye classes have low or no affinity for fibres. This is also the case for vat dyes. While the sodium leuco form has a high affinity for fibres, the vat acid (anthrahydroquinone) and normal pigment (anthraquinone) forms have no affinity. From this point of view, the "vat acid method", in which dyeing is started with the vat acid form of the dye, and the "pre-pigmenting" and "semi-pigmenting" methods, where dyeing is started with the pigment form, have been developed. These methods are described separately below.[1]

I. **Vat acid method:** In this method, the vat dye is first vatted in a concentrated form in another container (main vat), and when acetic acid is added to the dissolved (sodium leuco form) dye, sodium leuco form converts into vat acid (anthrahydroquinone), as shown in Figure 6.17. Since this is insoluble in water, it creates dispersion with the help of the appropriate dispersing agent in the medium.[1]

   Since the vat acid has no affinity for the fibre, when the textile material is treated with it, it is evenly distributed over the material. In the next step, when the material is contacted with caustic soda (it is beneficial to add some hydrosulphite), the vat acid turns into sodium leuco and is fixed upon penetrating the fibres. In the procedures according to the "vat acid temperature steps" method, the dye vatted in the main vat is placed in a cold liquor containing acetic acid and dispersing agent. After being treated at this temperature for 15–20 minutes, the temperature is increased and dyeing is continued for 20–30 minutes.[1]

II. **Pre-pigmentation method:** This method is used for dyeing the wrapped yarns with IW and IN class dyes.[1] In this process, the vat dye is distributed on the textile material in insoluble form, that is, before vatting, as evenly as possible.[8] The textile material to be dyed is treated with a liquor containing dye pigment and auxiliaries (wetting agent, dispersing agent, levelling agent) for 10–15 minutes at 60–80 °C (at 60 °C with dyes that are at risk of over reduction).[1] If necessary, salt is added to the liquor to support uptake. Alternatively, the HT process can be used. If a brighter colour is desired on a previously unbleached ground, pre-pigmentation can be combined with peroxide bleaching.[8] After adding the calculated amount of caustic soda, hydrosulphite and auxiliaries, dyeing is continued for another 45 minutes at 60–80 °C. However, the addition of sodium nitrate should not be omitted if dyeing will continue at temperatures above 60 °C with dyes that are at risk of over reduction.[1] Pre-pigmentation can be performed on any discontinuous dyeing device.[8]

   Then pre-pigmentation method is recommended if the dyeing depth is less than 3%, that is, for light and medium shades. A sample dyeing profile and dyeing recipe for this are given in Figure 6.18.

   In the dyeing of fabrics on a jigger, dyeing can be done according to a kind of pigmentation method. In the jigger-pigmenting method, the dry fabric is passed through

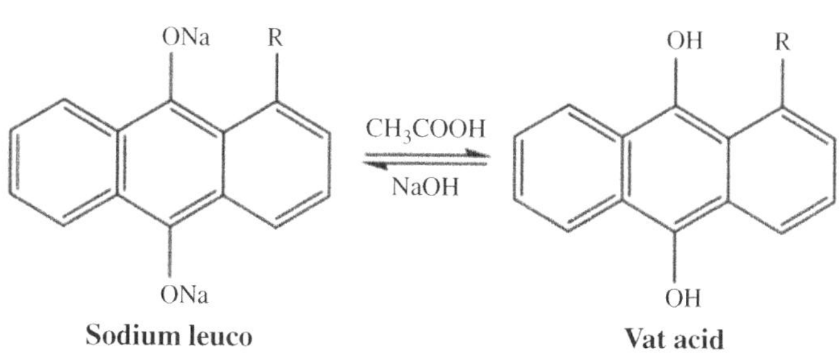

Figure 6.17 Conversion of the sodium leuco form of the vat dye to the vat acid. Ref. [1]/Ege Üniversitesi Mühendislik Fakültesi Çoğaltma Yayınları.

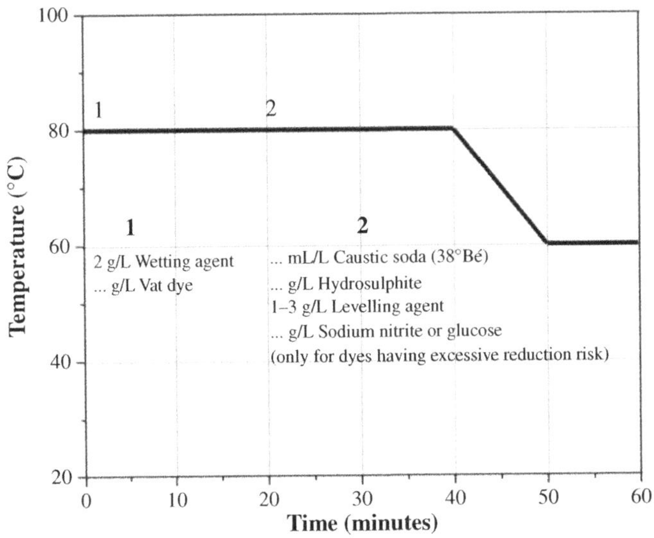

Figure 6.18 An example dyeing graph and recipe for dyeing according to the pre-pigmentation method used in bobbin dyeing at depth below 3%. Ref. [9]/Dystar.

the pigment dispersion in two passages at 20 °C and two passages at 40 °C (more in small lots) on the jigger. In the meantime, care should be taken to keep the liquor ratio as short as possible. After the pigmentation is finished in this way, the liquor level in the jigger is adjusted to the liquor ratio (1:3–1:5) and heated up to the dyeing temperature. Calculated amount of caustic soda and hydrosulphite are added. In order for the ends of the dyeing to be of the same darkness, the vatting of the dye remaining in the jigger liquor must be finished before the first passage. Therefore, after the addition of caustic soda and hydrosulphite, a waiting period of a few minutes is necessary before the fabric passage is started. If it is necessary to add salt, it is recommended to put half of it after the second and the other half after the third passage.[1]

III. **Semi-pigmentation method:** This process is based on the fact that the vatting rate is slower at lower temperatures. The dyebath is prepared at 15–20 °C with dyes and chemicals, and dyeing is started. The temperature is then gradually increased, and during this time the dye, which has not yet been vatted, disperses within the substrate in an insoluble form.[8] By raising the temperature, the vatting rate also increases, and the dye is vatted where it is deposited on the fibres. Thus, better penetration into the fibres and more uniform dyeing is obtained. In this method, dyeing process starts at 20 °C with a liquor containing dye, caustic soda, hydrosulphite and auxiliaries. After the process is carried out for 15 minutes at this temperature, the temperature is increased to 60 °C and dyeing is continued for further 20–30 minutes.[1]

The HT semi-pigmentation method is recommended especially for dyeing of very light and medium shades. According to this method, the dye pigment, caustic soda, HT stable reducing agent and all auxiliaries are added to a cold liquor and heated as quickly as possible to the dyeing temperature (95–115 °C) and dyeing is carried out at this temperature for 30–60 minutes.[1]

According to the pigmentation methods, the dyes to be used in the studies should be the brands that contain the dye in the form of the finest particles (such as colloisol, fine powder, microdisperse powder and ultradisperse powder). The uniform distribution of pigment or vat acid in the liquor depends, on the one hand, on the fineness and uniformity of their particles in the liquor, and on the other hand, on the uniformity of the dyeing machine and the material to be dyed. For example, in windings with different densities, the liquor will pass through the less dense parts where the resistance is the least, and therefore the pigments will accumulate in these areas more. This may lead to uneven dyeing. Therefore, it is necessary to pay attention to the aforementioned issues.[1]

6.5.1.2 Oxidation and Aftertreatments

After dyeing is finished, oxidation process is carried out in order to convert the dye in leuco form, which was taken up by the fibres, into water-insoluble vat pigment. However, it is of great importance that the excess liquor (containing reducing agent and alkali and unfixed dye) on the product is removed by squeezing, suction or rinsing before oxidation. Otherwise, the pigment particles as a result of oxidation that are mechanically bonded to the fibres will cause low washing and rubbing fastness values of dyeing. The oxidation process can be done in various ways depending on the dye used:

- By hanging in the air or air suction
- With running water
- With peroxides (hydrogen peroxide, sodium perborate and sodium percarbonate)
- With potassium bichromate and acetic acid
- With sodium chlorite and acetic acid
- With sodium hypochlorite.

It is stated in the catalogues, either as lists or on the page of each dye, which of these methods should be used to oxidise the dye and which should not.[1]

In order to remove the caustic soda residues on the fabric, vatting is done after the oxidation is completely finished, especially in fabric dyeing. However, when methods using acetic acid during oxidation are applied, no neutralisation with acid is required.[1] Because in this case, neutralisation is also achieved.[3]

Atmospheric oxygen is actually an oxidant sufficient to convert the reduced vat dye into its original pigment form. However, in practice, such an oxidation will proceed slowly, as it is necessary to oxidise the product quickly and uniformly, and it can be used in bobbins, wound fabrics, etc. A non-uniform oxidation may occur in the products due to the air infiltration problem.[3] For this reason, the use of chemical oxidising agents is generally preferred. In the past, oxidation could be done with sodium chlorite in an acidic medium or with sodium hypochlorite in an alkaline environment. However, nowadays, the use of chlorinated compounds is avoided due to the problem of AOX.[2]

For oxidation with sodium chlorite, a liquor containing 1.5–4 g/L sodium chlorite and 6–12 mL/L acetic acid (50%) is used. The pH of the liquor should remain below 5 during the entire oxidation period. Oxidation with sodium hypochlorite is done in cold and alkali medium. Immediately after the dyeing liquor is drained, oxidation begins with the liquor containing 1–3 g/L active chlorine and 1–3 g/L dispersing agent. Antichlorination is then performed. Dyeings oxidised in this way should not be treated with acid, and it is beneficial to improve their rubbing fastness by soaping them first at 60 °C and then in boiling liquor, as in naphthols.[1]

Today, the most ecologically suitable oxidising agent is hydrogen peroxide. However, for some vat dyes, the oxidation potential of hydrogen peroxide remains very high under certain conditions, and excessive oxidation may occur in this case. This is an oxidation beyond the oxidation of the dye to the pigment form, and in such a case the sodium salt of m-nitrobenzene sulphonic acid can be used as an alternative oxidising agent. Although this substance is a weaker oxidant, it is expensive, and it is possible to use it in an alkaline environment.[3]

Oxidation of IW and IN dyes does not present much difficulty. They can be rinsed before oxidation, even with dyes such as Indanthren Blau RS, an immediate rinsing after dyeing will greatly increase the brightness of the dyeing. After dyeing with IW and IN dyes in hank dyeing apparatus, the yarns are rinsed and oxidised by one of the other suitable oxidation methods.[1] After dyeing the packages in medium (1–3%) and dark (>3%) colours in the apparatus, firstly, reductive washing (with a liquor containing 2 mL/L caustic soda (38 °Bé), 2 g/L hydrosulphite and 1 g/L dispersing agent at 20 °C for 5 minutes) is carried out, followed by oxidation, soaping and rinsing. In light shades (<1%), oxidation, soaping and rinsing are done directly without a reductive washing step.[3] In the jigger, after dyeing is finished, the fabric is overflow rinsed and oxidised in two to four passages with a suitable oxidising agent.[1]

Oxidation of IK dyes is generally difficult. If they are rinsed before they oxidise, some of the dye will bleed because their affinity for cellulose fibres is reduced. As a result of this, the shade of the dyeing becomes lighter, and even the levelness of the dyeing may deteriorate since this lightening in the windings will be different between the outer and inner parts.[1]

After oxidation, the fabric is subjected to soaping process in order to develop the final shade of the dyeing and to ensure high fastness values. This process is done by boiling for half an hour in a liquor containing 1–2 g/L calcined soda and 2–6 g/L soap. Although it is easy to perform an adequate soaping in jets, hank dyeing machines and hank dyeing apparatus, it is more difficult in jiggers and open-width washing machines. It is recommended that the jiggers be enclosed and the number of passages be increased. It is stated that in soapings made continuously in open-width washing machines, after impregnating the fabric with soda ash and washing agent in a vessel, steaming in a steamer or in 1–2 closed vessels for 0.5–2 minutes at 102–105 °C gives good results.[1]

During the aftertreatment, the dye portion mechanically bound to the fibres washes out, while the dye particles within the fibres combine to form larger aggregates. For this reason, the colour shade may change somewhat and the fastness values may increase slightly. Since some dyes have a high ability to form large particles, if rayon is dyed with them, the fibres may become partially dull.[1]

6.5.2 Dyeing by Impregnation Method

A prerequisite for all continuous dyeing processes with vat dyes is that the fabric should have good absorbency so that it wets out rapidly and uniformly. Since the differences in substantivity of the leuco dyes have less influence in padding method, selection of the dyes is not as limited as in exhaust dyeing processes. The pad-steam method is the most reliable and popular continuous dyeing process for vat dyes. The fabric is impregnated with the dye dispersion and then immediately dried. In a second padding process, alkali and reducing agent (usually hydrosulphite) are applied, then steamed for 40 seconds with saturated steam at a wet temperature of 98 °C and a dry temperature of 103 °C to reduce the dye and fix it to the fibre. This is followed by oxidation, rinsing and soaping in the open-width washing machine.[8]

Semi-continuous dyeing methods involve at least one continuous process, namely impregnation of the material with a dye dispersion (usually combined with an intermediate drying process). Subsequently, processing is carried out in various discontinuous dyeing machines for the vatting of the dye and its fixation to the fibres. Most of the time jigger is used for this. These processes provide well-penetrated dyeings with good levelness.[8]

6.5.2.1 Dyeing by Pad-steam Method

It is the most used continuous dyeing method for vat dyes. This method also has different application forms in itself. These can be listed as:

- Two-bath pad-steam method
- One-bath pad-steam method
- Wet steam method
- Modified wet steam method
- Standfast method
- Williams unit method
- Continuous quick fixation method.[1]

Today, only the first of these has found widespread use. However (although some of them are not used at all today), information about all of these methods will be given below.

6.5.2.1.1 The Two-bath Pad-steam Method

The dyeing process in this method consists of the following steps:

a. Impregnation with dye dispersion
b. Intermediate drying
c. Impregnation with chemical liquor
d. Steaming
e. Aftertreatment (such as rinsing, oxidation and soaping)

Below is detailed information about these steps.

a. Impregnation with dye dispersion
Impregnation with dye dispersion is done on padding mangles.[1] In principle, any type of padding mangle is suitable, provided that it has a homogeneous squeezing effect. The coating of both rollers must be of the same hardness (60–70° Shore). Various designs are available to compensate for cavitation in the squeezing rollers under load. The most important prerequisite for faultless dyeing is a uniform squeezing between the squeezing rollers. This can only be achieved if there is no change in absorbency across the width and length of the fabric. For this reason, pretreatment processes should

also be carried out continuously in an open-width form. The dye dispersion should have a neutral pH.[8] Having a low pick-up value (60–80% in cotton, around 70–90% in regenerated cellulose) is beneficial in reducing the risk of migration in the intermediate drying step.[1] The smaller the dye particles, the greater the tendency to migrate.[8]

For dyeing by impregnation method, the brands in which the dye pigments are in the finest powder form should be used. Special fine powder brands are poured in water 10–20 times their weight at 40–60 °C, while colloisol liquid brands are mixed into water at 20–30 °C. With cold or warm water, the liquor is made up to a volume close to the main volume. Auxiliaries (wetting agent, protective colloid and antimigration agent) are diluted with cold or warm water in a separate container and poured into the liquor shortly before use. The liquor prepared in this way is filtered and filled into the padding mangle. Different temperatures (20–30 °C, 40–50 °C and even 80 °C) are reported by various dye factories as the appropriate impregnation temperature.[1]
temperatures

The fabric should be padded and dried in one continuous operation.[8]

b. Intermediate drying

Drying is usually carried out in two stages. The fabric is first pre-dried in one or more infrared (IR) channels to a moisture content of 25–30%. Hot-flue or cylinder dryers are generally used for the main drying process. Stenters are not used to dry fabrics that are still wet from the dye padding due to the risks of dye migration and contamination of the stenter chain.[8]

Pre-drying and main drying of fabric impregnated with dye dispersion are processes that can have a decisive influence on the dyeing quality. Irregularities in drying can affect the migration of insoluble dye particles and therefore cause two-sidedness or listing, even if the fabric is properly impregnated in the foulard. It is recommended to pass the liquor-impregnated material through an air passage of 15–20 seconds between the padding and the first drying unit. The temperature in the first compartment of the hot-flue should be around 100 °C and the air circulation should be minimal. Drying can then be accomplished with higher air circulation at higher temperatures (120–140 °C) in other sections.[8] If the temperature and humidity of the hot air used in hot-flues are too high, an unpleasant streaky dyeing may occur, as well as dyeing with different shades along the width of the fabric or on both sides due to the dye migration that may occur for various reasons.[1] It is recommended that the first six to ten guide rollers be coated with polytetrafluoroethylene (PTFE) to prevent staining.[8]

The risk of migration exists only in the first phase of drying (until the moisture content drops to 25–30%). During impregnation with dye dispersion, the lower the pick-up value, the less the amount of surface water to be removed in the first phase, thus reducing the risk of migration. The correct operation of the dryer used, the avoidance of excess drying, the use of dye brands containing a small amount of dispersing agent and the addition of an appropriate antimigration agent to the liquor are measures to reduce the risk of migration. Another measure taken to prevent migration is to place a 2-m pre-drying channel in front of the dryer. In this channel, generally IR rays are used, sometimes with hot air. The aim is to ensure that surface water is removed as quickly as possible, thereby reducing the time during which migration can occur. The pre-drying channel is essential when using roller dryers. Proper operation of the channel is critical, since a large amount of moisture will be removed in a short time and in a short working area, so the slightest error in the channel (such as temperature difference, uneven air circulation) will, on the contrary, cause an increase in migration.[1] To reduce staining, the first six to eight drying rollers should be PTFE coated and temperature controlled, and the temperature of the first cylinder should not exceed 100 °C. The risk of two-sidedness is greater at higher temperatures and especially at low processing speeds. However, cylinder dryers have proven to be very suitable for smooth fabrics and much more economical than hot-flues.[8]

c. Impregnation with chemical liquor
Impregnation with chemicals can be done in a padder with small vessel (20–30 L) and 2-roller. It is beneficial to have a very short distance (up to 1 m) between the padder and the steamer. Otherwise, some of the reducing agent applied to the fabric will be decomposed by the atmospheric oxygen. The process is typically conducted at an impregnation temperature of 20–30 °C and a pick-up value of 80–100%.[1] It is recommended to cool the fabric after the preceding drying. The pick-up should be as high as possible. This ensures that there is sufficient liquor available for the dyes and chemicals and that exposed parts of the fabric (e.g. twill lines) are always saturated with sufficient reducing liquor. This gives an improved fabric appearance and prevents frosting. The reducing agent consumption depends on the amount and type of dye applied and the processing conditions, since a certain portion of the hydrosulphite is consumed by atmospheric oxygen. Depending on the dye concentration and the amount of liquor pick-up, approximately 50–100 mL/L caustic soda (27% by weight) and 20–50 g/L hydrosulphite are used.[8]

d. Steaming
Steaming is usually done in horizontal steamers within 30–60 seconds, depending on the type of fabric and dye and the depth of the dyeing.[1] Steamers often have driven top rollers and speed control to ensure the correct fabric tension. The fabric capacity is between 15 and 60 m, so the steaming time varies between 30–60 seconds, depending on the speed.[8] As it is known, as the amount of fabric that the steamer can take increases, it is possible to keep the production speed high without decreasing the dwell time in the steamer. It should be noted that the diameter of the transport rollers in the steamers should not be less than 125 mm, or even 150 mm if working with heavy fab rics with a width of 150–160 cm, and the distance (upright) between the two transport rollers should not be more than 1 m.[1] Hanger steamers have proven to be suitable for fabrics that need to be treated with minimum tension, where necessary precautions are taken to prevent chemical liquor from accumulating at the bottom of the loops.[8]

The process requires saturated steam at normal atmospheric pressure, and the steam atmosphere should be as air-free as possible.[8] Before operation, the steamer should be heated for 30–60 minutes, or dyeing should be started after making sure that there is no air inside. During dyeing, it should be ensured that the steam temperature remains constant at 102–105 °C and the steam pressure does not change.[1] Moisture content is as important as temperature in the reduction process. Correct adjustment is necessary for a successful dyeing process. The steamer exit must be constructed in the form of a water seal to prevent air from entering the steamer. The water supply should be 3–5 L/kg of material, and the temperature in the water seal should not exceed 40 °C, otherwise, some of the vat dye may bleed. The water should be fed into the water seal across full width of the fabric but not sprayed onto the fabric. The water drains via overflow, again across the whole width. The design of the steamer should prevent the formation of condensation drops from the steamer roof and from the steamer throat entry. The air passage between the padder and the steamer entry is critical and should be as short as possible to prevent the hydrosulphite from breaking down. Likewise, the air passage between the steamer water seal exit and the first wash box should be kept short. This is especially true for Indanthrone blue dyes, which are susceptible to excessive oxidation by atmospheric oxygen at high alkaline concentrations and form a dull yellow-green azine derivative.[8]

e. Aftertreatment

Aftertreatments consist of rinsing, oxidising, soaping and final rinsing, and are carried out continuously immediately after steaming. The most suitable equipment is an open-width washer, preferably with eight compartments. The fabric capacity of each compartment is approximately 15 m, although this depends on the speed of processing. Efficient oxidation and especially soaping depend on the processing time and concentration of chemicals, so the fabric capacity of these compartments should be proportional to the processing speed. Oxidation is done with peroxide or perborate at pH 8–9 and 50 °C. The fastness of the dyeing and the final hue are affected by the soaping process. In the presence of water at an elevated temperature a change takes place in the state of the reoxidised dye on the fibre that is accompanied by a hue change. Dyes selected for continuous processing should be relatively insensitive in this respect, since it is essential for the final hue to be attained within the time available for soaping. Two factors are decisive for the treatment: the time should be not less than 30 seconds and the temperature should be 95–100 °C. To ensure good reproducibility of the dyeings, rinsing water and chemical additions should be controlled by flowmeters and specific temperatures maintained by automatic controllers.[8] An example is given below.[1]

1st compartment: Cold water overflow rinse

2nd compartment: Overflow rinse with warm (maximum 50 °C) water

3rd and 4th compartments: Oxidation with 3–5 mL/L perhydrol at 50–60 °C (if pH is rising, pH should be adjusted to 8–9 by adding acetic acid) or with 3–5 g/L sodium perborate at 65 °C

5th and 6th compartments: Soaping at boiling temperature with liquor containing;
1–2 g/L Washing Agent
1–3 g/L Soda ash (pH should be 10)
5–10 g/L Sequestering agent

7th compartment: Overflow rinsing with water at 70 °C

8th compartment: Neutralisation in cold water with acetic acid (fabric pH should be 6).[1]

6.5.2.1.2 *One-bath Pad-steam Method*

According to this method, which does not find much use, mostly non-dense woven fabrics can be dyed. This method consists of following steps:

a. Impregnation with liquor containing dye, caustic soda and reducing agent
b. Steaming
c. Aftertreatment.

It is advantageous over the two-bath pad-steam method in that it does not require intermediate drying.[1]

As the reducing agent, a substance that is inactive at room temperature but acts rapidly under steaming conditions is required. For this reason, hydrosulphite cannot be used. Good results are obtained when a sodium sulphoxylate-based reducing agent stabilised with acetaldehyde and ammonia and a suitable stabilising agent are used together. When preparing the liquor, dye dispersion, stabilising agent solution, reducing agent and auxiliaries should be added to the liquor, respectively. Steaming is done in an airless steam atmosphere within 1–2 minutes. High steam temperatures up to 110 °C have a positive effect, but in this case, care should be taken not to dry the fabric.[1]

6.5.2.1.3 Wet Steam Method

In this method, the fabric padded with dye dispersion enters the special steamer. Inside the steamer, at the bottom, there are 5–6 small vessels (booster) that can be cooled. These vessels contain a liquor containing caustic soda, hydrosulphite and some dye. By means of a pump mechanism, the liquor taken from the last vessel is evenly distributed across the full width of all vessels via a pipe system that goes all the way to the first vessel. Using a dosing apparatus, a calculated amount of caustic soda, hydrosulphite and salt solution is added to the liquor continuously. Since there is a cooling system around the vessels, it can be ensured that the temperature of the liquor in them does not exceed 70–80 °C.[1]

The advantage of this method is that no intermediate drying is required. Since the fabric enters and exits the caustic soda and hydrosulphite liquor several times in steam atmosphere, all the hydrosulphite in the liquor can be utilised. The amounts of hydrosulphite, caustic soda and salt to be added to the liquor can be found from the graphs provided by the dye manufacturers. Steaming (102–105 °C) time is 1 minute.[1]

This method, which is recommended for dyeing products such as workwear, military fabrics, dyed in large quantities, and products with a high risk of migration in intermediate drying, such as towels and velvet, lost its importance with the development of the modified wet steam method.[1]

6.5.2.1.4 Modified Wet Steam Method

This method, developed by the BASF factory in 1971, consists of following steps:

a. Impregnation with dye dispersion
b. Impregnation with chemical liquor
c. Steaming
d. Aftertreatment

It is more advantageous in that it does not require intermediate drying compared to the normal pad-steam method. It is especially recommended for dyeing fabrics with a high risk of migration due to the lack of intermediate drying. In addition, this method has advantages such as no need for hot-flue for intermediate drying, better levelness of the resulting dyeings and less contamination of the steamer. The disadvantage of the method is that it consumes an average of 15% more dye compared to normal pad-steam method and does not give satisfactory results in densely woven, mercerised fabrics.[1]

a. Impregnation with dye dipersion
Pick-up value during impregnation with dye dispersion should be as low as possible, since the fabric will enter the chemical liquor without being subjected to an intermediate drying (from wet to wet). The pick-up value will vary according to the previous pick-up value of fabric, and the fabric that has high pick-up value in the first padder will not be able to get enough of the chemical substance (such as caustic soda, hydrosulphite) from the liquor. It is beneficial to add an appropriate antimigration agent to the dye dispersion liquor, even though intermediate drying will not be performed. The purpose of this addition is to reduce the bleeding of the dye pigment on the fabric into the padder containing chemicals. Another point to be considered is that the impregnation auxiliary (wetting agent) to be used should be of the type that does not foam during steaming. The temperature of the impregnation liquor is about 25 °C. It is beneficial to pass the fabric impregnated with the dye through an air passage before entering the liquor containing chemicals. In the meantime, the swelling of the cellulose fibres progresses, and since some of the surface water will be used as the swelling water, the amount that can be taken from the chemical liquor increases.[1]

b. Impregnation with chemical liquor
The impregnation of the liquor containing caustic soda and hydrosulphite on the wet fabric is provided by passing the fabric through small vessels of special structure. The fact that the volume of the vessel does not exceed 20–30 L is beneficial in terms of replenishing the liquor in a short time.[1] In the wet-steam process chemical liquor impregnation is carried out so that there is a maximum difference in pick-up between wet fabric entering and leaving the liquor trough. This may be achieved either by applying an unweighted top roller or by lick-roll application. After leaving the nip or the lick roller, the fabric must pass directly to the steamer without passing over any guide rollers.[8] When the fabric impregnated with the dye dispersion enters the liquor of chemicals, some dye will tend to bleed into the liquor until an equilibrium is established. In order to prevent this, it is necessary to add 100 mL/L of dye dispersion liquor to the chemical liquor first placed in the vessel.[1]

c. Steaming
In this method, steaming with air-free steam for 30–60 seconds at 102–105 °C gives good results.[1]

d. Aftertreatment
It is similar to the normal pad-steam method.[1]

6.5.2.1.5 Standfast Methods

In these methods, a Standfast machine is used instead of a steamer. In these machines, the fabric is passed through the molten metal. An alloy that melts at 71 °C is used as a metal. Heating the metal up to 95 °C is done either indirectly with steam or with electrical energy. The distance passed by the fabric in the molten metal is usually 3 m and the speed of passage is 30–100 m/min and hence the residence time varies between 6 and 1.7 s. At the exit, the fabric passes through the sodium sulphite bath, and metal residues that may remain on the fabric are removed. Dyeing is performed using two methods with the Standfast machine.[1]

a. ***Pad-standfast method:*** The steps are as follows:
   - The fabric is impregnated with dye pigment in the padder.
   - The fabric is subjected to intermediate drying if necessary.
   - The fabric is impregnated with a chemical (hydrosulphite, caustic soda) bath in the vessel at the entrance of the Standfast machine (Pick-up: 120–150%).
   - The fabric is passed through molten metal at 95 °C and the dye is fixed at this time.[1]

b. ***Standfast-leuco method:*** In this method, which can be used mostly in dyes with good uniform dyeing properties, the vat dye is first vatted and then heated to 75–80 °C after it is sent to the vessel in the Standfast machine. The fabric (pick-up of 120–150%) passing through this liquor passes through the molten metal, where the dye is fixed.[1]

6.5.2.1.6 Williams Unit Method

The Williams unit is a vessel equipped with rollers and a special structure. Here, due to the heatable walls inside the vessel, although the liquor ratio is low, it is possible for the fabric to have a long path length within the liquor. A good liquor mixture and a high liquor temperature shorten the vatting and fixation time, increase levelling and allow the dye to penetrate into the product. After the fabric is impregnated with the dye dispersion in the padder, it is passed through two Williams units, usually placed one behind the other. In the Williams unit, there is a liquor heated to 85–95 °C containing caustic soda and hydrosulphite. This method can only be used for dyes with good stability and the consumption of caustic soda and hydrosulphite is high.[1]

6.5.2.1.7 Continuous Quick Fixation Method

The basis of the quick fixation method relies on the fact that the fabric impregnated with the dye dispersion can be vatted and fixed within 5–8 seconds in the vatting liquor, thanks to a special auxiliary (reaction accelerator) added to the liquor. Since the dwell time is shortened, the amount of padding liquor can be reduced, allowing the bath to be replenished more quickly, reducing the risk of excessive reduction, and, in short, enabling a dyeing with uniform shade at the centre and edges.[1]

6.5.2.2 Dyeing by Pad-batch Method

Dyeing can also be performed with vat dyes by the pad-batch method. The basis of the method consists of the following steps:

a. Impregnation with dye dispersion
b. Intermediate drying (if required)
c. Impregnation with vatting liquor
d. Batching
e. Aftertreatment.[1]

a. Impregnation with dye dispersion
   The fabric to be dyed according to this method should have good hydrophilicity. Not all vat dyes (especially Indanthren Blau RS and some other blue dyes) work well with this method. It is necessary to pay attention to this issue when choosing the dye to be used. When working with intermediate drying or dyeing fabrics with high liquor

pick-up ability, it is useful to add a suitable anti-migrating agent to both the dye dispersion liquor and the vatting liquor. It is beneficial to add alginate thickener with high viscosity to the dye dispersion liquor in pad-dyeing without intermediate drying (from wet to wet). It is beneficial to add a suitable protective colloid and complexing agent to the dye liquor. The impregnation temperature is 25 °C and the pick-up value is 50–60%.[1]

b. Intermediate drying
The process can also be applied to fabrics that can absorb large amounts of liquor (such as towel, corduroy) without intermediate drying. Intermediate drying is essential for fabrics made of mercerised satin, poplin and regenerated cellulose fibres. Intermediate drying can be done in a hot-flue at 80–120 °C.[1]

c. Impregnation with vatting liquor
In addition to the caustic soda and hydrosulphite, some amount of dye dispersion (2–10% of the vatting liquor put into the vessel, according to the pick-up value) is added to the vatting liquor during the first filling of the vessel. The pick-up value in the vatting liquor varies between 20 and 150% depending on whether intermediate drying is done or not and the type of fabric. The fabric coming out of the vatting liquor must be thoroughly wet. Therefore, it is not squeezed between the rollers. The control of liquor pick-up is provided either by passing it between two rollers on top of each other (squeezing is provided only by the weight of the top roller and no additional squeezing pressure is applied), as specified in the modified wet steam method, or by the help of a tension roller whose tension can be adjusted.[1]

d. Batching
The weakest part of the method, therefore, the step that requires the most attention, is the wet fabric being wrapped in rolls and batched. It is essential that the edges of the fabric overlap completely during wrapping. Otherwise, it is possible that the hydrosulphite on the edges is consumed more by the atmospheric oxygen than in the other parts of the fabric, and the dye fixation is insufficient here. The width of the polyethylene foil used should be greater than the width of the fabric. A 5 cm wide wet cloth should be tightly wrapped in 3–4 layers on the area where the foil will be tied at the ends. The edges of the foil should also be tightly tied onto this wrapping with thread or thin strips. The batching time is usually 2 hours and for some dyes it is 4–6 hours. It is unnecessary to increase the holding time above the recommended time in the dye catalogue.[1]

e. Aftertreatment
This process, which comprises rinsing, oxidation, soaping and rinsing, is carried out as described in the pad-steam method.[1]

6.5.2.3 Dyeing by Pad-jig Method

In this method, the fabric is impregnated with vat dye pigment (aqueous dispersion) in the foulard. The vatting of the dye on the fabric and its fixation to the fibres are carried out on a jigger in a second step. The calculated amount of dye is poured in 10–20 times its weight in water at 40 °C, and an aqueous dispersion is prepared and completed to the volume of the impregnation liquor. It is of great benefit to add a suitable impregnation auxiliary to the liquor.[1] A good wetting agent is required as well as an antimigration agent.[8] Generally, the impregnation temperature is 20–40 °C, and the pick-up value is 60–70%.[1]

The fabric, together with the end-cloth, is impregnated with the dye dispersion in the padder and wound onto a roll. If the fabric is not to be dried before treatment in the jigger, the wet batch must be rotated slowly and continuously, and the fabric should be wrapped in plastic film.[8]

The decision to perform intermediate drying depends on the operating possibilities and cost conditions, as well as the quality of the dyeing. For example, in the dyeing of mercerised fabrics by pad-jig method, an intermediate drying improves the appearance of the dyeing. If intermediate drying is to be done, an antimigration (reducing the mobility and displacement of the dye pigments in the fibres) auxiliary should be added to the impregnation liquor. The amount of hydrosulphite, caustic soda and salt to be put into the jigger is calculated according to the type of dye used (IK, IW and IN). They are added together with other auxiliaries (such as protective colloid, levelling agent, complexing agent and sodium nitrite) to the liquor in the jigger before the first passage.[1]

Since some of the dye impregnated in the textile material will tend to bleed into the liquor in the jigger until an equilibrium is established between the dye in the fibre and the dye in the liquor, some amount of the impregnation liquor in the pad is added to the liquor in the jigger. The formulas and principles suggested in various catalogues to calculate the amount of impregnation liquor to be added to the liquor in the jigger are different from each other.[1]

The amount of impregnation liquor required depends on the pick-up value of fabric and the liquor ratio in the jigger process. The condition of the vatting bath is checked from the selvedge of the cloth with vat yellow paper before the cloth is re-entered into the bath.[8]

While it is suggested that all of the impregnation liquor to be added to the jigger liquor should be added before the first passage, there are also those who suggest that 2/3 of it should be added before the first passage and 1/3 before the third passage. Generally, four to eight passages are required for the vatting and fixation of the dye in the jigger. In many cases, hydrosulphite and caustic soda need to be added to the liquor during dyeing. It should be noted that the controls with vat yellow paper are not made by inserting this paper into the liquor but by touching the edge of the fabric coming out of the liquor.[1]

6.5.2.4 Dyeing by Pad-fabric Beam Method

In this method, the fabric is first impregnated with a dye pigment in a padder. The completion of the dyeing is done in the apparatus by winding it on the fabric beams. It is a method used mostly for dark shades in tightly woven fabrics, mercerised products (e.g. mercerised poplin for trench coats) or linen fabrics.[1]

The fabric, which is impregnated with the dye dispersion in the padder, is carefully wound on the beams (by paying due attention to issues such as winding density, tension uniformity, levelness of the edges, covering the perforations in an appropriate extent and lot size), and the fabric beam is placed in the dyeing apparatus. The liquor containing caustic soda, hydrosulphite and auxiliaries is prepared in the preparation tank and heated to the dyeing temperature (75–80 °C) and filled into the dyeing apparatus in the direction from outside to inside without starting the circulation pump. The circulation direction is changed to be inside–out, and the circulation pump is started so that the pressure gradually increases. During the whole dyeing process, the circulation is from inside to outside. The dyeing temperature is 70–80 °C, and the duration is 45–60 minutes. In this method, if desired, dyeing can be done under HT conditions.[1]

6.6 Problems Encountered in Dyeing with Vat Dyes and Precautions to Be Taken

Various dyeing faults, causes and solutions are explained below.

6.6.1 Dark Selvedge

This occurs due to (i) thicker edge which absorbs more dye and (ii) in jigger dyeing the edge is exposed to air and causes rapid oxidation, partially dries and absorbs more dye while passing through the next bath. Spraying $Na_2S_2O_4$ and NaOH solution on the edge during dyeing stops the oxidation and reduces this problem.[19]

6.6.2 Light Spot

This occurs due to (i) improper preparation of the fabric, (ii) local oil stains which adhere to the material during spinning and were not effectively removed during pretreatment and (iii) partial conversion of cellulose into oxycellulose or hydrocellulose during washing. A mild scouring at the post-dyeing stage followed by re-dyeing reduces this problem.[19]

6.6.3 Oxidation Marks/Dark Stains

This is the most common fault observed in vat-dyed products. After dyeing, washing should be thorough and no alkali residue should remain on the fabric. The presence of alkali during oxidation with H_2O_2 enhances these marks since the decomposition of H_2O_2 to release oxygen occurs at around a neutral pH. The presence of alkali causes H_2O_2 to release perhydroxyl ions, not oxygen.[19]

When such problems are encountered, the dye should be treated with $Na_2S_2O_4$ and NaOH solution at the reduction temperature for 15–30 minutes, followed by thorough washing and re-oxidation.[19]

6.6.4 Streaky Dyeing

This may be due to (i) reed marks on fabric, (ii) variable tension in the warp and weft causing either one to become apparent and (iii) faulty mercerisation. To avoid this problem, dyes with good covering power should be used.[19]

6.6.5 Poor Rubbing Fastness

The main reasons are (i) dyeing at a very short liquor ratio, which results in the formation of several dye layers on the cellulose surface, (ii) prolonged exposure of the dissolved dye to air, which results in partial oxidation of the dye and subsequent surface accumulation of the dyes, (iii) incomplete reduction of the dye towards the end of dyeing, (iv) ineffective soaping after dyeing and (v) dyeing in hard water, where metal salts react with the anionic dye and cause this complex to precipitate on the dyed material.[19]

Most of the errors described can be corrected by treating the dyes with NaOH and $Na_2S_2O_4$ at the vatting temperature in the presence of levelling agents, but there is a possibility that some of the dye may be lost from the bath. The dye should be added to the liquor at a rate of 10–15% of the original recipe. A higher liquor ratio at all stages of dyeing and appropriate control over the vatting and dyeing temperatures is a solution to most of the errors occurring in vat dyeing.[19]

References

1 Tarakçıoğlu, I. Tekstil Boyacılığı I Teksiri. İzmir, Türkiye: Ege Üniversitesi Mühendislik Fakültesi Çoğaltma Yayınları, 1982.

2 Yurdakul, A., Atav, R. Boya baskı esasları. Bornova, İzmir: Ege Üniversitesi Tekstil ve Konfeksiyon Araştırma Uygulama Merkezi Yayını, 2006.

3 Akçakoca, E.P. *Küp ve İndigo Boyaları ile Boyama*. Ordenim Semineri, Kayseri, 2002.

4 Eyüboğlu, Ü., Okaygün, I., Yaraş, F. Doğal boyalarla yün boyama, uygulamalı ve geleneksel yöntemler. İstanbul, Türkiye: Arkeoloji ve Sanat Yayınları, 1983.

5 Öztürk, İ. Doğal bitkisel boyalarla yün boyama. Ankara, Türkiye: Ürün Yayınları, 1999.

6 Dystar. Küp boyalarının 100. yılı. IX. Uluslararası İzmir Tekstil ve Hazır Giyim Sempozyumu, Altınyunus/Çeşme-İzmir, Türkiye, 26–29 Ekim 2001.

7 Ünal, B., Chakraborty, A. Neden küp boyamacılığı? (140–143). X. Uluslararası İzmir Tekstil ve Hazır Giyim Sempozyum Kitabı, Çeşme-İzmir, Türkiye, 2004.

8 Latham, F.R. Dyeing with vat dyes. In Cellulosics Dyeing (Ed. J. Shore, Chapter 5, 246–278). Bradford, West Yorkshire, UK: Society of Dyers and Colourists, 1995.

9 Dystar. Indanthren dyes technical documents.

10 Akçakoca, P.E. Denim kumaşlar ve indigo boyamacılığı. Tekstil ve Konfeksiyon Dergisi, 1999; 79(2), 136–142.

11 Roessler, A., Dossenbach, O, Meyer, U., Marte, W., Rys, P. Direct electrochemical reduction of indigo. Chimia, 2001; 55(10), 879–882.

12 Carr, W.W., Michielsen, S., Beckham, H.W. Fundamental investigation of ultrasonic effects in textile wet processing. Link of access: https://www.researchgate.net/publication/242318912_Fundamental_Investigation_of_Ultrasonic_Effects_in_Textile_Wet_Processing, [Date of access: 19.02.2021].

13 Akçakoca, E.P., Atav, R. Tekstil boyacılığında alternatif indirgeme yöntemleri. Tekstil Maraton, 2004; 14(75), 61–66.

14 Roessler, A., Dossenbach, O, Meyer, U., Marte, W., Rys, P. Direct electrochemical reduction of indigo. Journal of Applied Electrochemistry, 2002; 32, 647–651.

15 Çağlar, E. Denim terbiye işlemlerinin araştırılması (Yıl içi projesi). Süleyman Demirel Üniversitesi, Mühendislik ve Mimarlık Fakültesi, Tekstil Mühendisliği Bölümü, 2005.

16 Bechtold, T., Burtscher, E., Turcanu, A., Bobleter, O. The reduction of vat dyes by indirect electrolysis. Journal of the Society of Dyers and Colourists, 1994; 110(1), 14–19.

17 Roessler, A., Dossenbach, O., Rys, P. Electrocatalytic hydrogenation of indigo, process optimization and scale-up in a flow cell. Journal of The Electrochemical Society, 2003; 150(1), D1–D5.

18 Comisso, N., Mengoli, G. Catalytic reduction of vat and sulfur dyes with hydrogen. Environmental Chemistry Letters, 2003; 1(4), 229–232.

19 Chakraborty, J.N. Fundamentals and Practices in Colouration of Textiles. Woodhead Publishing India Pvt Ltd, 2010.

7

Dyeing with Indigo Dye

Indigo is a dye in the vat dye class obtained from the *Indigofera tinctoria* plant, which naturally grows in tropical areas. As an alternative to this plant, indigo dye can be obtained from the plant called *Strobilanthes cusia*, which grows in colder tropical regions. *Indigofera tinctoria* plant is 1–2 m tall with light green leaves and dark pink-purple flowers. Indigo dye is obtained from the leaves of this plant by fermentation. The most characteristic usage area of indigo dye (Figure 7.1), which is also expressed as the "king of dyes", is denim products.[1]

The history of indigo dyeing is quite old. Indigo dye was used in dyeing in India and China in the second century BC. It is said to have been used since the 4000s. Indigo dye was obtained naturally from ancient times until the industrial revolution. Previously, indigo dye was used in India, where it was also produced. Used as a dye, indigo reached the regions of Ancient Greece and Rome, where it was later introduced as a luxury product. This dye, which is used in textile dyeing and printing, has been the most used dye in many Asian countries (China, India, Japan, etc.) for many years. Sources indicate that indigo was also used in ancient Mesopotamia, Egypt, Greece, Rome, England and even Peru.[1] Mummies from Egyptian tombs from about 2500 BC have been found covered by indigo dyed hemp fabrics. The association of indigo with India is reflected in the Greek word for the dye, *indikón* (ινδικόν, coming from India). The Romans Latinised the term to indicum, which passed into Italian dialect and eventually into English as the word "indigo".[3]

Production of indigo dye, which was obtained naturally until the industrial revolution, was investigated by synthetic means with the industrial revolution, and some important studies were carried out on it. With the development of the industry, the amount of naturally produced indigo dye in the world in 1897 was 19,000 tons, but this value decreased to 1000 tons in 1914. This is an indication that synthetic indigo dye was increasingly being used. Today, natural indigo dye has been replaced by synthetic indigo dye in the textile industry.[1] By 2011, 50,000 tons of synthetic indigo were produced worldwide.[4] Natural indigo is mostly used for special purposes and applications of textile handicrafts.[5]

Today, indigo dye is obtained synthetically using N-phenylglycine and N-phenylglycine-o-carbonic acid.[6] As shown in Figure 7.2, indoxyl is obtained from N-phenylglycine and N-phenylglycine-o-carbonic acid, and indigo dye is formed by oxidation of this with the help of air oxygen.

Indigo is a dye that is insoluble in water and organic solvents, has a very low affinity for cotton fibre, and is in the group of vat dyes. Indigo dyes are named as "CI Vat Blue 1".

Figure 7.1 Natural indigo dye. Ref. [2].

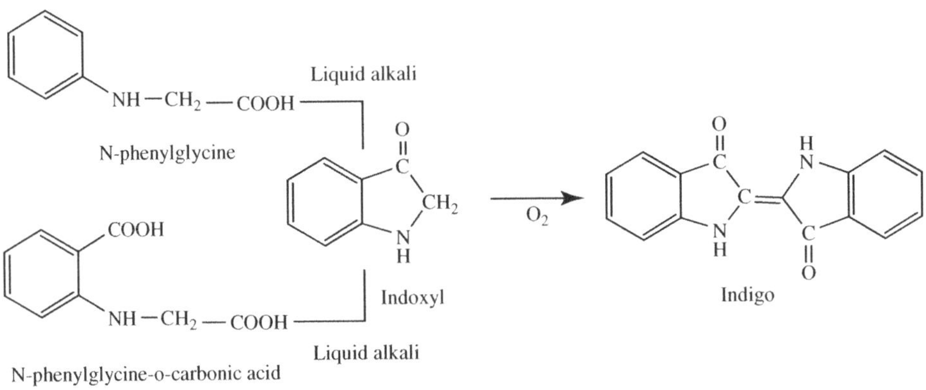

Figure 7.2 Preparation of indigo dye from N-phenylglycine and N-phenylglycine-o-carbonic acid. Ref. [6].

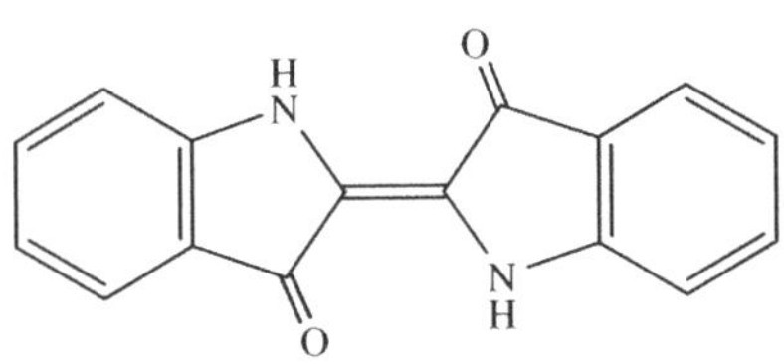

Figure 7.3 Chemical structure of indigo dye. Ref. [3]/ Woodhead publishing.

These dyes contain at least two oxygen atoms attached to the ring and conjugated with ring electrons[1] (Figure 7.3).

7.1 Fastness Properties of Indigo Dye

Although the washing fastness of indigo dyes is good, their rubbing fastness is low due to their weak affinity for cotton fibres. This situation, which seems like a disadvantage at first, actually creates the "easy fading" feature that reflects the main characteristic of denim fabric. The low rubbing fastness is due to the low affinity of the indigo dye for the cotton fibre, so it cannot penetrate into the fibre sufficiently. This ensures that the indigo dye can only adhere to the surface of the fibre and prevents it from reaching the core of the fibre. This situation is called **"Ring Dyeing"** in the literature[1] (Figure 7.4).

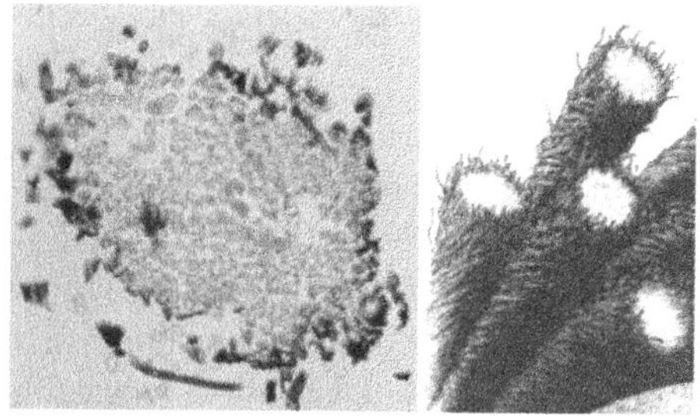

Figure 7.4 Cross-sections of indigo-dyed cotton fibre (left). Ref. [7]/CHT Group and cotton yarn (right). Ref. [6]/with permission of Tekstil ve Konfeksiyon.

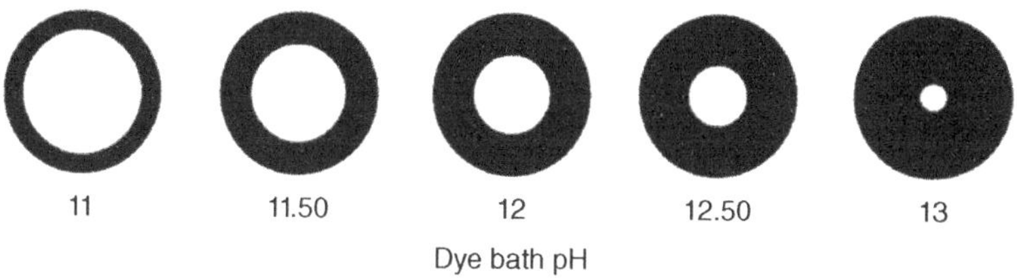

Figure 7.5 Effect of pH on ring dyeing. Ref. [8]/with permission of Elsevier.

Table 7.1 Effect of various anion forms of indigo dye on dyeing.

	Monoanion form	**Dianion form**
pH	11	>13
Fibre penetrating ability	Bad	Good
Condition of dye particles	Disperse	Dissolves
Ring dyeing feature	Much	Litte

Ref. [1]/MSc thesis.

As can be seen from Table 7.1 dyeing with indigo varies depending on pH. During vatting, if the pH is around 10–11, only one of the two conjugated oxygen groups of the indigo dye is decomposed, and the dye turns into monoanion form. Since this form cannot penetrate into the fibre, ring dyeing is high, and therefore the rubbing fastness of the dyeing is low.[1] The colour shade becomes easier to take effect in subsequent stone washing. Furthermore the colour shade becomes greener and duller.[8]

When the pH rises to 13 and above during vatting, the other conjugated oxygen group of the indigo dye also changes to the anion state and dianion form is formed. This form penetrates into the core of the fibre. In this case, ring dyeing is less (Figure 7.5), but rubbing fastness is higher.[1] Therefore, the colour shade becomes harder to take effect in subsequent stone washing. Moreover, the colour shade becomes redder and brighter.[8]

The scarcity or abundance of ring dyeing affects the fading properties of the denim fabric to be produced during the denim washing steps. A denim fabric with a high ring dyeing exhibits high fading, and the rubbing effects on the fabric surface are more vivid. On the other hand, a denim fabric with less ring dyeing has less fading and less friction effects on the fabric surface. Here, according to the customer's request, indigo dyeing can be done with more or less ring dyeing effect.[1]

7.2 End-uses of Indigo Dye

Indigo, which used to be widely used in every field, has lost its importance to a great extent with the invention of dyes with better fastness. Today, indigo dye, which cannot maintain its prominence in other fields, maintains its place as an indispensable dye class in denim dyeing.[9] Denim remains popular with men and women of all ages and social classes. Much of denim's popularity can be attributed to the appeal of the colour blue and the effects that occur when the fabric is washed repeatedly. The blue colour of the denim fabric is due to the cotton warp threads traditionally dyed with indigo.[8]

7.3 Forms of Indigo Dye

Indigo dyes are sold in the market in different forms, such as microgranule, paste and liquid. Indigo dyes are also available in partially reduced and water-soluble form.[10]

7.4 Dyeing with Indigo Dye

When dyeing with indigo dye is desired, dyeing methods belonging to the class of vat dyes should be used. According to this, the indigo dye is first converted to a water-soluble form with affinity for fibres with a reducing agent in alkaline medium. After dyeing with this form of the dye, the dyes taken by the fibres are subjected to oxidation process in order to finalise the dyeing and are converted back into a water-insoluble state.[1]

7.4.1 Reduction of Indigo Dye

The reduction of indigo dye with reducing agent in alkaline medium is called "vatting" (Figure 7.6), and the molecular form that emerges after vatting is called "leuco-indigo". Leuco form is called "Indigo red" in the industry. If the reducing agent used in the vatting is sodium dithionite, this form of leuco is also called "sodium leuco-indigo".[1]

In ancient times, arsenic trisulphide (As_2S_3) was used in the calcined solution for the reduction of indigo. By the twentieth century, arsenic trisulphide, which was poisonous for reduction, was replaced by hydrosulphite.[6] Today, sodium hydroxide or in other words caustic soda (NaOH) is used as alkali, and sodium dithionite or in other words hydrosulphite ($Na_2S_2O_4$) is used as reducing agent.[1] Below, after briefly mentioning the vatting

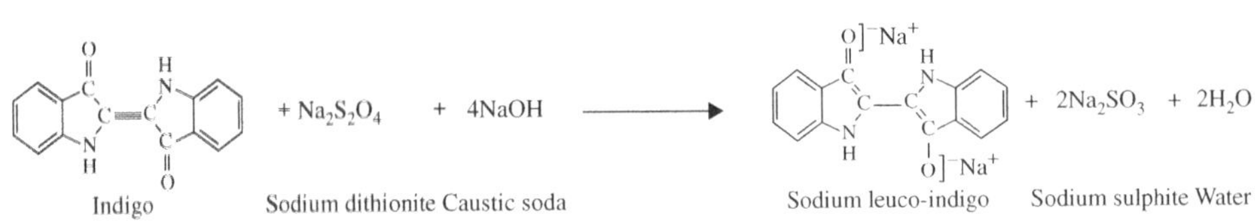

Figure 7.6 Reduction of indigo dye. Ref. [1].

methods such as fermentation vat, iron (II) sulphate-lime vat and zinc-lime vat, which have been used in the past and have lost their importance today, detailed information will be given about the caustic-hydrosulphite vat used today.

a. ***Fermentation vat:*** In this oldest vatting method, substances such as caustic soda, soda, lime and wood ash are used as alkali. The reduction is achieved by fermentation of starch- or sugar-containing products such as oily clay, flour, bread, rice, dates, grapes and figs. This method, which is difficult to implement, has completely lost its importance today.[11]
b. ***Iron vitriol-lime vat:*** In this method, lime is used as an alkali and iron (II) sulphate is used as a reducing agent. The main vat obtained in this way should be left for 4–6 hours before being used. Apart from that, it is easy to prepare, and the durability of the liquor is high. However, it has a significant drawback that it degrades 25–35% of indigo during vatting.[11]
c. ***Zinc-lime vat:*** In this method, lime is used as alkali and zinc salt is used as reducing agent. The main vat must be left for five hours before being used. The degradation of indigo in this method is considerably lower than in the iron vitriol method.[11]
d. ***Caustic-hydrosulphite vat:*** In this method, caustic soda is used as alkali and hydrosulphite is used as reducing agent. Vatting with hydrosulphite is advantageous in that it gives more specific dyeing and less indigo degradation than other methods.[11]

In indigo dyeing, as in dyeing with normal vat dyes, the main reducing agent used is sodium dithionite, which is called hydrosulphite or hydro. The reduction potential of hydrosulphite can meet all practical requirements. In the caustic-hydrosulphite method, which is widely used to reduce indigo dye, it is generally not possible to prepare indigo vats above 80 g/L due to the dissolution problem. To reduce 1 kg of indigo, 660 g of sodium dithionite is required stoichiometrically. In practice, approximately twice as much sodium dithionite is used to stabilise the dyeing system. Since sodium dithionite is sensitive to atmospheric oxygen, there must always be excess hydrosulphite in the medium.[6]

The reduction reaction takes place over a period of 20–30 minutes at approximately 50–60 °C. The reduction of the indigo dye can be followed by the colour change. When the chromophore groups (C=O) of indigo break down and oxygen atoms form phenolate, the blue colour of the indigo dye turns yellow. The appearance of yellow colour indicates that residual leuco-indigo is present in the environment.[1] To see if the vatting is complete, when a glass plate is inserted and removed from the liquor, the liquor should run clear and yellow, taking 12–15 seconds to turn green. If the colour is not quite yellow or turns green in a shorter time, some more hydrosulphite should be added. Because when enough hydrosulphite is not used, the indigo dye cannot be fully dissolved and an efficient dyeing cannot be made. An excess of hydrosulphite alone should also be avoided. Because, when working in excess of hydrosulphite, uneven dyeing occurs and ecological damage increases.[6] Meanwhile, if the caustic soda concentration in the liquor is insufficient, the dyeing becomes dull and uneven. The excess of caustic soda slows down the dye uptake.[12]

The indigo dye, which has turned into the leuco-indigo form, has affinity for cotton fibres and is water-soluble. Now dyes in leuco-indigo form, which have affinity for cotton, can penetrate cotton fibres and perform indigo dyeing.[1] The dyeing process is done by adding the vatted indigo dye to the liquor containing water, caustic soda, hydrosulphite and dispersing agent material. In some brands, salt is added to the liquor to increase the dye uptake. After the pigment layer deposited on the surface is removed, dyeing begins. Dyeing is carried out by immersing several times in a dye bath at 20–25 °C and air oxidation. After dyeing, a very good rinsing is done and the product is dried.[11,12]

Due to the ecological drawbacks in the reduction with the caustic-hydrosulphite system, various researches are carried out on new reduction systems in order to both reduce the cost of dyeing and to find an ecological dyeing method. These researches can be classified under six main headings.

I. The use of pre-reduced forms of indigo dyes that are suitable for less caustic-hydrosulphite use

There are also dye brands containing indigo in reduced state on the market. They do not need to be vatted in the main vat separately. They can be dissolved directly in the dyeing liquor. After adding 1.5 mL/L caustic soda (38 °Bé), 1.5 g/L hydrosulphite, 1.0 g/L auxiliary and 20.0 g/L table salt into the liquor at 20–25 °C, 5.0 g/L dye is dissolved in this liquor, and dyeing begins when the colour of the liquor becomes greenish-yellow.[11]

The 20% leuco-indigo solution produced by catalytic hydrogenation technique without harming the environment is transported in a nitrogen atmosphere, stored and dosed directly from the stock tank to the dyebath.[6] The 20% indigo solution, which was launched in 1994, does not contain any sulphur compounds in its structure. This is because its production takes place without the use of hydrosulphite.[9] With the use of 20% indigo solution, the concentration of sulphur compounds in the dyehouse wastewater can be reduced by approximately 50% compared to the conventional method.[6] The amount of sulphide wastewater, which poisons bacteria and prevents biological treatment in treatment plants, has decreased considerably. In reinforced concrete wastewater treatment systems, the corrosion hazard mainly caused by sulphide and sulphate during anaerobic treatment is considerably reduced.[9] Beyond that, it is possible to find a solution to the colouration problem of wastewater by recycling the indigo recovered from the rinsing water by ultrafiltration to the dyeing process.[6]

II. Less caustic-hydrosulphite use by vatting the indigo dye in a nitrogen atmosphere

In the indigo reduction system called Ben-Indigo, launched by the Beninger company, the vatting is done in a 160-L reactor, the amount of which can be adjusted according to the need, in an airless environment without excess hydrosulphite. In this case, the vatting efficiency rises above 95%. The reduction reaction takes place in the nitrogen atmosphere in the ultrasonically operated reactor. In this way, it is possible to reduce indigo within 2 minutes at 50 °C. Thus, it is possible to use 40% less caustic soda and hydrosulphite compared to conventional reduction. In addition, the indigo concentration, which was 100 g/L in the conventional reduction, was increased to 140 g/L.[6]

III. Less caustic-hydrosulphite use by vatting of the indigo dye in an ultrasonically operating reactor

In the case that indigo dyes are vatted in an ultrasonic reactor, less caustic soda and hydrosulphite are used as the vatting efficiency increases, and it is even possible to

Table 7.2 Amount of reducing agent required to reduce 1 kg of indigo.

	Conventional reactor		Ultrasonic reactor	
Reducing agent	**Consumption (kg)**	**Sulphite load (kg)**	**Consumption (kg)**	**Sulphite load (kg)**
Hydrosulphite	1.30	1.20	0.70	0.64
Sodium sulphoxide	—	—	0.45	0.33
Thiourea dioxide	—	—	0.45	0.33

Adapted from Ref. [6].

use organic reducing agents such as sodium sulphoxide or thiourea dioxide instead of hydrosulphite. In this case, the sulphite ($SO_3^=$) load in the wastewater also decreases. In Table 7.2, the amounts of reducing agents used in conventional and ultrasonic reduction to reduce 1 kg of indigo dye are given comparatively.[6]

As can be seen from Table 7.2, in case of using an ultrasonic reactor in the vatting process, the amount of hydrosulphite required to reduce 1 kg of indigo dye is reduced by approximately 50%. In this case, the sulphite load in the wastewater also decreases by half. During the vatting process in the ultrasonic reactor, the dye aggregates get smaller with the effect of ultrasound, and the vatting efficiency increases due to the decrease in the particle diameter and the increase in the active surface area where the reaction takes place.[13]

IV. Use of organic reducing agents instead of hydrosulphite
Although an organic reducing agent (glucose, α-hydroxy ketones, thiourea dioxide, etc.) whose oxidation products can be biodegraded can be used instead of sodium dithionite, complete success has not yet been achieved with these substances.[13] However, if 20% pre-reduced indigo is used, it is possible to successfully use weaker organic reducing agents.[6]

V. Electrochemical reduction of indigo dye
It is possible to reduce solid indigo microcrystals immobilised on the surface of several electrode materials in buffered aqueous solution, and the results are closely related to those obtained for indigo dissolved in solvents. However, indigo behaves markedly differently in aqueous suspension and cannot be reduced electrochemically under these conditions.[14] Unfortunately, it was not industrially feasible to reduce indigo electrochemically under dye bath conditions on a planar electrode. The most limiting factor seems to be the poor contact between indigo particles and the electrode. Direct electrochemical reduction of dispersed indigo for the dyeing process is somewhat complicated due to the electron transfer difficulties between solid electrode and solid particle of dye, and therefore indirect electrochemical methods have been suggested. In this system, the reduction of the dye occurs not through direct contact of the dye with the cathode but through the mediator that is reduced by contact with the cathode.[3]

Electrochemical reduction of indigo is used in the slasher dyeing machine of Sucker–Muller–Hacoba. In this system, with the help of electron transfer agents,

20% pre-reduced indigo is reduced at an adjustable redox potential in the electrolysis cell and regenerated at an integrated cathode. The portion of the electrolyte consumed by air oxidation is constantly replenished at the cathode. In this way, the formation of by-products such as sulphites is prevented. Furthermore, with electron transfer agents, it is possible to recover indigo and recycle the washing water.[6]

VI. Reduction of indigo dye by electrocatalytic hydrogenation
Studies on the conversion of indigo to water-soluble leuco form by electrocatalytic hydrogenation in a two-compartment current cell using Raney nickel electrodes in aqueous solutions have been carried out and the optimisation of the system conditions has been emphasised. It has been found that increasing the pH can increase the reduction rate and the maximum conversion can be achieved by increasing the concentration and temperature. By working in optimised conditions, the efficiency of 12.7% can be increased to 95%. The use of Raney nickel as an electrode provides low cost and resistance to alkaline environment. Since black platinum electrodes are unstable, their use in industrial applications does not seem possible.[13]

7.4.2 Oxidation of Indigo Dye

The last step in dyeing is the process of making the reduced indigo dyes that penetrate the cotton fibres insoluble in water again and thus finishing the dyeing. For this, oxidation process is applied. The cheapest method for oxidation is oxidation with air. Oxygen molecules in the air oxidise the reduced leuco-indigo dyes and bring them back to their normal indigo form. The dye, which returns to indigo form, has high washing fastness since it is now insoluble in water in the fibre.[1]

$$C_{16}H_{10}O_2N_2Na_2 + \frac{1}{2}O_2 + H_2O \rightarrow C_{16}H_{10}O_2N_2 + 2NaOH$$

The oxidation time directly affects the dyeing efficiency. In practice, air oxidation for less than 1 minute cannot oxidise sufficient leuco-indigo, while more efficient dyeing can be obtained in oxidations of 1 minute or more. The oxidation process can be followed by the colour change as in the reduction. While the colour of the reduced leuco-indigo is yellow, the colour of the leuco-indigo that starts to be oxidised changes from yellow to green and then to blue[1] (Figure 7.7).

As can be seen from Figure 7.7, while the indigo dye in the lower part of the yarn rope was still yellow because it is still in the leuco form, the indigos gradually turned green and then blue towards the top of the rope, turning back to the indigo form. Here, the oxidation process is done by the oxygen in the air.[1]

Figure 7.7 Colour change during oxidation of indigo. Ref. [7]/CHT Group.

7.4.3 Post-treatments After Indigo Dyeing

Unfixed dye and chemical residues are removed from the material by passing through two or three washing chambers after dyeing in several passages (dip in a dye bath and air oxidation) depending on the desired dyeing depth.[10] If it is worked according to the iron vitriol or zinc method, after the final oxidation, acidification is done with 1 mL/L conc. H_2SO_4 and then rinsed thoroughly. If it is worked according to the hydrosulphite method, there is no need for acidification. After the final oxidation, a good rinse with cold water is sufficient.[11]

7.5 Indigo Dyeing Methods and Machines

Since indigo dyes have low affinity for cellulose fibres, materials in wound form, such as bobbins, cannot be dyed in dark shades by exhaust method.[10] Figure 7.8 shows the techniques that can be used to dye cotton with indigo. Among these, the hank dyeing technique is mostly applied as a craftsmanship and is not a technique used in industrial-scale production. Contrary to continuous yarn dyeing methods, ring dyeing does not occur in hank dyeing, the dye also penetrates into the yarn. The continuous fabric dyeing technique consists of passing the fabric through successive leuco-indigo baths. After each dipping, the fabric is squeezed between rollers and then exposed to air for oxidation. Then the fabric is washed, passed to the drying section and wrapped in rolls. This method has not become industrially widespread for technical reasons. When using this technique, it is extremely difficult in practice to achieve good dyeing uniformity and good colour depth as well as acceptable dyeing fastness values.[8] Therefore, the indigo dyeing process is mostly carried out continuously by passing rope or open-width yarns through vats containing vatted indigo dye. Due to the low affinity of vatted indigo to cellulose fibres, even if the process is carried out for a long time in a single bath, it is not possible to actually make the dye absorbed on the fibre, and only dyeing in light shade can be achieved.[10]

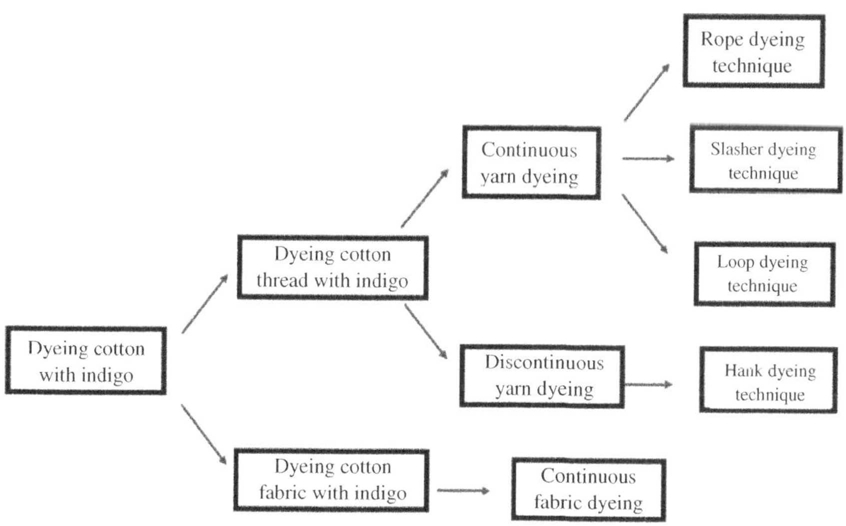

Figure 7.8 Techniques for dyeing cotton with indigo. Ref. [8]/with permission of Elsevier.

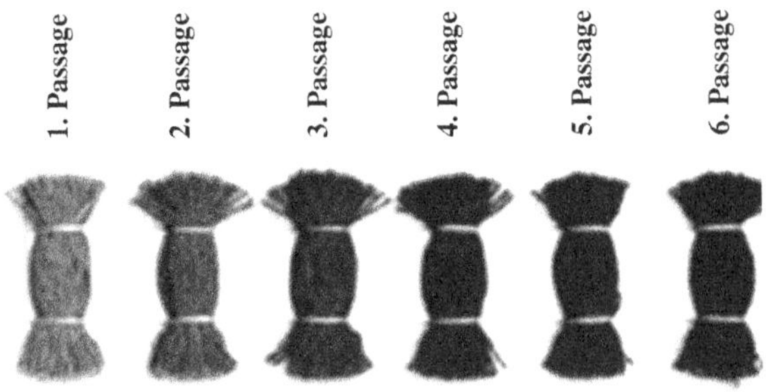

Figure 7.9 Colour obtained depending on the number of passages in indigo dyeing. Ref. [7]/CHT Group.

If dyeing in dark shades is desired, increasing the amount of dye in the bath does not increase the efficiency; on the contrary, it reduces the rubbing fastness too much. The methods used for obtaining dark shade dyeing can be listed as follows:

- Overdyeing on several vessels.
- Intermediate squeeze at each vessel exit.
- Air passage between exiting the vessel and diving into the new vessel.
- Adding a small amount of salt to the dyebath.[1]

Dye uptake in a single bath is between 0.5 and 0.8%. As the number of baths is increased, overlapping is made and the colour darkens. For this reason, the number of passages is increased according to the desired dyeing depth[10] (Figure 7.9).

Intermediate squeezing at each vessel exit ensures that the yarns entering the next vessel absorb the new liquor better and thus obtain a darker dyeing. Thanks to the air passage between each vessel, leuco-indigo oxidises and turns into blue-coloured pigments that are insoluble in the fibre. In this way, the yarns can take up the leuco-indigo again in the next vessel, and the colour becomes darker.[1] The addition of sodium chloride to the dyeing liquor decreases the negative zeta potential of the cellulose and increases the uptake of the dye by the fibres by reducing the ionic repulsion between the fibre and the anionic leuco-indigo. Thus, obtaining dark dyeing becomes easier.

For the reasons explained above, denim fabrics are dyed in warp yarn form in more than one passage, except for very light shades. The passage means dyeing with vatted indigo and oxidising. If each stage is to be classified simply, it consists of the immersion part, through which the material can pass and take up the dye, then the squeezing part, and then the air passage to ensure the oxidation of the dye. Yarns with already fixed and oxidised dyes are dipped into subsequent baths, and the same process is repeated. Only a small amount of indigo dye is displaced in the baths after the first bath. With this continuous process, the depth of colour increases. After the last air passage, unfixed dye and chemical residues are passed through two or three washing chambers and are removed from the material.[10]

Today, there are basically two different dyeing machines in which the warp yarns are dyed with indigo dye in more than one vessel. These are as follows:

- ***Rope dyeing:*** This dyeing technique dates from 1915, when the first rope dyeing machine appeared.[8] It is the most preferred dyeing method in the market, enabling the denim fabric to show its characteristic features in the best way. Here, first the warp threads are turned into ropes, and after the warp ropes are dyed by passing through

6–12 dyeing vats, the threads in the form of rope are opened and wrapped in a beam. After that, sizing is done. In the rope dyeing machine, 12–36 ropes, each consisting of 350–450 warp threads, can be dyed at the same time. The ropes are passed through the machine separately and properly. In this technique, higher quality dyeings are obtained compared to other techniques. A homogeneous dyeing is achieved because the risk of right–left differences is less due to the fact that the yarns in the form of rope are then distributed in the width of the warp. However, the cost of this technique is higher due to additional rope preparation and unwinding.[1]
- ***Open-width dyeing (slasher or sheet dyeing):*** Slasher dyeing was introduced in 1970 to be more rapid and flexible in response to changes on the denim market. This technique works particularly well on lightweight denim. These dyeing machines take up less space and allow for smaller quantities to be produced.[8] It is a less preferred method than rope dyeing. Here, the warp threads wrapped in the warp beam are directly dyed by passing through 6–8 dyeing vats in open-width, and then they can enter directly into sizing.[1]

In addition to passing the yarns through more than one dyeing vessel in order to obtain a dark colour in indigo dyeing, another possibility is to pass the yarns through the same dyeing vessel a certain number of times (4–10 times) until the desired colour depth is achieved. **The loop dyeing machine** in the market works according to this principle.[1] This dyeing technique was developed in 1980 by Eckhardt Godau and his company, Looptex.[8]

References

1 Bağıran, İ.C. Denim yıkamada karşılaşılan sorunlar ve bunlara yönelik çözüm önerileri (MSc Thesis). Ege Üniversitesi, Fen Bilimleri Enstitüsü, İzmir, Türkiye, 2011.
2 Link of access: https://en.wikipedia.org/wiki/Indigo_dye, [Date of access: 20.02.2021].
3 Chavan, R.B. Indigo dye and reduction techniques. In Denim: Manufacture, Finishing and Applications (Ed. R. Paul, Chapter 3, 37–67). Sawston, Cambridge, Birleşik Krallık: Woodhead Publishing, 2015.
4 Link of access: https://tr.wikipedia.org/wiki/%C4%B0ndigo_a%C4%9Fac%C4%B1, [Date of access: 20.02.2021].
5 Aslan, M. Denim yıkama prosesinde enzim kullanımı (MSc Thesis). Ege Üniversitesi, Fen Bilimleri Enstitüsü, İzmir, Türkiye, 2004.
6 Akçakoca, P.E. Denim kumaşlar ve indigo boyamacılığı. Tekstil ve Konfeksiyon, 1999; 79(2), 136–142.
7 CHT Group Technical Document.
8 Meksi, N., Mhenni, M.F. Indigo dyeing technology for denim yarns. In Denim: Manufacture, Finishing and Applications (Ed. R. Paul, Chapter 4, 69–105). Sawston, Cambridge, UK: Woodhead Publishing, 2015.
9 Yazmuradov, A., İşkabulov, C. Denim kumaş terbiyesi (Lisans tezi). Ege Üniversitesi Tekstil Mühendisliği Bölümü, İzmir, Türkiye, 2001.
10 Çağlar, E. Denim terbiye işlemlerinin araştırılması (Yıl içi projesi). Süleyman Demirel Üniversitesi, Mühendislik ve Mimarlık Fakültesi, Tekstil Mühendisliği Bölümü, 2005.

11 Tarakçıoğlu, I. Tekstil Boyacılığı I Teksiri. İzmir, Türkiye: Ege Üniversitesi Mühendislik Fakültesi Çoğaltma Yayınları, 1982.

12 İnce, H. Tekstil endüstrisi atıksularında indigo boyanın geri kazanımı (MSc Thesis). Fırat Üniversitesi, Fen Bilimleri Enstitüsü, Çevre Mühendisliği Anabilim Dalı, Türkiye, 2014.

13 Akçakoca, E.P., Atav, R. Tekstil boyacılığında alternatif indirgeme yöntemleri. Tekstil Maraton, 2004; 14(75), 61–66.

14 Roessler, A., Dossenbach, O., Meyer, U., Marte, W., Rys, P. Direct electrochemical reduction of indigo. Chimia, 2001; 55, 879–882.

8

Dyeing with Sulphur Dyes

Despite their long history, little is known about the chemical structures of sulphur dyes. The first commercial sulphur dye was synthesised in France in 1873 by Croissant and Bretonniere. These researchers prepared brown dyes for cotton by heating various organic materials of animal or vegetable origin with aqueous sodium sulphide (Na_2S) or polysulphide.[1] Sulphur dyes are composed of high molecular weight compounds formed as a result of the reaction of sulphides with amines and phenols. There are many colourants containing sulphur in their molecules in the market, but only those that become water-soluble as a result of a reaction with Na_2S under alkaline conditions are called sulphur dyes. Because they are mixtures of highly complex molecules, their exact chemical structure is not fully known exactly.[2] Since little is known about their structure, sulphur dyes are classified according to the chemical structure of the precursors used in their preparation.[1] Amino derivatives, nitrobenzenes, nitro and aminobiphenyls, substituted phenols, substituted naphthalenes, condensed aromatic compounds, indophenols, azines, oxazines, thiazole, azine and thiazine rings can be part of these compounds.[2] While colours such as yellow, orange or brown are generally obtained from thiazole, thiazone or thianthrene, purple, bluish-green and also black dyes are produced from indophenol and indamines and form the basis of burgundy and reddish-brown colours[3] (Figure 8.1).

Sulphur dyes have a polymeric structure in which coloured polymer units are linked by disulphide bonds.[3] Sulphur dyes contain sulphur both as part of the chromophore and in the polysulphide side chains.[2] Compounds or other aromatic species (containing nitro, amino and/or hydroxyl groups in their structure) whose structures are given in Figure 8.1 form sulphur dyes by connecting with disulphide or disulphoxide bridges.[1,4] Figure 8.2 shows the structure (structure number 53185) of colour index (CI) Sulphur Black 1, a sulphur dye known as Sulphur Black T.[1,5]

Sulphur dyes are dyes that have different chemical structures, but are studied as a group because their production and dyeing methods are similar. Another common point is that they all have sulphur atoms in their molecules.[6]

The basis of dyeing with sulphur dyes is similar to that of vat and indigo dyes. The disulphide bridges in the structure of the sulphur dye, which has no affinity for fibre and is insoluble in water, are converted into ionised mercaptan groups (−S ~) by reduction (with Na_2S) in a basic medium during the dyeing process, and thus the dye is turned into a water-soluble form, which has affinity for fibres (leuco form). After dyeing, disulphide

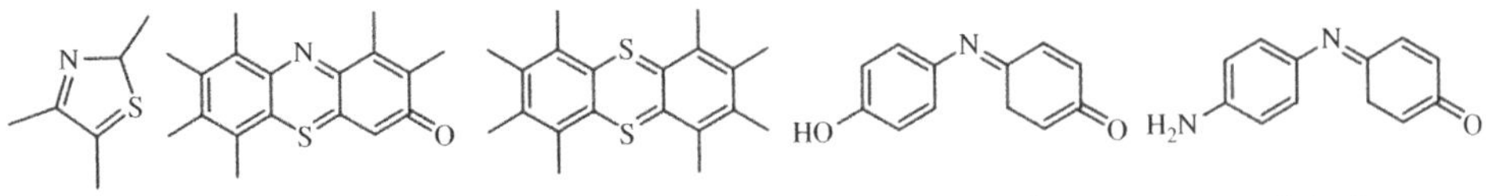

Figure 8.1 Various precursors that can be found in the structure of sulphur dyes. Ref. [3]/Springer Nature.

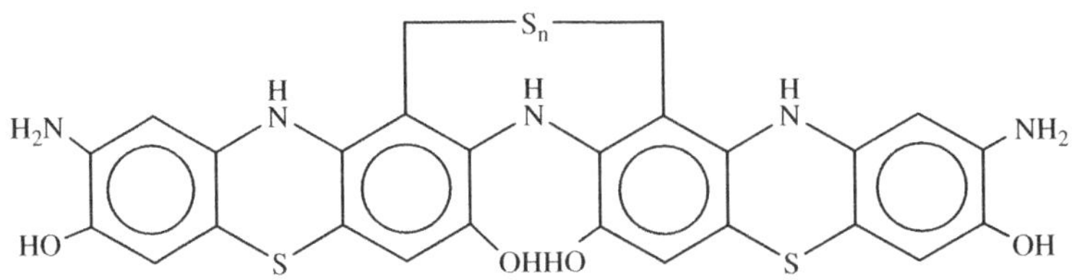

Figure 8.2 CI Sulphur Black 1. Ref. [5]/European Patent Office.

bridges are re-formed again in the dye molecules taken up by the fibres by oxidation, allowing the dye molecules to convert to their original water-insoluble state and thus show good fastness properties.[1,3,6] Finally, the reducing agent, alkali, salts and unfixed dye are removed from the fibres in rinsing and washing processes.[2]

One of the most important requirements today is ecology. Sulphur dyes do not contain heavy metals or adsorbable organic halogens (AOX). Therefore, sulphur dyes should be able to meet ecological demands. However, the issue that creates ecological problems is not the dye itself, but the application process. To put it more clearly, these are the reduction and oxidation processes.[7] Although sulphur dyes lost their popularity in the past due to the high sulphur content of the wastewater, the increase in the wastewater load of the reduction and oxidation agents, and the excessive chromium content of the textile material they later regained a certain share of usage. This is both because the dyeing process has become more ecological due to the developments in dyes and the emergence of alternative reducing substances, and because this dye group gives good results in obtaining ageing effects such as fashionable **"antique dyeing"**.[8]

8.1 Classification of Sulphur Dyes

Sulphur dyes can be classified in different ways. According to the CI, they are classified into four groups:

1. CI sulphur dyes
2. CI leuco sulphur dyes
3. CI soluble sulphur dyes
4. CI condensed sulphur dyes.[8]

8.1.1 Conventional Sulphur Dyes

The dyes in the first group, which can be called conventional sulphur dyes or normal sulphur dyes, are dyes that do not dissolve in water and have no affinity for fibres.[4] Since

Conventional sulphur dyes are insoluble in water and therefore case do not have affinity for cellulose fibres, they are reduced to water-soluble form during dyeing. Na_2S is generally used to reduce these dyes.[6] During reduction, disulphide bridges in the structure of the dye molecule are broken.[4] These water-soluble (reduced) forms have a high affinity for cellulose fibres. After treatment with these, the reduced dye is oxidised in the fibres and converted back to the original water-insoluble form.[6]

$$\text{D-S-S-D} \underset{\text{Oxidation}}{\overset{\text{Alkali reduction}}{\rightleftarrows}} \text{2D-SH} \underset{\text{Hydrolysis}}{\overset{\text{Alkali}}{\rightleftarrows}} \text{2D-SNa}$$

Insoluble in water
Low affinity

Soluble in water
High affinity

These dyes are generally used in dyeing by the exhaust method.[9] In brands where the dye is in powder form, the dye is first made into a paste with a suitable anion-active wetting agent, after adding the required amount of Na_2S, it is diluted with 10–30 times its weight in hot water and dissolved by boiling for 2–3 minutes. In the case of dissolving large amounts of dye, 1/3–1/2 of the dye liquor is dissolved by boiling dye and sodium sulphide in the liquor, and the temperature of the liquor is reduced to 40 °C by adding cold water to the liquor. In brands where the dye is in the form of granules, it is recommended to pour the dye into hot water by stirring, without transforming it into paste.[6]

Some special, large molecules and black sulphur dyes containing anthraquinone group in their structure provide dyeing with superior fastness values compared to all other sulphur dyes. Generally, the chlorine fastness of sulphur dyes is not good, but the chlorine fastness of these specific dyes is good, and there is no danger of damage to the material due to the formation of sulphuric acid during storage. The label "Indanthren" can be used for dyeing with these dyes. Dyeing with these dyes can be done at 60–70 °C by reducing with Na_2S (requiring crystalline Na_2S three times the weight of the dye is required), as in Conventional sulphur dyes, or by using caustic soda and hydrosulphite.[6]

8.1.2 Leuco Sulphur Dyes

Sulphur dyes are also produced in reduced or partially reduced form in powder or liquid form.[1] This second group of dyes, called leuco sulphur dyes or pre-reduced/ready-to-use sulphur dyes, have the same CI number as those in the first group, but they are in the form of water-soluble leuco and generally contain Na_2S or hydrosulphite in their structure. Such a structure is sufficient for the direct application of the dye.[2,4] However, sometimes a small amount of reducing agent may need to be added depending on the dyeing method used and dyeing depth. Compared to the dyes in the first group to which they belong, their dye content is between 25 and 50%.[4] In addition, these are the most suitable dyes for the one-bath pad-steam method.[9]

8.1.3 Water-soluble Sulphur Dyes

Water-soluble sulphur dyes have a different CI number than conventional sulphur dyes because they are water-soluble thiosulphuric acid derivatives of Conventional sulphur dyes.[4] Making sulphur dyes soluble in water is similar to rendering vat dyes water-soluble (obtaining vat leuco ester dyes). Sulphur dye, which is reduced in an alkaline medium, is subjected to a thiosulphation process to obtain a water-soluble sulphur dye form that is not oxidised under normal conditions.[6] These dyes have no affinity for cellulose, but they are reduced during dyeing and converted to an alkaline soluble thiol form.[4]

D-S-S-D	$\underset{\text{Oxidation}}{\overset{\text{Alkali reduction}}{\rightleftarrows}}$	2D-SNa	$\xrightarrow{\text{Thiosulphatation}}$	$D\text{-}S\text{-}SO_3^-Na^+$	$\xrightarrow{\text{Reducing agent/alkali}}$	$D\text{-}S^-$
Water-insoluble Low affinity		Water-soluble High affinity		Water-soluble Low affinity		Water-soluble High affinity

Sulphur dyes rendered water-soluble in this way are suitable for continuous dyeing processes, because, on the one hand, they are highly soluble in water without adding any reducing agents, and on the other hand, they do not show an affinity for cellulose fibres in this state. Thus, the first advantage of these dyes compared to Conventional sulphur dyes is the ease of continuous dyeing with them and obtaining good results. The second advantage is that good results are obtained when using Sulfhydrate F (Sodium hydrogen sulphide [NaHS]) instead of Na_2S as a reducing agent, unlike Conventional sulphur dyes, in dyeing performed with them. As a result of hydrolysis in solutions of Na_2S in water, 90% NaHS and 10% caustic soda are formed as shown below.[6]

$$Na_2S + H_2O \rightleftarrows NaHS + NaOH$$

When NaHS (Sulfhydrate F) is used directly instead of Na_2S, caustic soda is not formed in the environment; this has several benefits. Since the alkalinity value of the medium will be lower, the danger of undesired oxidation events is reduced. This means better rubbing fastness, less bronzing and less variation between the edges and the middle parts of jigger dyeing. In this case, the rinsing of the material becomes easier and thus the rinsing time is shortened.[6]

The dyeing of regenerated cellulose fibres using Conventional sulphur dyes and Na_2S causes various difficulties due to the caustic soda formed in the environment. These types of dyes are preferred for dyeing regenerated cellulose fibres, since these difficulties are eliminated in dyeing using water-soluble sulphur dyes and NaHS. Due to the low alkalinity value of the medium, the swelling of the fibres is less and thus the difficulties in dyeing the wound yarn (compression of the windings, circulation difficulties) will be eliminated, and the embrittlement of the material's handle is prevented.[6]

Although the brightness of the mercerised cotton material decreases in dyeing using Na_2S, it is not damaged by dyeing using NaHS. In addition, there is no danger of saponification of acetate fibres when dyeing acetate/cellulose blends using NaHS.[6]

8.1.4 Condensed Sulphur Dyes

Condensed sulphur dyes are in the form of sodium-S-alkyl or S-arylthiosulphate. Although they contain sulphur in their structure, their production methods differ significantly from those of conventional sulphur dyes. Na_2S or polysulphide is required for dyeing, but conventional dyeing methods are not suitable for these dyes.[4,9]

In addition to the four sulphur dye classes described above, there is also a small group of dyes similar to sulphur dyes. These are sulphurised vat dyes, or in other words, sulphur-vat dyes obtained by the sulphurisation process.[1] Sulphur-vat dyes are not classified separately in the CI. These dyes are selected from the CI sulphur and CI vat classes (e.g. CI Vat Blue 42, CI Vat Blue 43, CI Vat Green 7, CI Vat Green 20, CI Sulphur Red 10, CI Sulphur Brown 96, CI Sulphur Black 11) and have very good wet fastness properties. Although their chlorine fastness is not good, their bleaching fastness is better than that of conventional sulphur dyes.[4,6] They can be easily reduced, therefore, starch is not recommended as a finishing agent when dyeing with them. This is because during alkali washings, starch will partially decompose and show reducing properties, causing some dye to bleed by being reduced. The first of these dyes was launched by Cassella in 1909 under the name Hydron Blue R (CI Vat Blue 43; CI 53630). This dye gives reddish-blue hues and has been seen as an inexpensive alternative to indigo since it has better fastness properties than most blue sulphur dyes.[1] They are used especially during the dyeing of inexpensive goods (overalls, etc.), since they are cheaper compared to vat dyes and their dyeing processes are easier compared to indigo. They are between the sulphur and vat dyes in terms of dyeing technique. These dyes cannot be completely reduced with Na_2S. For this reason, they are usually vatted with hydrosulphite and caustic soda, as in vat dyes. In some cases, Na_2S can be used to replace some of the hydrosulphite. The amounts of hydrosulphite, caustic soda and Na_2S required for dyeing are calculated from the diagrams in the catalogues. With sulphur-vat dyes, dyeing can be done according to both discontinuous (apparatus, jigger) and continuous methods (pad-roll, pad steam). In dyeings based on impregnation (semi- and full-continuous), specially prepared "stabilisol" or "sol" form dyes are used instead of powdered dyes. Stabilisol dye is available as a ready-made dispersion as a very fine pigment. They can be impregnated at high temperatures. In its sol form, the dye is in a partially reduced state and forms a finely dispersed dispersion. They are only suitable for impregnation at low temperature (30–40 °C). This is because at high temperatures, the very fine distribution may be deteriorated as a result of oxidation.[6]

8.2 Fastness Properties of Sulphur Dyes

The fastness properties of sulphur dyes are between those of direct and vat dyes.[1] Wet fastness values (such as washing, water, perspiration) are generally good, but bleaching (chlorine and peroxide) fastness values are poor.[1,9] With some exceptions (such as CI Sulphur Black 1 and its equivalents), light fastness values are moderate, rarely above 5.[1] Although there are wide colour gamuts, sulphur dyes are mostly used for dyeing dark shades, since light shades have poor light fastness.[2] Rubbing fastness is at medium levels.[9] The main reason for the demand for sulphur dyes is their good fastness properties and low prices.[1]

8.3 End-uses of Sulphur Dyes

Although sulphur dyes are used especially for cotton and viscose materials, they can also be used for dyeing blends of cellulose fibres with synthetic fibres (including polyamide and polyester). They are rarely used for dyeing silk. Sulphur dyes play almost no role in textile printing, with the exception of black shades.[2] Since the shades of the dyeings obtained with sulphur dyes are quite dull, these dyes are mostly used for dyeing of cheap, ordinary materials (such as cord fabric, workwear, warp thread for denim fabric, sewing thread) that are not mercerised, particularly in dark and dull colours such as blue, black, navy blue, brown, olive and khaki. There is no true red colour, only dull burgundy colours are available.[1,2,4,6,9] In addition to the cost advantage, the dyed materials also have good fastness properties (albeit lower than that of vat and naphthol dyes).[6,7]

8.4 Forms of Sulphur Dyes

Today, sulphur dyes are available in many different forms such as powder, pre-reduced powder, granule, disperse powder, disperse paste and liquid, water-soluble form.

a. ***Powder form:*** They are highly durable. Therefore, they are preferred in long-distance transportation.[4] Amorphous powders are insoluble or partially soluble in water. They are made soluble by boiling with Na_2S and water.[2]
b. ***Pre-reduced powder form:*** Na_2S or hydrosulphite (dithionite) is added to obtain the leuco form.[4]
c. ***Granular form:*** These are pre-reduced dyes. The amount of Na_2S or hydrosulphite and electrolyte is adjusted to form granules.[4]
d. ***Disperse powder form:*** It is based on forming microparticles in the presence of a disperse system. The drying process is controlled and is done in the presence of a coagulation inhibitor to prevent aggregation.[4] The dispersible pigments can be used in dyeing by impregnation methods by adding a dispersing agent. These may contain a certain amount of reducing agents in their formulations, and in this case they are called "partially reduced pigments".[2]
e. ***Disperse paste form:*** This is the form that has not been dried. Therefore, they are cheaper than the disperse powder form.[4]
f. ***Liquid form:*** Liquid forms are obtained by adding caustic soda and sodium hydrosulphite. These additions are made directly to the thiolating melt. Sometimes hydrotropic substances are added to preserve the reduced dye in solution. Partially reduced liquids are more concentrated than fully reduced liquids. This also reduces the transportation cost. An additional reducing agent is required for the reduction of these dyes. However, this reducing agent may be glucose or dithionite to reduce the sulphur content in the waste.[4]
g. ***Water-soluble form:*** They can be in powder or liquid form. Liquid brands are less expensive than powder brands, as there is no drying cost. These are thiosulphuric acid derivatives of sulphur dyes.[4] They exist in the structure of Bunte salt ($D\text{-}S\text{-}SO_3Na$). They are soluble in hot water but have no affinity for fibres. The addition of alkali and reducing agents to the liquor makes them substantive towards the fibres.[2] In addition, the bronzing problem, which will be explained later, is very low with these dyes.[4]

8.5 Dyeing with Sulphur Dyes

With sulphur dyes, both discontinuous (apparat, jigger, jet, etc.) dyeing by the exhaust method and continuous dyeing by the impregnation method (pad-batch, pad-roll, pad steam, etc.) can be performed.

8.5.1 Chemicals Used in the Dyeing Process and Their Functions

The chemicals and auxiliaries used during the dyeing process with sulphur dyes are as follows:

- ***Reducing agents:*** Na_2S, NaHS and thiourea dioxide are most commonly used. As alternative reducing agents, binary mixtures of glucose and sodium dithionite, glucose and hydroxyacetone, or glucose and formamidine sulphinic acid are also used.
- ***Alkalis:*** Soda ash or caustic soda is used as alkali.
- ***Salts:*** Sodium chloride or sodium sulphate is used as salt.
- ***Complexing agents:*** In some cases, especially dyeing processes where the liquor circulates, EDTA and polyphosphates are used to protect against the negative effects of alkaline earth metals.[2]
- ***Antioxidants (stabilising agents):*** Na_2S, NaHS and sodium polysulphide are used as stabilising agents for glucose. However, sodium polysulphide is not compatible with hydrosulphite.[4]
- ***Dispersing agents:*** Generally, naphthalenesulphonic acid-formaldehyde condensates, ligninsulphonates and sulphonated oil-based products are used.[7]

The functions of chemicals used in dyeing are explained below.

a. ***Reducing agents***

Reduction of sulphur dyes can be performed at lower potentials (400 to 500 mV) compared to those for vat and indigo. Therefore, the most commonly used reducing agents are Na_2S and NaHS. Sulphur compounds are used as reducing agents in 90% of the sulphur dyes used in the world. In some cases, they are included in the formulation of the dye, and sometimes they are added to the dyeing liquor. While a part of the compound put into the liquor is consumed, the remaining part is found in the wastewater and causes an increase in chemical oxygen demand (COD), as well as bad odour and toxicity issues. There are other ecological and toxicological problems caused by sulphur. Above pH 9, they cause an unpleasant odour. They release hydrogen sulphide (H_2S) gas at neutral and especially acidic pH values. This situation manifests itself in the form of the smell of rotten eggs and creates a toxic problem when inhaled. Therefore, sulphur-containing compounds should be eliminated.[7] For this purpose, studies have been carried out on the usability of different reducing agent alternatives in sulphur dyeing in recent years.

- **Na_2S:** It is a reducing agent that is most commonly used in the dyeing process with conventional sulphur dyes, is cheap and has sufficient reducing power. Since Na_2S is available in various forms in the market, knowing the type of Na_2S used is essential for preparing a correct recipe.[6] They are found as crystals (30–35%) or in sequin form (usually 60–62%, but may be 50% in some parts of the world).[7]

  Generally, calcined Na_2S is less expensive than crystalline Na_2S, but it is not as clean as the latter. Especially during the dyeing of the wound yarns (such as cross bobbin, warp beam), the windings act as a filter and collect the dirt on them. One recommendation to partially prevent this unpleasant phenomenon is to dissolve calcined Na_2S in a concentrated form (1:2–1:5) first, wait and use after the impurities have collapsed. Since Na_2S will damage copper, aluminium and their alloys, devices made of these should not be used during dyeing with sulphur dyes.[6]

The amount of Na_2S to be used depends on the dye and percentage of dyeing. It is generally recommended to use crystalline Na_2S 1–3 times the weight of the dye. The exact ratio is indicated for each dye in the catalogues. It is recommended to use more in yellow, orange, brown and khaki colours. The amount specified in the catalogue may not be sufficient, especially if light shades are dyed in long liquor ratios. In this case, it should be ensured that the concentration in the liquor is at least 4–5 g/L by increasing the amount of Na_2S. For example, if 100 kg of material is to be dyed at a 1:20 liquor ratio with 1% sulphur dye, and if 1.5 times of crystalline Na_2S should be used for this dye, for 100 kg of material, 1 kg of dye and 1.5 kg of Na_2S will be required. Since this amount will be put into $100 \times 20 = 2000$ L liquor, the Na_2S concentration will be 1500 g/2000 L, that is, 0.75 g/L. This amount is insufficient and a concentration of at least 4 g/L. is required. Therefore, it is necessary to use 8–10 kg of crystalline Na_2S, not 1.5 g/L, for this dyeing.[6,7]

- **NaHS (Na_2S, sodium hydrosulphide):** It has a wide range of uses, especially in dyeing with water-soluble sulphur dyes. It has now largely replaced Na_2S and is available in liquid, powder, or many different concentrations. The amount of use is similar to Na_2S, but the 35% form of NaHS is equivalent to the 60% sequin form of Na_2S.[7] A critical point to be considered in dyeing made using Sulfhydrate F is the addition of sufficient soda ash to the medium. This is because acid is released during the reduction of water-soluble sulphur dyes. If this acid is not neutralised by adding soda ash, the acid and NaHS will react with each other to produce toxic and very foul-smelling H_2S gas. For this reason, 3–4 g/L soda ash should be added to the liquor in dyeings made at long liquor ratios, and in dyeing done at short liquor ratios, soda ash should be added at the rate of 1/5–1/4 of the weight of Sulfhydrate F used. Care should be taken to add soda ash to the liquor before Sulfhydrate F.[6] If desired, 5 grams of caustic soda can be added for every 20 grams of NaHS (35%) instead of soda ash.[7]
- **Hydrosulphite (Sodium dithionite) ($Na_2S_2O_4$):** Hydrosulphite alone is not preferred because the system is difficult to control and does not yield stable results. However, hydrosulphite can be used for dyeing with CI Sulphur Black 11.[4]

  The caustic/hydrosulphite ($NaOH/Na_2S_2O_4$) system has been the only sulphur-free reduction system for sulphur dyes for many years. However, the difficulty of the controllability of the system and its lack of reproducible results prevented it from becoming widespread. Sodium carbonate/hydrosulphite ($Na_2CO_3/Na_2S_2O_4$) is only suitable for water-soluble sulphur dyes because it is a weak alkali system for water-insoluble sulphur dyes. Attention should be paid to the risk of poor colour formation as a result of excessive reduction.[6] For example, CI Solubilised Sulphur Red 6, CI Solubilised Sulphur Brown 12, CI Solubilised Sulphur Brown 15, CI Solubilised Sulphur Green 3, CI Solubilised Sulphur Green 26 are not suitable for hydrosulphite reduction. However, some dyes such as CI Solubilised Sulphur Red 11, CI Solubilised Sulphur Brown 93 give better results with hydrosulphite than with Na_2S or NaHS.[4]
- **Glucose (dextrose monohydrate)/hydrosulphite:** The use of glucose with sodium carbonate, caustic soda or a mixture of both is increasing due to environmental conditions. However, there are some problems. Since it is difficult to stabilise

glucose at high temperatures, not all dyes can be reduced. For this reason, binary systems such as glucose/hydrosulphite have been tried and it has been determined that they can be used in the application of sulphur dyes. In an alkaline environment, glucose loses its reducing ability by forming various degradation products more or less rapidly. However, if hydrosulphite is added to the system, more stable products are formed as shown in Figure 8.3.

Figure 8.4 shows the comparison of the redox potential of the glucose/hydrosulphite system with conventional reducing agents. As can be seen from the figure, glucose alone

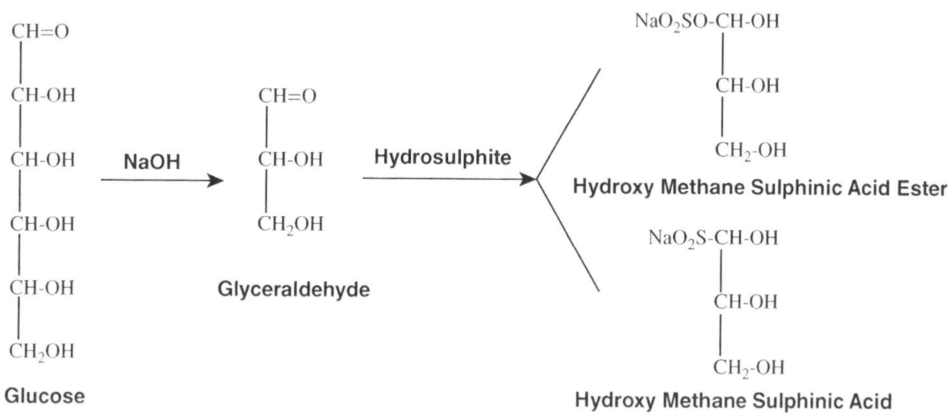

Figure 8.3 Glucose/hydrosulphite reduction system. Ref. [7]/with permission of Tekstil ve Konfeksiyon.

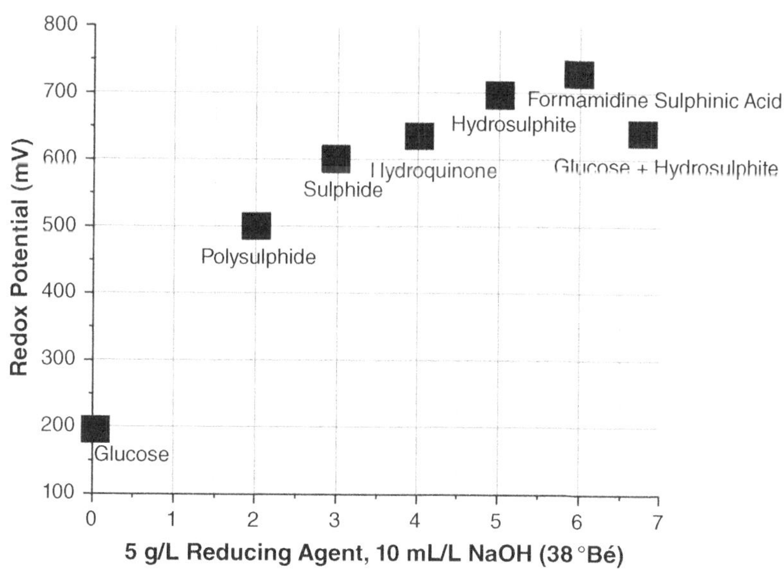

Figure 8.4 Redox potentials of various reducing agents. Ref. [7]/with permission of Tekstil ve Konfeksiyon.

cannot provide sufficient potential for the reduction of the sulphur dye, whereas the glucose/hydrosulphite reduction system provides a sufficient reduction potential.[7]

The combination of glucose and hydrosulphite results in lower COD levels compared to using Na_2S. Glucose and hydrosulphite react together in the dyebath and quickly form addition products with optimum reducing properties for sulphur dyes. The redox potential of this binary system is lower than that of the hydrosulphite alone because glucose lowers the redox potential of the hydrosulphite. This ensures a good dye yield and eliminates the risk of excessive reduction and damage to the sulphur dyes. The advantages of the glucose/hydrosulphite system can be listed as follows:[7]

✓ There is no danger of excessive reduction of the dye.
✓ Dyeing efficiency is almost the same as using sulphur.
✓ It provides good reproducibility.
✓ It gives lower COD values in wastewater compared to dyeing with hydrosulphite alone.
✓ It provides 10–15% higher dyeing efficiency compared to dyeing with glucose only.
✓ Undesirable odour formation is not an issue due to the breakdown products of glucose.
✓ Reductive washing is not necessary when dyeing PES/CO blends.[7]

Other than glucose/hydrosulphite, other binary reducing systems provide similar benefits. Examples of these are glucose/hydroxyacetone, glucose/sulphinic acid.[7]

- **Thioureadioxide:** Thioureadioxide, which is used together with caustic soda, provides limited use. Although the wastewater has a lower COD/BOD value compared to the caustic soda/hydrosulphite system, its use has not become widespread since reproducible production results could not be obtained.[7]
- **Thiogleucol (2-mercapta ethanol):** This reducing agent, which is not suitable for conventional sulphur dyes, can be used for dyeing for water-soluble sulphur dyes by the exhaust and one-bath pad-steam method. Although it provides lower colour yield than Na_2S when combined with caustic soda, it can be used to reduce most water-soluble sulphur dyes. The most important advantage is that the waste is sulphur- and odour-free. The most important disadvantage is the high cost of dyeing.[7]
- **Hydroxyacetone:** As with thioglycol, the colour yield is lower and the cost is higher compared to the use of Na_2S. In addition, hydroxyacetone has the ability to ignite. Dye baths have a characteristic odour similar to acetone. However, when used with soluble sulphur dyes as in thioglycol, the wastewater does not contain sulphur. The advantages of hydroxyacetone can be listed as follows:
  ✓ less chemical use,
  ✓ lower COD value due to biodegradation and
  ✓ wastewater does not contain sulphur, sulphite or sulphate.[7]

The main environmental benefit from using sulphur-free reducing agents is the minimisation of the sulphur content in wastewater. In order to obtain the maximum environmental and economic benefit from the proposed technique, the condition of primary importance is to consume only the amount of reducing agent required for the reduction of the dye. Therefore, consumption of the reducing agent by the oxygen inside the machine should be prevented as much as possible. An effective method to achieve this is to use nitrogen to remove oxygen from the liquor and air in the machine.[2]

In general, in all application methods of sulphur dyes, a certain amount of reducing agent is required. Since all reducing agents are non-regenerative chemicals, studies have also been carried out to find systems that can be regenerated. Indirect electrolysis, which can be used successfully for vat and indigo dyes, has also been tried for the reduction of sulphur dyes. Indirect electrochemical reduction is preferably used in dyeings made by the exhaust method, which does not have the problem of dilution of the dyebath and is easy to regenerate. Studies on cathodic reduction have shown that this method provides better reduction and that the colour depth and hue are similar to dyeings done under standard conditions.[7]

b. ***Alkalis***
In the sulphur dyeing process, dye, Na_2S and, if desired, a complexing agent, as well as soda ash are added to the dyeing liquor. The purpose of adding soda ash is to stabilise the liquor and prevent precipitation of calcium salts. Generally, calcined soda ash is used 2–4% of the weight of the material to be dyed, depending on the depth of the dyeing.[6]

c. ***Salts***
During dyeing with sulphur dyes, Glauber's salt or table salt must be added to the liquor. The purpose of the salt addition is to facilitate and accelerate the uptake of dyes by the fibres. Addition of 5–20 g/L salt is sufficient according to the depth of the dyeing and the liquor ratio. Using too much salt causes bronzing. If it is difficult to obtain a uniform dyeing, it is beneficial to add salt towards the end of the dyeing, and it would be appropriate to completely abandon the addition of salt in light-shade dyeing and when dyeing mercerised cotton.[6]

d. ***Complexing agents***
It is essential to use soft water during dyeing with sulphur dyes. This is because most of the sulphur dyes form precipitates with the substances that cause the hardness of the water. Especially in wound yarn dyeing, these precipitates are filtered by the windings and cause the quality of dyeing to decrease. For this reason, it is useful to add some complexing agent (sequestering agent) to the liquor.[6]

e. ***Antioxidants (stabilising agents)***
In order to prevent premature oxidation and thus stabilise the dyebath, sodium polysulphide should be added to the dyebath as a stabiliser for Na_2S, NaHS and glucose. Thus, the risk of bronzing effect and reduced rubbing fastness is reduced. However, sodium polysulphide is not compatible with hydrosulphite.[4] Polysulphides should also be considered when talking about the wastewater load. They can cause the same problems as Na_2S or NaHS. Furthermore, acidification of polysulphides leads to elemental sulphur. This can turn into SO_2 form on dyed materials and cause bad odour.[7]

f. ***Dispersing agents***
Naphthalenephonic acid-formaldehyde condensates, ligninsulphonates and sulphonated oil-based products are used as dispersing agents.[7] It is necessary to use a dispersing agent, especially in the process steps where the dye has not yet been reduced or is converted back to its initial form by oxidation.[2]

8.5.2 Chemicals Used in Post-processes and Their Functions

The first step after dyeing is rinsing. After rinsing, the dyeing is always finished by an oxidation process, forming water-insoluble dye molecules within the fibres.[6] Oxidation can be carried out using a variety of chemicals such as,

- air
- potassium or sodium bichromate/acetic acid
- sodium chloride
- potassium iodate/acetic acid
- sodium bromate
- sodium perborate
- hydrogen peroxide and peroxide compounds
- special oxidising agents.

Detailed information about each of them is provided below.

a. ***Oxidation by air:*** This method of oxidation, achieved by rinsing the material with abundant amounts of cold and then hot water and suctioning air through the wound yarns after rinsing on the wound yarns is not always recommended. It should never be applied as air oxidation in some blue colours will cause an uneven appearance. When working with other dyes, it should be kept in mind that the full colour shade of the dyeing will be obtained at the end of air oxidation and the colour shade will change during storage.[6]
b. ***Oxidation with potassium or sodium bichromate/acetic acid:*** Chromium (VI) compounds rapidly and completely oxidise the sulphur dye in the reduced state.[2] Although the colour shades obtained as a result of bichromate oxidation are a slightly dull, it is a widely used oxidation method especially in piece dyeing due to its good wet fastness and repeatability. In wound yarn dyeing, this oxidation method is not recommended as the chrome residues that will remain on the yarns will result in a harsh handle. Oxidation is carried out in a liquor containing 1–1.5% potassium bichromate and 3% acetic acid (30%) (pH 4–5), first in the cold, then at 60 °C for 15 minutes. Then rinsing is performed with hot and then cold water to remove chromium salt residues. If 0.5–1% copper sulphate is added to the liquor during oxidation with potassium bichromate, there is an increase in the light fastness of the dyeing.[6] An important point is not to add copper sulphate until all the waste Na_2S is broken down and the pH is below 6. This is because this leads to a harder handle and is not recommended for black dyeing.[7] This method is not preferred much today, since it only causes a chromium load in wastewater.[6]
c. ***Oxidation with sodium chlorite:*** Chlorite oxidation gives dyeings good wet fastness values and a soft handle. However, it is not sufficient for the complete oxidation of some red-brown sulphur dyes. Moreover, it is environmentally unsuitable for use, as it causes a load of adsorbable organo halogen (AOX) in wastewater.[7]
d. ***Oxidation with potassium iodate/acetic acid:*** The use of chromium compounds is currently restricted. A substitute for dichromate is potassium iodate. However, this oxidation method is environmentally unsuitable for use, as it causes an AOX load in wastewater.[7]

e. ***Oxidation with sodium bromate:*** It is not sufficient for the oxidation of sulphur dyes alone or together with acetic acid. The properties of the oxidation system approach those of sodium bichromate/acetic acid when catalytic amounts of metavanadate are added. However, metavanadate causes pollution in wastewater.[9] In addition, an AOX load is generated in the wastewater due to bromine.[7]

f. ***Oxidation with sodium perborate:*** Vivid colour shades are obtained as a result of oxidation with sodium perborate. It gives a slight reddish hue compared to hydrogen peroxide. The wet fastness values of dyeing are slightly lower than those obtained by bichromate oxidation. The thoroughly rinsed material is treated initially with a liquor containing 1% sodium perborate in the cold, and the temperature is increased up to 40 °C. The oxidation time is 15 minutes, and after oxidation, it is thoroughly rinsed with cold water.[6]

g. ***Oxidation with hydrogen peroxide and peroxide compounds:*** The results obtained as a result of oxidation with hydrogen peroxide are similar to those of perborate. It has the drawbacks of slightly lower wet fastness values (one or more grey scale digits) and low repeatability compared to oxidation with potassium bichromate, despite yielding vivid hues. If the oxidation with both perborate and hydrogen peroxide is carried out an acetic acid environment, the colour shades become slightly duller, but wet fastness values and reproducibility properties improve.[6,7] Oxidation with hydrogen peroxide can be done in two different environments:
   - ***Oxidation with hydrogen peroxide in acidic medium:*** Oxidation with hydrogen peroxide in an acidic medium is a safe method. The obtained fastness values and shades are comparable to those of other classical oxidising agents.[7] The well-rinsed material is treated for several minutes with a liquor containing acetic acid. The pH of the liquor should be around 5. 1–1.5% sodium perborate or 0.7–1% perhydrol is added and oxidation is carried out for 15–20 minutes by heating in cold liquor up to 50–60 °C.[6]
   - ***Oxidation with hydrogen peroxide in alkaline medium:*** Oxidation with hydrogen peroxide at 40 °C in alkaline medium (pH around 10) is a simple method. In this method, lighter shades and lower wet fastness values are obtained. This oxidation method is mostly used in the dyeing of wound yarn.[7]

h. ***Oxidation with special oxidising agents:*** Especially factories producing sulphur dyes put special oxidising agents on the market under various names. These commercial products, which contain additives such as dispersing agents in addition to the oxidising agent, can provide slightly better results in some cases compared to other oxidising agents.[6]

The oxidation process after dyeing used to be done with bichromate and acetic acid. Today, however, the use of chromium salts has decreased due to environmental pollution.[8] Oxidising agents such as bromate, iodate and chloride cause the formation of organic halogen compounds in wastewater that give high levels of AOX.[7] Therefore, while deciding which oxidising agent to choose and whether to perform the oxidation in an acidic or alkali environment, besides the changes in fastness and colour shade, economic and ecological criteria should also be taken into account. Especially in terms of ecology, the product that has found the most common usage area in recent years is hydrogen peroxide.[8]

8.5.3 Dyeing by Exhaust Method

Most of the sulphur dyes give the best results in dyeing at 80–90 °C. Some red, brown, blue and green dyes give the best results (fuller and more vivid colours) in dyeing at 60 °C. The most suitable dyeing temperature for each dye is specified in the catalogues.[6]

Since a large amount of dye remains in the waste liquor in dyeing with sulphur dyes, this waste liquor can be reused by adding an amount of dye equivalent to that consumed (dyeing in a standing bath). This system is especially suitable for dyeing with a single dye such as black, where the same colour will be dyed many times. After dyeing, the waste dye liquor is taken into the tanks and the dye solution prepared separately and the calculated amount of Na_2S, soda ash and salt are added. However, dyeing in a standing bath is not recommended due to possible differences in the affinity and fixation properties of each dye in mixed dyeing.[9]

A sample recipe and dyeing graph of yarn dyeing in apparatus with conventional sulphur dye are given in Figure 8.5.

After dyeing with sulphur dyes, rinsing, oxidation and hot washing are performed. Then, neutralisation with soda ash is performed to adjust the pH of the material, followed by soaping using neutral soap and soda ash at boiling temperature, followed by hot rinsing at 85 °C. If desired, post-treatment with a cationic fixing agent and then softening treatments are performed. Finally, washing is done with a weak alkali (2–3 g/L soda ash) or an alkali salt (2–3 g/L sodium acetate) at weak alkaline pH in order to prevent the strength losses that may occur as a result of the formation of sulphuric acid on the fabric during storage.[9]

Instead of the conventional sulphur dyes used in classical dyeing, newly developed sulphur dyes, in which the glucose/hydrosulphite system can be used for reduction, are used as previously explained in ecological dyeings. The use of the glucose/hydrosulphite

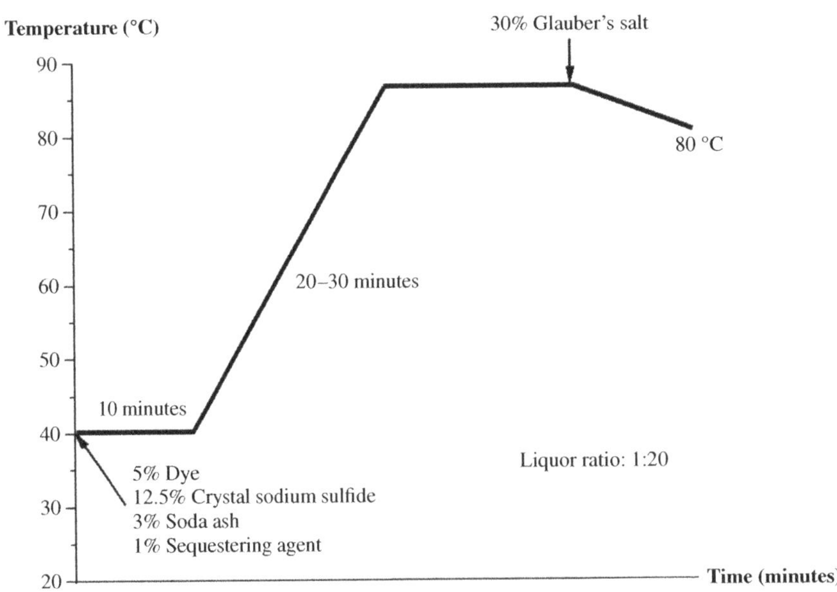

Figure 8.5 An example recipe and dyeing graph for yarn dyeing in the apparatus with Conventional sulphur dyes. Ref. [6].

system instead of Na_2S in conventional dyeing helps to keep the COD value at lower levels compared to applications with Na_2S. On the other hand, the oxidation process after dyeing has great importance on AOX compounds and heavy metals in wastewater. For this reason, chlorinated compounds that cause high AOX values in wastewater and chromium compounds such as potassium bichromate, which are directly harmful to health, have been replaced by the more environmentally friendly hydrogen peroxide.[7]

With CI soluble sulphur dyes, cationic pre-treated cellulosic materials can be dyed without the use of reducing agents. In this special process applied in piece dyeing, the fastness values are slightly lower than in classical dyeing. The material is first treated with a suitable cationic resin, and then dyeing is done by increasing the dyeing temperature. Cationic fixation can also be done continuously by the pad-dry-pad-steam method.[4]

8.5.4 Dyeing by Impregnation Method

Generally, impregnation methods are not recommended for conventional sulphur dyes. Because these dyes have limited solubility in water after reduction with Na_2S. In order to increase the solubility in water to some extent, it is necessary to heat the liquor. On the other hand, as the temperature increases, the penetration of the dye into the fibres will increase, making it difficult to obtain batches with the same depth from start to finish. However, colours such as black and navy blue, where colour differences cannot be noticed quickly, can be dyed continuously. Water-soluble sulphur dyes, on the other hand, are suitable for continuous dyeing as they do not show affinity for cellulosic fibres. This is one of the most important advantages of water-soluble sulphur dyes over conventional sulphur dyes.[6] By the impregnation method, dyeing can be done according to one of the pad-batch, pad-roll, pad-jig, pad-dry, pad-thermosol and pad-steam methods. Among these, the method that finds the most common usage area today is the pad-steam method. Each of these methods is described separately below.

a. ***Pad-batch method:*** After padding in the cold (20–25 °C) with dyes and chemicals, the dye is fixed by keeping it at room temperature for 2 hours. Rinsing, oxidation and post-treatments are then carried out.[6]
b. ***Pad-roll method:*** It is also possible to dye by the pad-roll method with sulphur dyes. In this method, after impregnation of the material with dye and chemicals, it is necessary to keep the material in a hot batching room at 100 °C for 1 hour. Then, rinsing, oxidation and post-processing are carried out.[6]
c. ***Pad-jig method:*** In this method, the material is treated in a jigger with a liquor containing chemical substances after impregnation with dye in the padder. Then rinsing, oxidation and post-treatments are carried out.[6]
d. ***Pad-dry method:*** The material impregnated in the padder is dried with a liquor containing dyes and chemicals. In the meantime, the fixation of the dye is also ensured. Then rinsing, oxidation and post-treatments are carried out.[6]
e. ***Pad-thermosol method:*** This method can only be used to dye light shades evenly. The material, which is impregnated with the liquor containing dyes and chemicals, is treated with dry heat for the fixation of the dye, followed by rinsing, oxidation and post-treatment.[6]

f. ***Pad-steam method:*** The material is impregnated with dye, reducing agent and wetting agent in one-bath or two-bath processes. In the one-bath pad-steam method, the reducing agent and dye are added to the liquor at the same time. In the two-bath pad-steam method, the material is first impregnated with a liquor containing dye and wetting agent, and if necessary, the reducing agent is applied in the second step after an intermediate drying. The material is then exposed to fully saturated steam. Afterwards, rinsing, oxidation and rinsing processes are applied.[2]

Below are the sample recipe and process steps for dyeing by the two-bath pad-steam method.

30 g/L dye (dissolution of dye is done by mixing with hot soft water and boiling for 3–4 minutes) and impregnation at 50 °C in a padder with pick-up value of 70% with a liquor containing 1 mL/L wetting agent → Intermediate Drying → Air Passage → Impregnation at 30 °C in the padder with pick-up value of 80% with the liquor containing 70 mL/L NaHS, 17.5 g/L soda ash, 20 g/L sodium sulphate and 20 mL/L first dyeing liquor → Steaming (at 102–105 °C 30–60 seconds) → Air Passage → Rinse → Oxidation → Drying.[6]

8.6 Problems Encountered in Dyeing with Sulphur Dyes and Measures to be Taken

8.6.1 Bronzing Effect

If care is not taken during dyeing with sulphur dyes, the dye in the reduced state but not penetrated into the fibres is oxidised on the fibres and mechanically bonded to the fibres. This phenomenon, which causes the rubbing fastness to decrease and the dyeing to gain a metallic appearance, is called **"Bronzing Effect"**.[6] The bronzing effect usually occurs in navy blue and blue tones. The reasons can be listed as follows:

- due to insufficient amount of reducing agent in the dyebath, the formation of superficial dyeing and its rapid oxidation,
- excessive dye on the material due to high dye or salt concentration, and
- delay between draining the liquor, and rinsing and oxidation steps after dyeing.[9]

The first precaution to prevent bronzing is to use a sufficient amount of reducing agent and to ensure that the dye dissolves completely.[1] The second most important precaution is that after dyeing, before oxidation, the excess dyeing liquor is quickly and effectively removed from the material, that is, effective rinsing is carried out.[6] In addition, the surfactants, complexing agents, dispersing agents, dyebath stabilisers and antioxidants added to the liquor also partially prevent bronzing.[6,9] In addition, special auxiliaries to be added to the liquor act in a way to prevent premature oxidation of the reduced dye. Thus, the part of the dye that is oxidised and mechanically bonded on the surface of the material is reduced. These substances also prevent the edges from being dyed differently compared to the middle in the dyeings made in the jigger. On the other hand, it should not be forgotten that the use of too much salt in dyeing will increase the risk of a bronzing effect.[6] Therefore, care should be taken to use the appropriate amount of salt. In order to eliminate this situation in the product that has a bronzing effect, treatment with 2–3 g/L Turkey red oil and 1–2 g/L ammonia solution at 50–60 °C for 10–15 minutes or treatment with soap solution at boiling temperature can be useful.[9]

8.6.2 Strength Loss During Storage

Some of the sulphur of the dye can turn into sulphuric acid with the effect of heat, humidity and air during storage in materials dyed with sulphur dyes in dark shades such as navy blue and black, in which a large amount of dye is used. This causes damage to the fibres.[6,9] Therefore, if the material is subjected to post-treatment with alkali substances such as soda ash (5 g/L) or sodium acetate (10 g/L) after dyeing, the soda ash or sodium acetate on the material will immediately neutralise the sulphuric acid that may be formed during storage. However, as the soda ash or sodium acetate in the material will be removed after washing, their protective effect disappears.[6] Therefore, rinsing should not be done after this process in order to ensure that the final pH of the fabric is slightly alkaline.[9] Special auxiliaries have also been introduced to the market. These protect the material against sulphuric acid that may be formed during storage, and their effects are resistant to washing.[6]

8.6.3 Low Washing and Rubbing Fastness Values

Low washing and rubbing fastness values are usually caused by the dye not being dissolved well enough, dye precipitation, low solubility of dyes, poor or insufficient washing of unfixed dyes after dyeing, and poor or insufficient soaping process. In order to obtain good fastness properties,

- the dye must be completely dissolved and filtered before being added to the dyebath. If there are undissolved dye particles, they adhere to the surface of the material and cause uneven dyeing and low washing and rubbing fastnesses.
- considering the maximum solubility of the dye, a sufficient amount of water should always be used for dissolving the dye.
- water and salts should not contain calcium and magnesium ions. Because if these ions are present, they form salts that are insoluble in water and precipitate in the dyeing machine in the form of sludge.
- washing after dyeing before proceeding to the next process steps such as oxidation and neutralisation should be effective in removing waste dye and chemicals.[9]

Beyond what is explained above, if dyeing is done on fabrics containing starch size before the starch is completely removed, the size will be dyed as the starch size has a slightly reducing property, which causes the rubbing fastness to decrease. For this reason, it must be ensured that the desizing of fabrics containing starch size is carried out completely and effectively.[6]

8.6.4 Low Peroxide Fastness Values

Since the peroxide fastness of sulphur dyes is generally low, special auxiliaries that increase peroxide fastness should be used.[8] This is because most of the washing detergents in the market contain perborate type peroxides. Therefore, various auxiliaries have been developed by various dye factories to improve the fastness of sulphur dyes. For example, the fastness of the material to washing at the boil, peroxide, soda ash boiling and mercerisation increases in the post-treatment with an alkylating agent-based auxiliary. This is because such substances provide the improvement of fastness by alkylating and oxidising the reduced dye as shown below.[6]

$$2D\text{-}SNa + {}^{-}Cl^{+}[R]^{+}Cl^{-} \rightarrow D\text{-}S\text{-}R\text{-}S\text{-}D + 2NaCl$$

Since oxidation also occurs during such a process, there is no need for peroxide, perborate or bichromate oxidation. Air oxidation is required before this process only in quinonimine type dyes. According to the exhaust method, the thoroughly rinsed material is treated with a liquor containing 2–3% auxiliary, 1–1.5% calcined soda ash, 1–2 g/L dispersing agent in cold temperatures for 10–15 minutes, then the temperature is increased to 80–85 °C and the material is treated at this temperature for 5–10 minutes. It is then rinsed with hot and cold water. In the processes based on impregnation, the material is rinsed in the first three vessels of the open-width washing machine and then thoroughly squeezed, and in the fourth vessel, filled as low as possible (enough to cover the lower transport rollers), it is impregnated with the above-mentioned auxiliary (20–30 g/kg) and soda ash (10–15 g/kg) and steamed for 1 second at 100 °C. Then it is rinsed with hot and cold water.[6]

8.6.5 Obtaining Low Colour Yield

Low colour yield may occur due to reasons such as an insufficient amount of reducing agent, presence of calcium salts in water and salt, excessive reduction of the dye, excessive oxidation of the dye, etc.[1] For this reason, attention should be paid to the reduction and oxidation steps, and care should be taken to work under optimal conditions.[9]

8.6.6 Uneven Dyeing

If unevenness has occurred as a result of dyeing, partial removal and levelling of the dye can be achieved by treating with a blank liquor containing Na_2S, disperse agent, complexing agent and wetting agent at 80–90 °C.[9]

References

1 Mansoor, I. Textile Dyes. Karaçi, Pakistan: Rehbar Publishers Karachi, 2008.
2 Türkiye Tekstil Terbiye Sanayiciileri Derneği. IPPC Best Available Techniques (BAT) reference document for the textiles industry, 2002.
3 Rattee, I.D. The Colouring of textiles. In Chemistry of the Textiles Industry (Ed. C. Carr, Chapter 9, 276–332). Hollanda: Springer Science, 1995.
4 Senior, C. Dyeing with sulphur dyes. In Cellulosics Dyeing (Ed. J. Shore, Chapter 6, 280–319). Bradford, West Yorkshire, UK: Society of Dyers and Colourists, 1995.
5 Kaisha, N.K.K. E.U. Patent No. 0325172B1. Tokyo: European Patent Office, 1994.

6 Tarakçıoğlu, I. Tekstil Boyacılığı I Teksiri. İzmir, Türkiye: Ege Üniversitesi Mühendislik Fakültesi Çoğaltma Yayınları, 1982.
7 Akçakoca, E.P., Atav, R. Kükürt boyaları ile ekolojik boyama. Tekstil ve Konfeksiyon, 2004; 14(1), 42–46.
8 Yurdakul, A., Atav, R. Boya baskı esasları. Bornova, İzmir: Ege Üniversitesi Tekstil ve Konfeksiyon Araştırma Uygulama Merkezi Yayını, 2006.
9 Mahapatra N.N. Textile Dyes. Yeni Delhi, Hindistan: Woodhead Publishing India, 2016.

9

Dyeing with Reactive Dyes

Initially, studies on reactive dyes were initiated by Imperial Chemical Industries (ICI) Limited in 1951. In fact, the initial aim was to develop reactive dyes for wool fibres, which are known to be highly reactive to various chemicals. For this purpose, a wide variety of dyes containing active halogen atoms were studied. In 1953, Dr W.E. Stephen synthesised the 2,4-dichloro-s-triazine-6-amino dye by reacting cyanuric chloride with an amino dye (Figure 9.1).[1]

Later, he handed this dye over to Ian Rattee to study the application technique and dyeing mechanism of the dye. Rattee found that this dye was not suitable for wool but began working on the dyeing behaviour of this dye for cotton, considering that cyanuric chloride is used in the modification of cellulose in xylene solvent. It was determined that the dye fixed to cellulose in a weakly alkaline solution with excellent efficiency. After this successful result in the initial studies, research commenced on the development of dyes containing new reactive groups. Thus, the use of reactive dyes in the dyeing of cellulosic fibres has received increasing attention.[1]

Today, reactive dyes constitute the most important dye class used in dyeing cellulosic fibres and their blends, due to their advantages such as high wet fastness, wide colour palette, vivid and lustrous colour shades and compatibility with many dyeing methods. More than half of cellulosic fibres are dyed with reactive dyes.[2] In terms of application technique, 53% of reactive dyes are applied via the exhaust method (40% cold brand and 13% hot brand); 26% in dyeing by impregnation method 15% with pad-batch and 11% with continuous dyeing, and 21% in printing. Therefore, it can be said that more than half of reactive dyes are applied by the exhaust method.[3]

The most distinctive feature of reactive dyes, which are so important in cellulose dyeing, compared to other dye classes is that they contain groups (reactive groups) that can react with fibre macromolecules. Depending on the reactivity of these groups, the dyeing conditions (temperature, time and alkali) of various reactive dyes vary significantly.[4]

The advantages of dyeing with reactive dyes can be listed as follows:

1. Their light fastness is generally very good; wet fastness is high. Their washing fastness ratings can be increased by cationic aftertreatment.
2. They allow lustrous and vivid colours to be obtained.

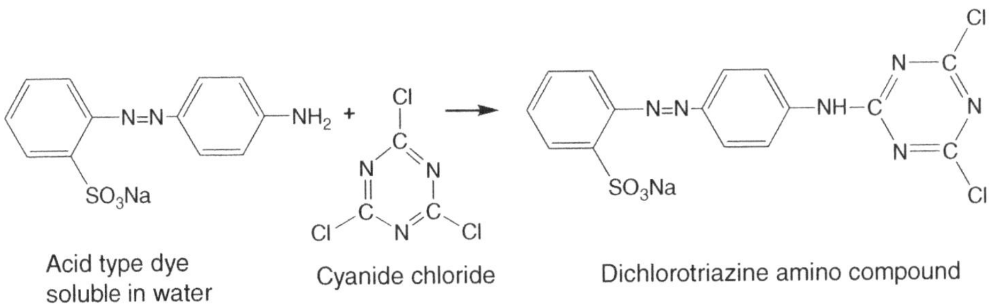

Figure 9.1 Synthesis of dichlorotriazine-based reactive dyes. Ref. [1]/ICI.

3. Their colour gamut is complete.
4. They show good compatibility in combination dyeings.
5. They are suitable for all dyeing methods (both exhaustion and impregnation). Their dyeing methods are simple.
6. They show high reproducibility.
7. They have high solubility.
8. Since reactive dyes containing azo groups can be easily cleaved, they can be easily used as a ground colour in discharge printing.
9. Their prices are lower than those of vat dyes.[5]
10. It is possible to level the dyeing prior to fixation (addition of alkali).[3]

The disadvantages of dyeing with reactive dyes are as follows:

1. Their resistance to basic solutions (mercerisation, boiling basic washing, etc.) and their chlorine fastness are not good. In alkaline processes, especially at high temperatures, some of the dye that is covalently bonded to the fibres is hydrolysed and the bond is broken, converting the dye into a non-reactive form.[5]
2. Perborate wash fastness is not very good with some vinylsulphone types. Over time, colour fading occurs.[5]
3. Their light fastness, especially in red colours, is limited.[2]
4. Generally, their wet light (perspiration + light) fastness is low.[2]
5. Since some of the dye is hydrolysed by reacting with water, the fixation yield is low (60–80%).[2]
6. Washing and post-processing steps after dyeing with reactive dyes are long and time-consuming. If attention is not paid, wet fastness values will be low.[5]
7. Post-dyeing treatment costs are high, and there are also significant water consumption and wastewater issues.[5]
8. They have low affinity for unmercerised products.[5]

In addition to the advantages and disadvantages of dyeing with reactive dyes described above, it should be noted that the swelling property of fibres dyed with reactive dyes and their dissolution in cuoxam are reduced, since the —OH groups of cellulose macromolecules are partially blocked by dye molecules, and even crosslinking occurs between macromolecules when working with some dyes.[4]

9.1 Structure and Classification of Reactive Dyes

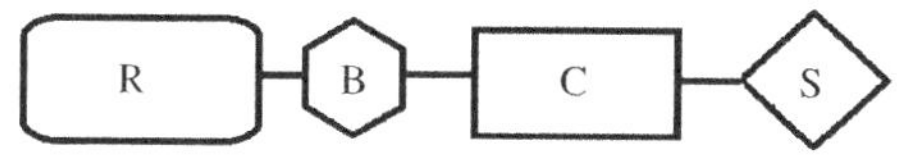

Figure 9.2 Structure of reactive dyes.

The basis of reactive dyes is based on the ability of a chromophore group (azo, anthraquinone, phthalocyanine, etc.) to be chemically bonded directly to the fibres with the help of a reactive group. The reactive group in the reactive dye molecule forms a covalent bond with the hydroxyl groups in the cellulose fibre, and thus the dye chromophore is covalently bonded to the fibres through the reactive group. At least the reactive group in the structure of the dye must be bifunctional (which can react with two molecules), so that it can bind to the chromophore group on the one hand (during dye production) and to fibres (during dyeing) on the other.[4] The structure of reactive dyes is shown schematically in Figure 9.2.

While the chromophore group influences the colour, fastness, levelling, substantivity and application temperature of the dye, the reactive group determines fastness, dischargeability, fixation and application temperature.[3] The groups in the structure of reactive dyes and their properties are as follows:

- ***Water-solubilising group (S):*** In the structure of reactive dyes, there are one or more sulphonic acid groups acting as the solubilising group.[6] Reactive dyes are soluble in water, as they contain groups that impart water solubility, and dyeing is done directly with their aqueous solutions, as in direct (substantive) dyes.[4]
- ***Colouring (Chromophore) group (C):*** It is possible to find a wide variety of chromophores as colouring groups in the molecule of reactive dyes.[6] However, to generalise, it can be said that yellow, orange, red and dark blue colours are azo and bisazo; brown, purple, burgundy and olive colours are metal azo complexes parliament blue and bright blue shades are anthraquinone and triphenodioxazine turquoise and bright green shades are phthalocyanine; and combination blues are copper formazan derivatives.[7]
- ***Bridge group (B):*** These are groups such as –NH–, –CO– and $–SO_2–$ that connect the chromophore group and the reactive group. They also have effects other than acting as a bridge. For example, they influence the reactivity of the reactive group. Dissociation of an amino bridge can reduce reactivity by a factor of 10. In such a case, the substantivity and, accordingly, the dye binding rate decreases. In addition, the most important feature of bridge bonds is that they prevent the cleavage of the dye and the fibre bond.[6]
- ***Reactive group (R):*** Reactive dyes may contain heterocyclic rings and/or vinylsulphone groups as reactive groups.[3] The reactive group is the group that forms covalent bonds with the functional groups in the fibres. The functional groups that can react with the reactive group are hydroxyl groups in cellulose.[6] Since water is present in the dyeing medium, the reactive group of the dye can react not only with the hydroxyl groups of cellulose macromolecules, but also partially with the hydroxyl groups of water (hydrolysis), and the reactive group, which has reacted with water (hydrolysed), loses its ability to react with cellulose. The hydrolysed part is lost for dyeing. On the other hand, some of the hydrolysed dye is mechanically attached to the fibres, causing a decrease in the wet fastness of the dyed material.[4] Since the fibre-dye binding

reaction and the water-dye hydrolysis reaction are in competition with each other, conditions must be adjusted to favour the bonding reaction.[6] Since reactive dyes are always applied in an aqueous medium, it is impossible to completely prevent reactions with water. Efforts are being made to reduce this reaction as much as possible and to allow most of the dye in the liquor to react with cellulose.[4] The success of dyeing with reactive dyes depends on the stability of the covalent bond between the fibre and the dye. It is important that this bond is resistant to hydrolysis in washing and finishing operations.[5]

It is possible to classify reactive dyes according to the structure of the reactive group, the structure of the chromophore group, the type of reaction with the fibre and the control parameters.[5]

9.1.1 Classification According to Reactive Groups

Today, a wide variety of reactive groups are known and used in the structure of reactive dyes.[4] About 2/3 of the reactive dyes contain heterocyclic rings, and 1/3 contain protected vinylsulphone-based reactive groups.[8] Reactive dyes can be divided into three groups according to the reactivity of their reactive groups as follows:

- dyes that are able to dye cellulose in the cold (high reactivity),
- dyes that are able to dye cellulose in the warm (moderate reactivity) and
- dyes that are able to dye cellulose in the hot (low reactivity).[4]

However, since reactive dyes that dye at truly low temperatures (at 25 °C) are no longer used, reactive dyes are currently classified into two groups as follows:

- dyes having high reactivity that are able to dye cellulose in the cold,
- dyes having low reactivity that are able to dye cellulose in the hot (around 80–95 °C).

Advantages of cold-dyeing reactive dyes are as follows:

- Since they have high reactivity, that is, they react easily, it is possible to dye more rapidly.
- Fewer chemicals are consumed. Since their reactivity is high, less alkali is needed for their fixation. On the other hand, since the dyeing temperature is low, the substantivity is higher than in hot dyeing processes, and therefore slightly less salt is used compared to hot-dyeing dyes.
- Since dyeing is done at low temperature, energy consumption is lower.
- Since dyeing is done using less salt, it is easier to remove the hydrolysed dye from the material by washing. This results in water and energy savings and better wet fastness.
- Because of their high reactivity, fixation can be achieved under milder conditions in semi- and full-continuous methods.
- Since dyeing is done at a low temperature, the dye yield (the amount of dye taken up by the fibres at equilibrium) is generally higher.[4]

Advantages of hot-dyeing reactive dyes are as follows:

- When dyeing is performed at high temperatures by the exhaust method, it is easier to achieve better dye penetration and migration, and thus levelling is better.
- Since their reactivity is low, impregnation liquors prepared with them have good stability.
- The transport of the woven fabric or knitwear in rope form in the machine can be facilitated better at high temperatures.
- Especially in regenerated cellulose, by starting dyeing directly at high temperature, swelling of the fibres and shrinkage of the material can be reduced. Therefore, in dyeing regenerated cellulose fibres, it is advantageous to use hot-dyeing reactive dyes compared to cold dyes.
- Raw or insufficiently pretreated fabrics can also be dyed with hot-dyeing dyes. Even after dyeing, bleaching can be done in the same liquor.
- Since their reactivity is low, they can be added directly to the liquor when the colour is off-shade, that is, adjusting the colour nuance is easy.
- Since the risk of uneven dyeing in hot conditions is lower, cheaper sodium chloride can be used instead of sodium sulphate.[4]

As can be understood from the points mentioned above, dyes in both groups have strengths and weaknesses.[4] However, since energy consumption is a very important criterion today, it is worth noting that cold-dyeing reactive dyes that dye in cold are mostly used, while hot-dyeing dyes are preferred for dyeing materials that are difficult to dye uniformly, especially regenerated cellulose fibres.

Reactive dyes are a group of dyes that are used in different applications such as exhaust dyeing, impregnation dyeing and printing. Different levels of reactivity are needed depending on the application process to be used. For example, while dyes with high reactivity are required in dyeing by pad-batch and low-temperature exhaustion methods, in printing processes, it is necessary to work with dyes with very low reactivity to ensure the stability of the printing paste. Today, there are reactive dye groups with different levels of reactivity produced by various companies that meet all these requirements available on the market.[9]

Figure 9.3 shows the order of reactive dyes containing various reactive groups according to their reactivity.[4,8]

The reactivity of the various groups does not change linearly but logarithmically. For example, the reaction rate of the dyes containing the most reactive dichlorotriazine group is 1000 times the reaction rate of the dyes containing the trichloropyrimidine group with the lowest reactivity. There may be significant reactivity differences between dyes with the same reactive group within a dye brand, depending on the structure of the chromophore and bridge group and the location of the reactive group, as well as among dyes with different reactive groups *within* a separate group. For this reason, the reactive groups are shown in the figure not as a point, but as a line segment of varying length (region of reactivity). To give an example, although the reactivity of the dichloroquinoxaline group is generally higher than that of the vinylsulphone group, the reactivity of some commercially available

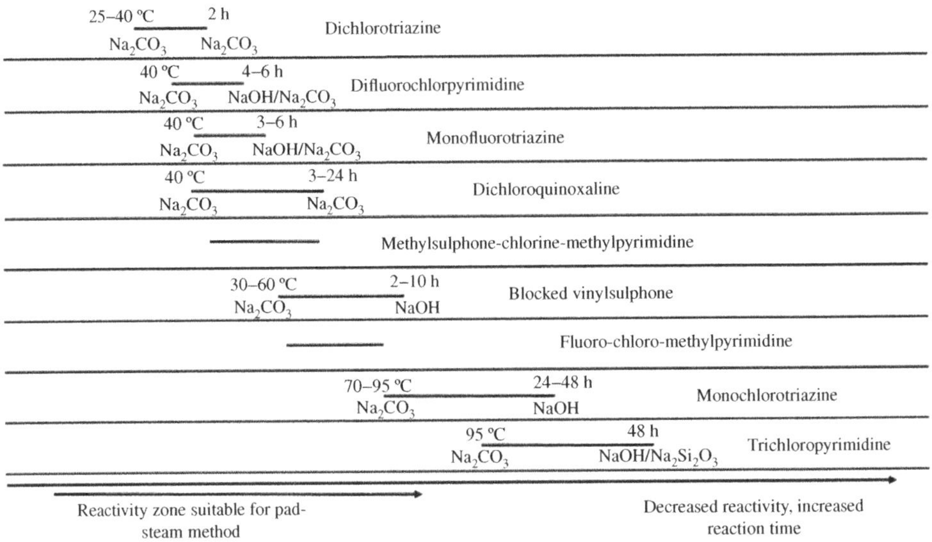

Figure 9.3 Relative reactivity of reactive dyes containing various reactive groups. (The line – indicates the reactivity region; the temperature and alkali on the left of the line indicate dyeing conditions for the exhaust method, while the time and alkali on the right of the line indicate dyeing conditions for the pad-batch method.) Ref. [4].

reactive dyes containing vinylsulphone groups may be higher than that of some reactive dyes containing the dichloroquinoxaline group.[4,8]

Since the chromophore and bridge group do not have a significant effect on the reactivity of the dye, the reactivity of heterocyclic ring-based reactive dyes primarily depends on the following factors:

a. ***The structure of the heterocyclic ring:*** For example, as the number of electron-withdrawing N atoms in the ring increases, the reactivity also increases. Therefore, the reactivity of groups with a triazine ring is higher than that of groups with a similar pyrimidine ring.[4,8]
b. ***Characteristics of the substituent group:*** The higher the electron withdrawing ability of the substituent, which is displaced during the substitution reaction, the higher the reactivity. For example, heterocyclic rings containing fluorine or methylsulphone ($-SO_2CH_3$) groups instead of chlorine atoms have much higher reactivity.[4,8]
c. ***Properties of other substituents attached to the heterocyclic ring:*** Electron-donating substituents decrease reactivity, while electron-withdrawing substituents increase it. For example, the reactivity of the dichlorotriazine group containing two electron-withdrawing chlorine atoms is much higher than that of the monochlorotriazine (MCT) group containing one chlorine atom. An important point is that if the displaced group is the same, the resistance of the formed dye-fibre covalent bond to hydrolysis generally decreases as the reactivity of a reactive dye increases. However, in reactive groups in which fluorine or methylsulphone groups are substituted instead of chlorine, good resistance to hydrolysis of the dye-fibre covalent bond can be achieved

despite a high reactivity. For this reason, many dye manufacturers have introduced reactive dye types with both high reactivity and good hydrolysis resistance to the market.[4,8]

In the structure of reactive dyes, there can be different numbers of reactive groups such as single (monofunctional), two (bifunctional) and multiple (polyfunctional). The best-known monofunctional reactive dye group is cyanuric chloride (2,4,6-trichloro, 1,3,5-triazine) derivatives. The three chlorine atoms in the triazine ring show different reactivity. Reactive dyes based on dichlorotriazine with high reactivity are obtained by condensation of cyanuric chloride with a chromophore containing an amino group. These highly reactive dyes are susceptible to hydrolysis.[5]

If two of the three chlorine atoms in the triazine ring are replaced by amino or alkoxyl groups, MCT-based reactive dyes are obtained, and dyes with this structure exhibit lower reactivity. They can dye cellulosic fibres at high temperature (80 °C) by the exhaust method. The reactivity of monochlorotriazine dyes is increased by the use of fluorine atoms instead of chlorine atoms. With monofluorotriazine dyes obtained in this way, dyeing can be done at a lower temperature (40 °C) by using the exhaust method.[5]

In addition to 1,3,5-triazines, other heterocyclic classes have also found significant use in cotton dyeing. In particular, monochlorotriazine (e.g. monochlorotriazine) dyes can be cited as an example. The reactivity of these dyes is lower than that of triazines, as previously described. The list of economically acceptable heterocyclic groups also includes dichloroquinoxaline-based reactive dyes. The reactive group with the greatest influence in the dye market is the 2-sulphoxyethylsulphone group. When this group is treated with alkali, sulphuric acid is removed from the structure and a vinylsulphone group is formed, which allows the fibre and dye to bond.[5] The chemical structure of all the mentioned reactive groups is given together in Figure 9.4.

As explained before, the reactive groups in the structure of reactive dyes can react with the hydroxyl groups of cellulose, as well as with the hydroxyl groups in the water, resulting in hydrolysis. The fixation degree of monofunctional reactive dyes containing a single reactive group in their structure is around 60%, that is, 40% of the dye used does not participate in the dyeing process and is hydrolysed by water. As a solution to this problem, reactive dyes with two reactive groups (bifunctional) were developed in the mid-1980s, thus enabling high fixation values.[5]

As can be seen from Figure 9.5, 40% of the dye undergoing hydrolysis in bifunctional reactive dyes still contains a second reactive group that is still intact. Assuming that this second reactive group also reacts with the fibre at a rate of 60%, 60% of the remaining 40% (i.e., 24% of the total dye) will have the chance to bind to the fibre. In this case, the percentage of dye bound to the fibre increases to 84%.

According to the calculations based on the fixation efficiency, the ratio of the fixation (reacting with cellulose) and hydrolysis (reacting with water) rates of reactive groups in exhaust dyeing varies between 3:1 and 10:1. Accordingly, 75–90% of the reactive groups undergo fixation and 10–25% undergo hydrolysis reaction during dyeing. Therefore, if all the reactive groups of the dye molecules are active at the start of the dyeing, process, under optimal conditions the fixation of the dye should not fall below 75% as a result of an optimized process. However, since the reactive groups in the reactive dye molecules can react

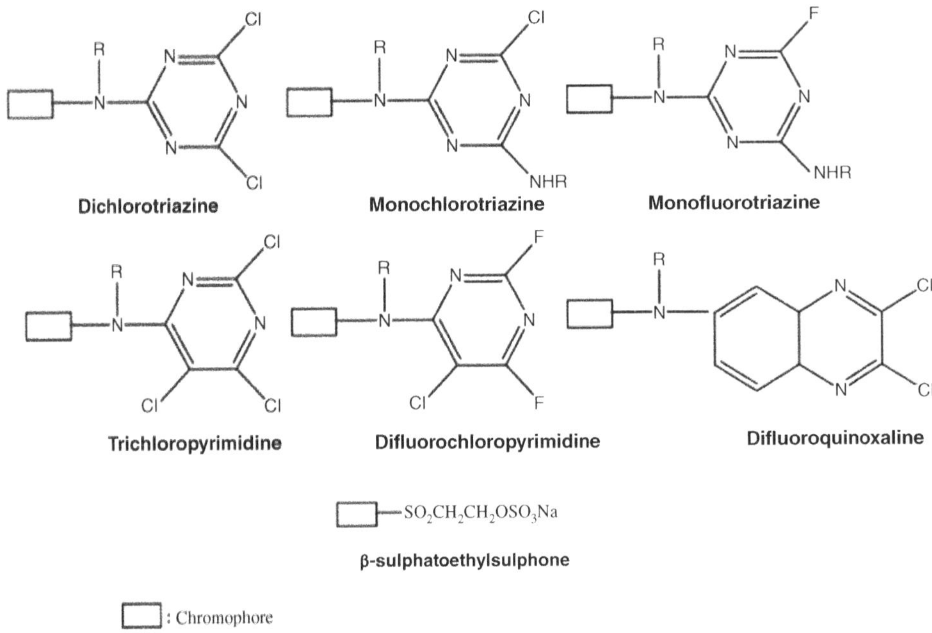

Figure 9.4 Various reactive groups that can be found in the structure of reactive dyes. Ref. [9]/Sumitomo Technical Documents.

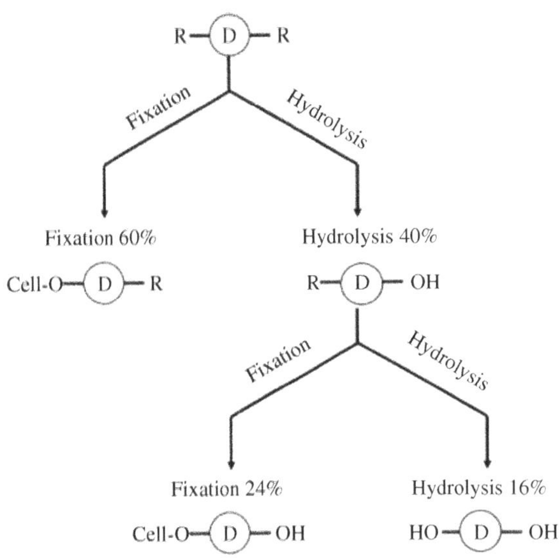

Figure 9.5 Reaction proportions of bifunctional reactive dyes with fibre and water. Ref. [10]/Ciba Technical Documents.

(hydrolyse) with water not only during dyeing but also during dye production in an aqueous medium, some dye molecules may not contain any reactive groups, and in this case, the fixation efficiency may even fall below 50%. However, thanks to computer-controlled production techniques, the average content of active reactive groups in the reactive dyes produced today has increased to 95%, and this situation is no longer a problem. However, it is not possible to increase the dye fixation efficiency above 90% with conventional reactive dyes containing a single reactive group. For this, a second reactive group must be attached to the dye molecules. In this way, the increase in yield is closely related to the location and reactivity of the reactive groups.[8]

Dyes with two reactive groups are divided into two groups within themselves. These are **homobifunctional** dyes that have the same two reactive groups and those that contain two different reactive groups, that is, **heterobifunctional** dyes. In Figure 9.6, the structure of MCT/MCT-based bifunctional reactive dye, in which two MCT groups are linked by a suitable bridge group, is given.[5]

This concept has created the opportunity to attach different chromophore groups to reactive dyes. In this way, the possibility of obtaining more distinctive colour shades has been achieved.[5] Figure 9.7 shows the different attachment ways of reactive and chromophore groups in bifunctional reactive dyes.

In the category of dyes with the same two reactive groups, dyes containing two vinylsulphone or 2-sulphoxyethylsulphonyl groups are also included. Besides the high fixation values of these dyes, their easy production and being cost-effectiveness are also important advantages.[5]

Reactive dyes containing different reactive groups were first described in 1959. These bifunctional dyes were first introduced to the market in the 1980s. These products are dyes consisting of two groups with different reactivity, such as the sulphatoethylsulphonyl group, which has a higher reactivity, and the monochlorotriazine group, which has a lower reactivity. Reactive dyes with two reactive groups, which are applied at warm and cold temperatures, are prepared with halotriazine and vinylsulphone groups. The bond between the triazine ring and the fibre is stable under alkaline conditions, while the bond between the vinylsulphone group and the fibre is resistant to acidic environments. Therefore, the good fastness properties of the material dyed with this dye can be maintained over a wide pH range.[5]

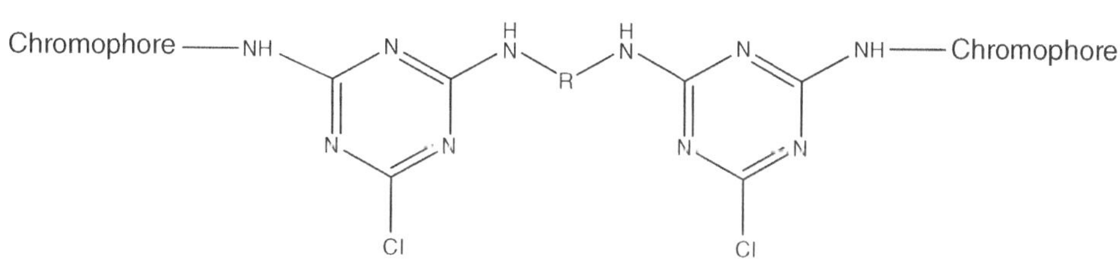

Figure 9.6 Homobifunctional reactive dye containing double monochlorotriazine group. Ref. [5].

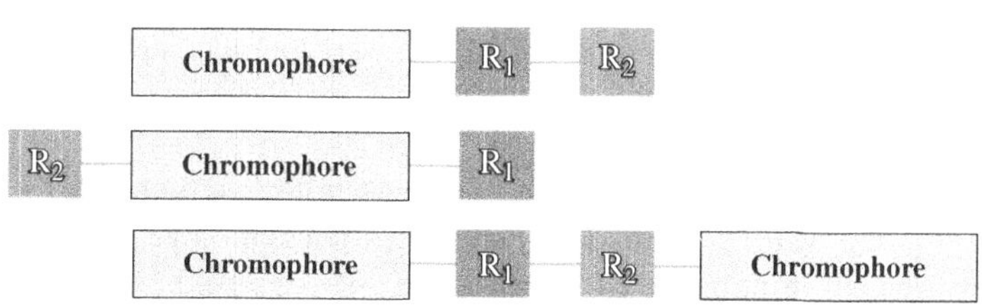

Figure 9.7 Different attachment ways of reactive and chromophore groups in bifunctional reactive dyes. Ref. [10]/Ciba Technical Documents.

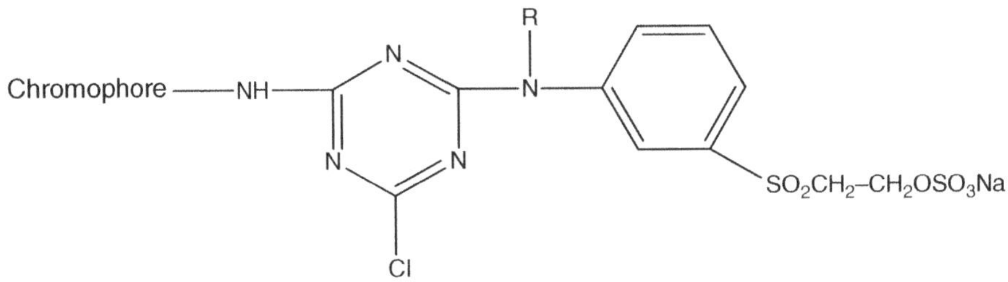

Figure 9.8 Heterobifunctional reactive dye containing monochlorotriazine and β-sulphatoethylsulphone group. Ref. [9]/Sumitomo Technical Documents.

Table 9.1 Variation of fixation efficiency of mono- and bifunctional reactive dyes with various reactive groups depending on the liquor ratio.

	Liquor ratio				
Reactive group	**1:3.5**	**1:5**	**1:10**	**1:20**	**1:40**
Dichloroquinoxaline	78	76	74	67	56
Difluoropyrimidine	80	78	76	72	63
Monofluorotriazine	79	77	75	71	64
Vinylsulphone	72	70	67	62	54
Vinylsulphone/Vinylsulphone	92	92	90	84	71
Difluoropyrimidine/Vinylsulphone	98	98	97	92	82

Ref. [9]/Sumitomo Technical Documents.

In Figure 9.8, the chemical structure of a heterobifunctional reactive dye containing monochlorotriazine and β-sulphatoethylsulphone groups is given.

Among the heterobifunctional reactive dyes, the most common are monochlorotriazine/vinylsulphone and monofluorotriazine/vinylsulphone-based dyes,[5] dichlorotriazine/vinylsulphone, trifluoropyrimidine/vinylsulphone, fluorochloropyrimidine/vinylsulphone and monofluorotriazine/trifluoropyrimidine-based dyes are also commercially available.[7]

Table 9.1 summarises the change of fixation efficiency of mono- and bifunctional reactive dyes with various reactive groups depending on the liquor ratio. As can be seen from the table, the fixation efficiency of bifunctional reactive dyes can exceed 90% depending on the liquor ratio.

Reactive dyes with more than two reactive groups (e.g. MCT[7]) were found to have only a minor effect on dye fixation. Therefore, the share of polyfunctional reactive dyes in the dye market is of secondary importance.[5]

Drimarene HF dyes are believed to incorporate a vinylsulphone group and a 2,4-difluoropyrimidine-6yl group in the dye molecule. Other tri-functional dyes capable of giving high fixation efficiencies from long-liquor dyeing processes approaching 90% in medium colour depths have been developed. Thus, Everlight has a tri-functional range (Everzol ED) and it is believed that some of the Procion HE-XL+ dyes are tris-monochloro-s-triazines; the latter may be prepared by the reaction of three moles of a dichloro-s-triazine dye with one mole of a tri-functional amine. As early as 1975, Hoechst launched the tri-functional dye, CI Reactive Red 181 (Remazol Red SBB), which is shown in Figure 9.9, and patented dyes of this type.[11]

Tetra-functional reactive dyes have been claimed in the patent literature. For example, Ciba has covered chromogens containing two vinylsulphone/monohalo-s-triazine residues. A patent describes tetra-functional reactive dyes prepared from the reaction of dichloro-s-triazine dyes (or difluoro-monochloro-pyrimidine dyes) with two moles of cysteamine and subsequent reaction of the two pendant primary amines with cyanuric chloride; the preparation route is shown in Figure 9.10.[11]

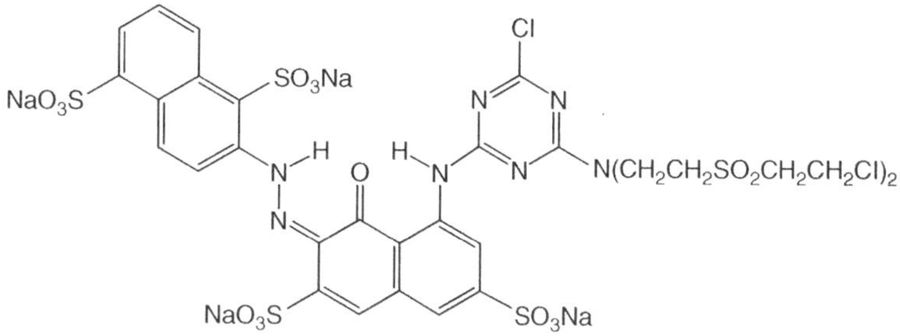

Figure 9.9 CI Reactive Red 181 dye. Ref. [11]/John Wiley & Sons.

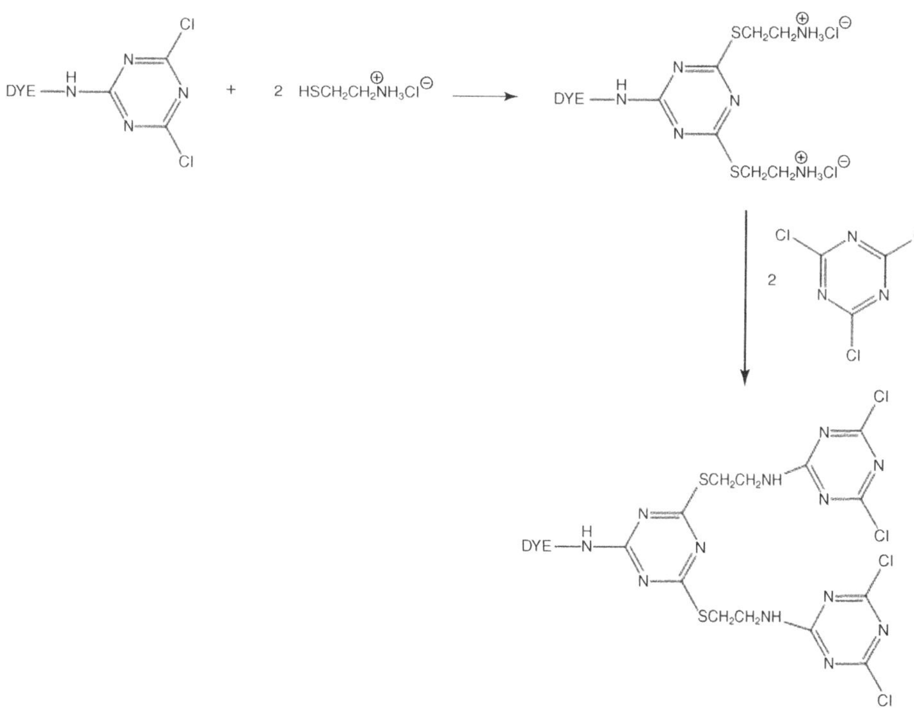

Figure 9.10 Synthesis of a tetra-functional chloro-s-triazine dye. Ref. [11]/John Wiley & Sons.

9.1.2 Classification According to Chromophore Groups

Various known chromophore groups can be used in the synthesis of reactive dyes. These are:

- **Azo and bisazo:** Most of the reactive dyes are in this group, and they contain an azo (—N=N—) group in their chromophore structure.[5] Generally, monoazo chromophores are found in yellow, orange, red and dark blue colours,[5] while bisazo chromophores are found in blue, brown and black colours. Light and bleaching fastness values vary depending on the chemical form of the dye.[7] CI Reactive Orange 1, CI Reactive Red 8 and CI Reactive Red 12 can be given as examples of azo dyes.[5]
- **Metal complex azo:** Especially colours with high light fastness are obtained with metal complex azo dyes.[5] The perspiration fastness of these dyes can be problematic in some cases.[7] Azo dyes containing copper complexes have a wide colour range (yellow, purple, brown, green, black). CI Reactive Red 23 is an example of metal complex azo dyes.[5]
- **Anthraquinone:** Anthraquinone-based reactive dyes are important for their brightness, good light fastness and chromophore stability in acidic and alkaline mediums.[5] These dyes have high light fastness as well as high bleaching and very good washing fastness.[7] Colour shades of commercial anthraquinone-based reactive dyes range from purple to blue. CI Reactive Blue 19 falls into this category.[5]
- **Phthalocyanine:** Phthalocyanine-based reactive dyes provide bright turquoise and green colours that cannot be obtained with other dye classes. The most important phthalocyanine-based reactive dye has copper or nickel as the central atom.[5] They have high light fastness but limited washing fastness.[7] CI Reactive Blue 15 and CI Reactive Blue 21 fall into this group.[5,7]
- **Triphenodioxazine:** Parliament blue and bright blue shades can be obtained with anthraquinone-based reactive dyes as well as with triphenodioxazine-based dyes.[7] However, the colour strength of these chromophores is much higher, especially when compared to anthraquinone-based dyes. Therefore, when working with them, the amount of dye that must be used to obtain the same colour depth is considerably lower than that required for anthraquinone-based dyes. Triphenodioxazine-based dyes, which provide great advantages, especially in lustrous blue reactive dyes,[8] show a high degree of agglomeration and increase their substantivity due to their planar arrangement.[5] The most important drawbacks are difficulty in removal during washing due to their high substantivity, and also their low bleaching fastness,[8] limited washing fastness in some cases[7] and sensitivity to acids and bases.[5] However, they generally have high light fastness.[7] CI Reactive Blue 163, CI Reactive CI Blue 172, CI Reactive Blue 187, CI Reactive Blue 198, CI Reactive Blue 204 and CI Reactive Blue 224 fall into this group.[5]
- **Formazan:** Different colours from red to greenish blue can be produced with copper complexes of dyes based on the formazan structure.[5] They have high light fastness, good bleaching fastness and very good washing fastness.[12] Sample dyes belonging to this group are CI Reactive Green 15, CI Reactive Blue 70, CI Reactive Blue 83, CI Reactive Blue 84, CI Reactive Blue 104 and CI Reactive Blue 160.[5]

In Figure 9.11, the chemical structures of the above-mentioned chromophore groups that can be found in the structure of reactive dyes are given.[5]

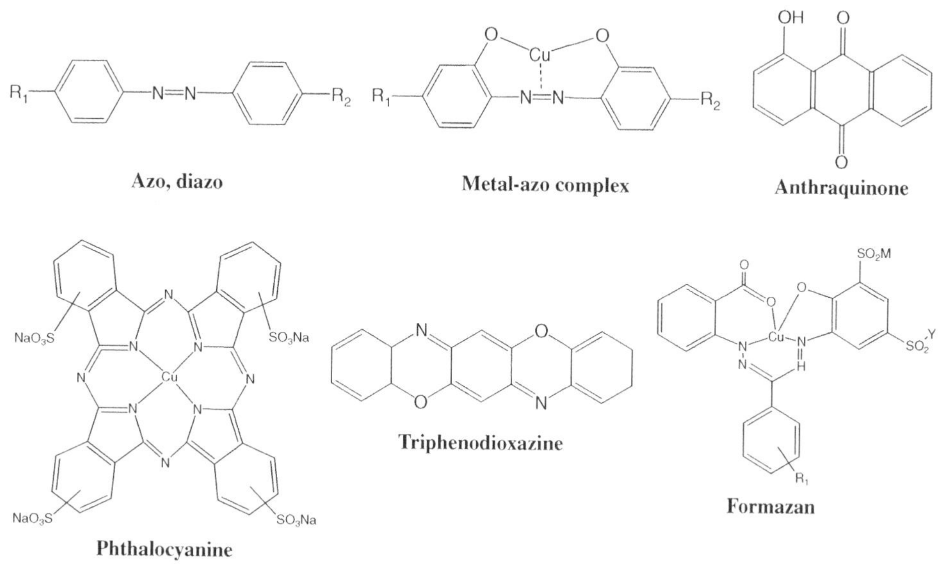

Figure 9.11 Various chromophore groups used in reactive dyes. Ref. [13].

9.1.3 Classification According to Reaction with Fibre

Reactive dyes can be divided into two groups according to their binding mechanism to the fibre:

- Heterocyclic ring-based reactive dyes that bond via nucleophilic substitution reaction and
- Vinylsulphone-based reactive dyes that bond via nucleophilic addition reaction.[2]

In bifunctional reactive dyes containing two different types of reactive groups in their structure, both reactions occur simultaneously or separately.[9]

As can be seen in Figure 9.12, each glucose unit in the structure of cellulose has three hydroxyl groups, one primary (attached to the sixth carbon atom) and two secondary (attached to the second and third carbon atoms). Studies have shown that the reactive dye is bonded to the cellulosic fibres preferably (75%) through the primary hydroxyl group at the sixth carbon atom, but it can also be bonded to the secondary alcohol groups (25%) at the second and third carbon atoms in case of reactive dyes that react via both substitution and addition reactions.[9]

As explained before, reactive dyes can react not only with hydroxyl groups in fibres but also with hydroxyl groups in water. However, dye molecules preferentially react with cellulosate ions, although the concentration of hydroxyl groups in water is considerably higher than that of cellulose (cellulose alcoholate) ions. Studies on the relative rate of dyeing have shown that the reaction rate of the dye with cellulose in triazine-based dyes is several thousand times higher than the reaction rate with water. In sulphatoethylsulphone-based dyes, the reaction rate of the dye with cellulose is several hundred to several thousand times higher than the reaction rate with water. In a study, it was revealed that although

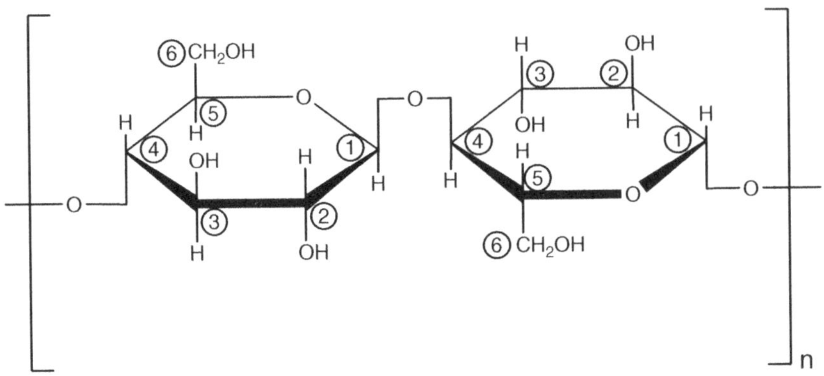

Figure 9.12 Structure of cellulose. Ref. [9]/Sumitomo Technical Documents.

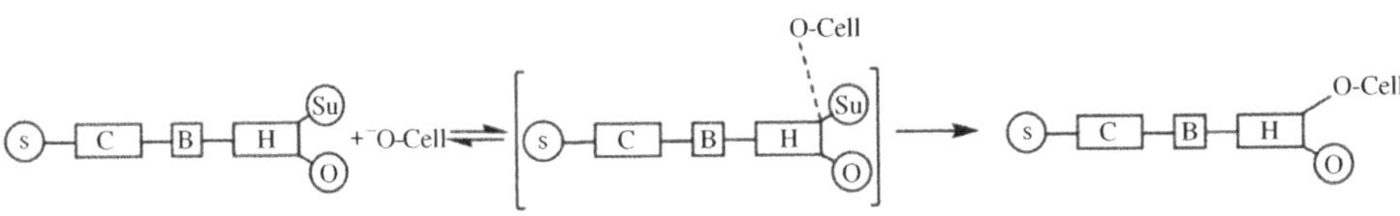

S: Solubilising group (–SO_3Na)

C: Chromophore

B: Bridge group (-NH,-CH_2,-N-,-NH,-C-)

H: Heterocyclic ring (triazine, pyrimidine, quinoxaline, quinazoline, pyridazone)

Su: Substituent group (-Cl, -F, -SO_2CH_3)

O: Other substituents attached to the ring

Figure 9.13 Covalent bonding of heterocyclic ring-based reactive dye to cotton fibre by substitution reaction. Ref. [8]/Tekstil & Teknik.

the activation energy required for the dye to react with the fibre is 9.2–15.8 kcal/mol, the activation energy required for the dye to react with the water is 16–26 kcal/mol, so the reaction with the fibre proceeds more easily.[9]

9.1.3.1 Reactive Dyes Undergoing Nucleophilic Substitution Reaction

The nucleophilic substitution reaction proceeds in two steps as shown in Figure 9.13. In the first step, the nucleophilic functional groups (—OH groups) of the fibre are activated in the alkaline medium and attack the electrophilic centre of the reactive group. In the second step, the elimination of a substituent, which is bonded to the ring, takes place.[5] In other words, in this type of dye, the covalent bond between dye and fibre is formed as a result of the replacement of a substituent, which is bonded to the ring, with functional groups in fibre macromolecules (—OH in cellulose).[4]

For example, in MCT-based reactive dyes, the electron-withdrawing nitrogen atoms of the triazine ring change the electron density of the ring and negatively charged substances such as Cell-O^- and OH^- are attracted by the ring. Then the chlorine atom attached to the

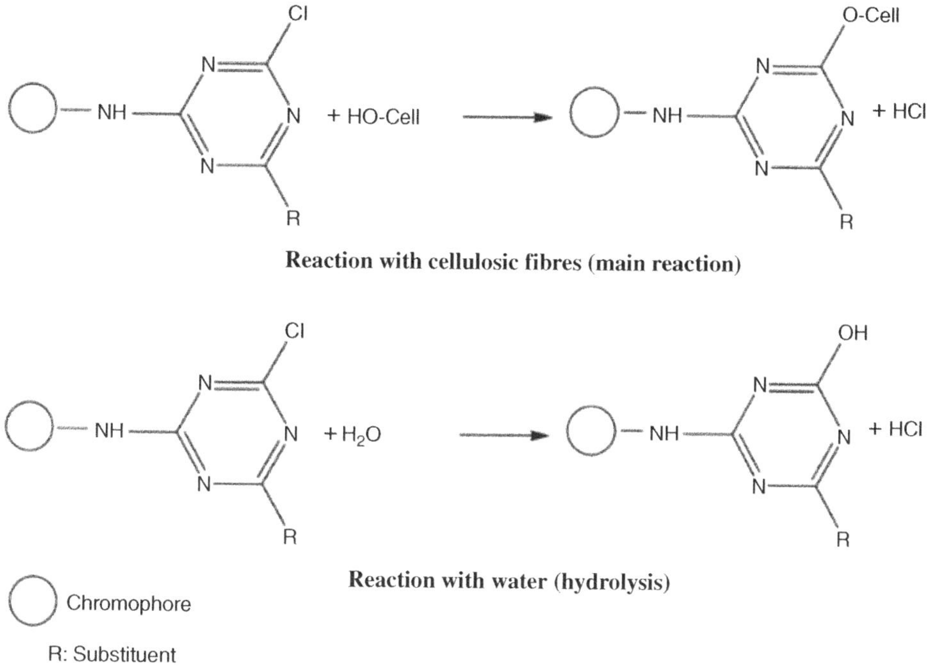

Figure 9.14 Reactions of monochlorotriazine-based reactive dye with cellulosic fibres and water. Ref. [9]/Sumitomo Technical Documents.

ring is replaced by them. Thus, the reactive dye is attached to the fibres by covalent bonds. Besides this main reaction, dye molecules also react with water and undergo hydrolysis as a side reaction. Figure 9.14 shows the reactions of MCT-based reactive dye with cellulosic fibres and water.[9]

9.1.3.2 Reactive Dyes Undergoing Nucleophilic Addition Reaction

Reactive dyes that react with nucleophilic addition reaction generally contain groups such as vinylsulphone ($-SO_2-CH=CH_2$), sulphatoethylsulphone ($-SO_2-CH_2-CH_2-OSO_3Na$) or chloroethylsulphone ($-SO_2-CH_2-CH_2-Cl$).[5] The reaction of vinylsulphone-based reactive dyes with -OH groups in cellulose macromolecules occurs in two steps. In the first step, the sulphateethylsulphone group transforms into a highly reactive vinylsulphone group in alkaline medium, and in the second step, it gives an addition reaction with cellulose to form an ether bridge in alkaline medium. The addition reaction occurs to the functional group of the textile fibre.[5,9] With the strong ionic effect of the sulphonyl group in alkaline medium, the carbon atom in the α position releases H^+, while the carbon atom in the β position tends to separate the $-OSO_3H$ group as an anion. This leads to the formation of the vinylsulphone group. The π electrons in the double bonds of the vinylsulphone group are activated by the electron interaction of the sulphonyl group ($-SO_2-$), and therefore the double bond enters the addition reaction with cellulose. Besides this main reaction, dye molecules can also react with water and even with each other as a

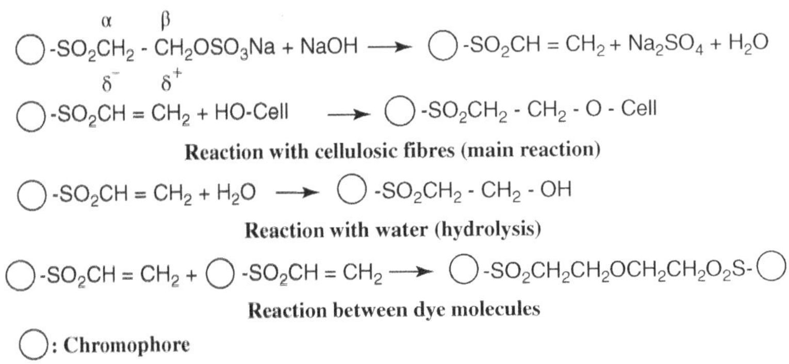

Figure 9.15 Reactions of sulphatoethylsulphone-based reactive dye with cellulosic fibres and water. Ref. [9]/Sumitomo Technical Documents.

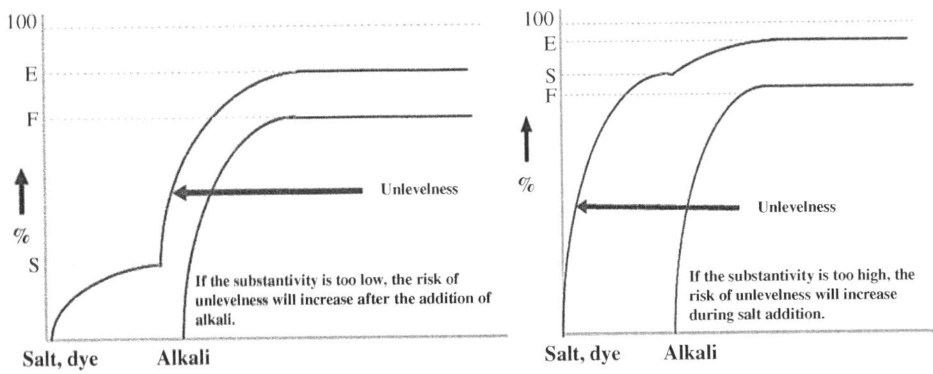

Figure 9.16 An example of the exhaustion (E) and fixation (F) curves of alkali-controlled (on the left) and salt-controlled (on the right) reactive dyes. Ref. [3]/Benar Kimya.

side reaction. Figure 9.15 shows the reactions that the sulphateethylsulphone-based reactive dye gives off with cellulosic fibres, water and each other.[9]

This reaction becomes easier as the strength of the acid that breaks off as a result of the elimination reaction increases. For example, the group eliminated from the structure of the dye in Figure 9.15 is the anion of a strong acid such as sulphuric acid ($-OSO_3H$). As the electron-withdrawing properties of the β substituents increase in dyes containing reactive groups of this type, both the formation of the vinyl group by elimination of the acid radical (elimination reaction) and the tendency of the resulting vinyl compound to undergo a nucleophile addition reaction with the —OH groups of cellulose increase.[8]

9.1.4 Classification According to Control Parameters

Reactive dyes can be classified into three groups depending on the parameter that needs to be kept under control for obtaining uniform dyeing.

- ***Group 1: Alkali-controlled dyes:*** The optimum temperature for these dyes is between 40 and 60 °C.[14] Figure 9.16 shows an example of the characteristic exhaustion and fixation curves of dyes in this group.[3] As can be seen from the figure, these dyes show low exhaustion in neutral salt solution before the addition of alkali. They have high reactivity, and alkali addition should be done carefully (with controlled dosage rate) for proper dyeing. Dichlorotriazine, chlorodifluoropyrimidine, dichloroquinoxaline and vinylsulphone-based reactive dyes can be cited as examples of reactive dyes in this group.[14]

- ***Group 2: Salt-controlled dyes:*** The optimum fixation temperature for dyes in this group is between 80 °C and boiling temperature.[14] Figure 9.16 shows an example of the characteristic exhaustion and fixation curves of dyes in this group.[3] As can be seen from Figure 9.16, these dyes show high exhaustion (i.e. high substantivity) in neutral salt solution before the addition of alkali. Therefore, salt addition should be done carefully to obtain uniform dyeing. Addition of salt should be done in portions and preferably at a controlled dosage rate. Reactivity of these dyes is low. Trichloropyrimidine, aminofluorotriazine, aminochlorotriazine and bisaminochlorotriazine-based reactive dyes can be cited as examples of reactive dyes in this group.[14]
- ***Group 3: Temperature-controlled dyes***: Dyes in this group react with cellulose in the presence of alkali at temperatures above boiling temperature. If desired, dyeing can be done under the conditions of Group 2. Since they have the ability to be taken up properly, there is no need for the use of auxiliaries in dyeing. The best levelness is achieved by controlled temperature rise.[14]

9.2 Fastness Properties of Reactive Dyes

Fastness properties may vary depending on the processes applied before dyeing, dye structure, colour depth and conditions (time, temperature, liquor ratio), type, amount and method of addition of chemicals used, characteristics of the water used, washing protocol and also the post-treatments at the end of dyeing. Today, the most emphasised fastness properties are wet fastness and light fastness. Fastness in dyeing with reactive dyes is closely related to the structure and properties of the dye used. In particular, the chromophore and reactive group of the dye have an important effect on the fastness values. As seen in Figure 9.17, the reactive group

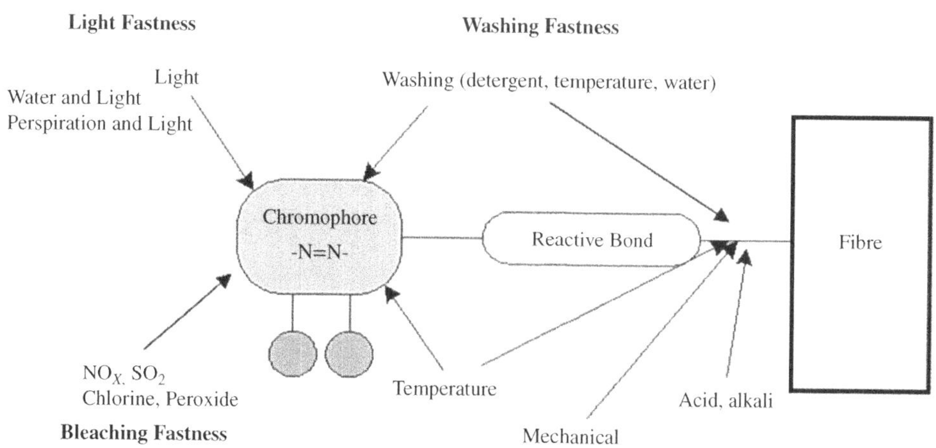

Figure 9.17 Factors on which the fastness of reactive dyeing depends. Ref. [7]/Dystar.

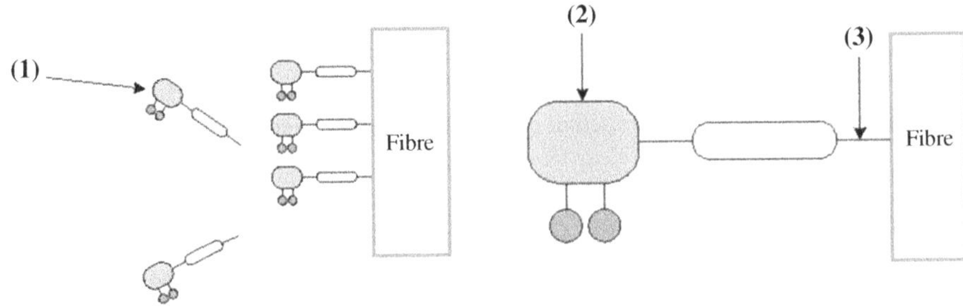

Figure 9.18 Factors affecting washing fastness of reactive dyes. Ref. [7]/Dystar.

especially affects the washing fastness, and the chromophore group closely affects the colour nuance, as well as the light fastness, wet light fastness, chlorine and peroxide fastness.[15]

Light fastness depends on the chromophore in reactive dyes, and it can be a problem especially in some shades (orange, red, dark blue, etc.) of non-metal complex azo dyes, while Cu and Ni metal complex chromophores provide high light fastness. Although the washing fastness is very good since there is a covalent bond between the fibre and the dye, problems may be encountered due to the following reasons:

1. Hydrolysed dye that is not removed effectively (insufficient washing, high-substantivity dye, etc.).
2. Factors that degrade the chromophore (bleaching agents and detergents containing these substances, etc.).
3. Factors that cleave the bond between the dye and the fibre (acid, alkali and temperature, etc.). This situation is shown schematically in Figure 9.18.[7]

Despite the development of the use of new chromophore groups in reactive dyes, new production techniques, the production of bifunctional dyes with high fixation efficiency, and new methods such as low salt dyeing, etc., in recent years, the problem of obtaining uniform, ecological, economical, reproducible dyeing with sufficient fastness values has not been fully resolved. Especially in dark colours, their wet rubbing fastness is not sufficient. The wet rubbing fastness of fabrics dyed in dark shades with reactive dyes is around 2–3. Although good washing fastness values can be obtained even in dark shades with the use of appropriate dyes, an exception is observed especially in wet rubbing fastness values. Insufficient wet rubbing fastness values, obtained especially in dark colour shades, are primarily caused by the contamination of the rubbing fabric with coloured fibre particles that break off from the tested fabric, and the post-treatment materials used after dyeing generally do not affect the wet rubbing fastness of the fabric.[16,17]

9.3 Forms of Reactive Dyes

Reactive dyes have been on the market only as powder dyes for many years. Liquid dye production started with the introduction of special reactive dyes developed by ICI in the early 1980s, which can dye cellulose fibres in a weakly acidic medium (pH 5–6). Today, liquid forms of normal reactive dyes are also available on the market. Some manufacturers have also introduced reactive dyes in granular form to the market.[8]

Liquid dyes, which provide significant convenience in the colour kitchen, especially during the dissolution of large amounts of dye, are also advantageous in terms of health due to the elimination of the risk of dusting. In contrast to powder dyes, granular dyes provide dye savings of up to 2%, since the dyes in the barrel can be emptied completely and without dusting. Elimination of the risk of remaining undissolved dye particles remaining in the liquor is one of the important advantages of liquid or granular dyes.[8]

9.4 Dyeing with Reactive Dyes

9.4.1 Dyeing by Exhaust Method

9.4.1.1 Steps of Dyeing and Factors Affecting These Steps

Exhaust dyeing processes with reactive dyes, whether performed with cold-dyeing or hot-dyeing types, basically consist of three steps as follows:

- Exhaustion of dye molecules by fibres **(migration step)**
- Reaction of dye molecules with fibres **(fixation step)**
- Removal of hydrolysed dye that is not fixed to the fibre from the material **(washing step)**.[2]

The first of these steps is determined by the substantivity and diffusion ability of the dye for the fibres; the second step (fixation) by the reactivity of the reactive group, pH and temperature; the third step (washing) by the fixation degree, solubility, substantivity and diffusion ability of the dye.[4] Examining these steps more closely:

- The first step of dyeing, the uptake of the dye by the fibres, is related to the substantivity of the dye used for the fibres and its diffusion ability into the fibres.[2] Although the substantivities of reactive dyes show significant differences among themselves, even the substantivity of reactive dyes with high substantivity is actually moderate. The substantivity of the reactive dye for the fibres (the ability to be taken up from the liquor by the fibres) and diffusion depend on factors such as the structure of the dye molecule, the type of cellulose fibre, the pretreatment processes applied to the fibre, the liquor ratio and dye concentration in the liquor, electrolyte (salt) concentration in the liquor, liquor temperature and liquor pH.

High-molecular-weight dyes have higher substantivity but lower diffusion abilities. For example, dyes containing phthalocyanine groups as chromophore (turquoise colours) have lower diffusion abilities than other reactive dyes. Therefore, these dyes are applied at higher temperatures to facilitate diffusion.[9] Generally, regenerated cellulose fibres have a higher substantivity (affinity) for dyes than cotton fibres.[4] On the other hand, diffusion is higher in cotton and linen fibres compared to viscose fibres. This is because the pore size of the amorphous regions in cotton and linen is larger than in viscose.[9] In addition to the type of cellulose fibre, the pretreatment processes also affect substantivity and diffusion. For example, both substantivity and diffusion increase in mercerised cellulose fibres.[4]

Addition of electrolyte (salt) to the liquor increases substantivity in dyeing performed by the exhaust method.[4] As the salt concentration increases, the dye aggregation increases, but in case of excessive use of salt, the problem of precipitation and unlevelness arises due to the decrease in the solubility of the dye.[9] On the other hand, increasing the dyeing temperature decreases substantivity and accelerates diffusion.[2]

Reducing the liquor ratio increases both substantivity and fixation efficiency. On the other hand, the increase in the liquor ratio supports the reaction of the dye with water, and thus the proportion of hydrolysed dye increases. Therefore, when the liquor ratio changes, the amount of salt and alkali to be used in the dyeing process should be corrected by consulting the dye catalogues.[2]

Another factor affecting substantivity is the pH of the liquor. When the pH is increased, the concentration of hydroxyl ions in the dye liquor increases and these ions are absorbed by the fibres. As a result, cellulose fibres become ionised and negatively charged. Therefore, the substantivity of cellulose fibres towards anionic dyes such as reactive dyes decreases in an alkaline medium.[9]

- The rate of reaction of reactive dyes with fibres (the second step of dyeing) varies depending on the type of reactive group, pH and temperature. As a general rule, increasing the **"pH-value by 1 unit"** or **"Temperature by 20 °C"** results in an approximate 10-fold increase in the reaction rate. Of course, with the increase in temperature and pH value, not only the rate of fixation (reacting with —OH groups in cellulose macromolecules) but also the rate of hydrolysis (reacting with —OH groups in water) increases.[2] In fact, as can be seen from Figure 9.19, the increase in the hydrolysis rate with the increase in temperature is slightly higher than the increase in the fixation rate.[8]
- The success of the washing process, which is the third step of dyeing, largely depends on the affinity of the hydrolysed dye for the fibre. The low-affinity hydrolysed dye

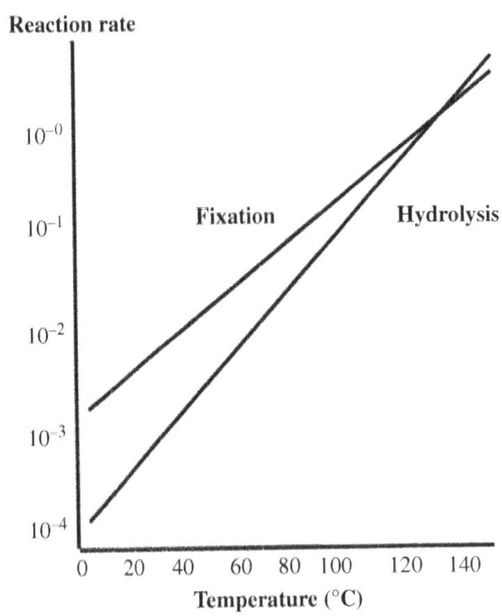

Figure 9.19 Variation of fixation and hydrolysis rates in reactive dyeing with temperature. Ref. [8]/with permission of Tekstil & Teknik.

is largely removed by cold water, while the high-affinity hydrolysed dye can only be removed by hot water.

In washing processes after reactive dyeing, the washing process becomes easier when:

- the amount of unfixed dye to be removed is lower (i.e., the fixation efficiency is higher),
- the substantivity of the dye for the fibres is lower,
- the diffusion and solubility of the dye are higher.[2]

The most effective way to reduce the substantivity of the dye and accelerate its diffusion during washing is to increase the washing temperature.[2] Since a significant part of the unfixed reactive dyes mixed with the wastewater during washing cannot be removed from the water at the end of normal physicochemical and biological treatments, increasing the dye fixation efficiency has gained great importance in recent years not only to save dye but also to reduce the wastewater load. As stated earlier, the development of dyes with two reactive groups is among the most important studies in this field. Thanks to bifunctional reactive dyes, the amount of hydrolysed dye remaining in the bath after dyeing can be significantly reduced in order to obtain a certain colour depth.[8]

9.4.1.2 Factors Affecting Dyeing Result in Dyeing by Exhaust Method

In exhaust dyeing with reactive dyes, it is necessary to increase the amount of dye taken up by the fibres as much as possible in order to increase the dye fixation. Dye exhaustion depends on factors such as substantivity of the dye, dye bath temperature, liquor ratio, type and amount of salt, type and amount of alkali.[9] So, the factors affecting dye uptake and fixation and thus dyeing efficiency in reactive dyeing can be listed as follows:

- Type of cellulose fibre and yarn/fabric properties
- The pretreatment processes of the fibre
- Type and molecular structure of the dye
- The liquor ratio and the dye concentration in the liquor
- Salt (type, amount, method of adding into the liquor)
- Alkali (type, amount, method of adding into the liquor)
- Water quality
- Auxiliary chemicals (type and amount)
- Dyeing temperature and time
- Characteristics of the dyeing device used.

These factors, on the one hand, affect the substantivity and diffusion ability of the dye, and change the dye uptake; on the other hand, they affect the degree of fixation. Accordingly, they also affect the washing process, which is the third step of dyeing. The effects of these factors on the dyeing result are explained in detail below, taking into account their effects on all steps of dyeing.

9.4.1.2.1 Effect of Cellulose Fibre Type and Yarn/Fabric Properties on Dyeing

When cellulosic fibres are mentioned, natural cellulose fibres such as cotton, linen, hemp, jute and regenerated cellulose fibres such as viscose, modal and lyocell come to mind. The dyeing properties of all cellulosic fibres are basically the same. However, there may be differences between them depending on the fibre type.[2] For example, while ramie and jute from cellulosic fibres have a very high orientation, cotton fibres have a lower orientation. This situation causes ramie and jute to take up dye with more difficultly than cotton. On the other hand, the average degree of polymerisation of flax fibres is higher than that of cotton. This causes the dye uptake behaviour of linen to be different from cotton. It should not be forgotten that even cotton fibres of different origin (Aegean cotton, Egyptian cotton, Greek cotton, etc.) will have different properties such as fineness, length and maturity, and therefore there will be differences in dye uptake between them.

Since regenerated cellulose fibres are cellulosic fibres, they can be dyed with dyes, auxiliaries and dyeing methods used in dyeing cotton in practice. However, there are significant differences in the adsorption, diffusion and fixation of dyes with respect to cotton fibres. The morphological structure of the fibre is of great importance in these differences. It is not recommended to use vat and sulphur dyes because these fibres are not very resistant to alkali. The dye class that gains importance in dyeing regenerated cellulose fibres is reactive dyes, since sufficient fastness values cannot be obtained with direct dyes. Since the amount of amorphous region of these fibres is quite high compared to cotton fibres, generally the ability of regenerated cellulose fibres to absorb dye from the liquor, that is, their substantivity (affinity) for dye, is higher than that of cotton fibres. On the other hand, since regenerated cellulose fibres swell excessively, the diffusion ability of dye molecules into the fibres decreases. Especially in tightly woven fabrics and packages, it becomes difficult to obtain dyeings that penetrate well into the fibres.[4]

Although substantivity is higher in regenerated cellulose fibres with more pore surface, the initial dye uptake rate is higher in the dyeing of cotton. This is followed by a very slowly advancing state of equilibrium. On the other hand, the initial dye uptake rate in viscose fibres is not high. The main factor that determines this speed is the degree to which the outer mantle is penetrated by diffusion. As is known, although the amount of amorphous region in viscose fibres and the ability to swell in water are high, dye uptake by the fibres is slow. The reason for this is the presence of the mantle layer on the outside of the fibres, where macromolecules are more tightly and uniformly located. This outer layer has a restraining effect on dye uptake. However, normal viscose can absorb dye a little faster and in greater amounts than modal fibre. This can be explained by the less porous structure of modal fibre and the fact that its average pore diameter is smaller than that of normal viscose fibres.[18]

Cuprammonium rayon, on the other hand, has higher exhaustion and fixation values than viscose, cotton and even mercerised cotton.[9] Perhaps the most important one among the various reasons for the different dyeing properties of cuprammonium and viscose rayon is that although there is a denser part outside the viscose fibres compared to the inner part, the density of each side is the same in cuprammonium rayon fibres.[4]

Although viscose, modal and lyocell are regenerated cellulose fibres, it should not be forgotten that there will be significant differences between their dyeing behaviours. Unlike viscose, the absence of a core/mantle layer in modal and lyocell, differences in the amorphous region, the pore structure and morphological features between the fibres change the dyeing properties of these fibres. If the dyeing properties of cotton and regenerated cellulose fibres are compared, it can be said that the colour depth increases in the direction of cotton → modal → viscose → lyocell.[2]

The reason why viscose can be dyed a little faster than modal fibre and can take up more dye can be explained by the less porous structure of modal fibre and the smaller average pore diameter of viscose fibres.[2] In addition, in modal fibres, the fibre elements are placed more evenly and tightly in the fibre cross-section, they do not show any skin-core difference, and the average polymerisation degree is higher compared to viscose fibres.[19]

One of the important factors in dyeing is the diffusion rate of the dye into the fibres. The determining factor in diffusion is the pore structure of the fibre. In addition, pore distribution and pore shape are also influential factors. Lyocell fibres have, on the one hand, a highly porous structure, and on the other hand, relatively large pores like viscose. Such a fibre structure makes the diffusion of dye into the fibres extremely convenient. Therefore, lyocell fibres can absorb more dye than viscose. Studies have shown that when dyeing lyocell fibres with reactive dyes, the fixation efficiency of both mono- and bifunctional reactive dyes is generally higher than cotton, viscose and modal fibres. In addition to the feature of lyocell fibres being dyed darker and faster than other cellulosic fibres, it has also been observed that the fastness values obtained are around the levels of cotton and viscose. In the literature, it is stated that some polyfunctional reactive dyes have a positive effect on the fibrillation state of the fibres when dyeing lyocell fibres with reactive dyes. In the examinations carried out, a significant decrease in fibrillation was observed as a result of the formation of cross-links between adjacent cellulose chains and fibrils by such reactive dyes.[18]

Considering that the fabrics produced from cellulosic fibres often have elastane content, it is thought that it would be useful to specify some points that should be considered in the dyeing of cellulosic fabrics containing elastane.

If the properties of the elastane fibre part are desired to be preserved in cellulose/elastane blends, the following points should be considered:

- Knitting of cellulose/elastane fabric should be done at minimum tension, and the winding mechanism of the fabric should be adjusted to stretch the fabric the least. Even on normal standard machines the elongation is greater than the recommended 10% elongation for raw fabric storage. For this, the fabric needs to be controlled and re-wound to allow for the release of internal tensions. If the fabric is not re-wound, and stored in this way, it becomes "cold-fixed" and it is not possible to reach the desired properties later on. The fabric should be stored in a roll form free from wrinkles. In practice, the lower the tension, the more it is possible to eliminate wrinkles.[20]
- Raw cellulose/elastane is prone to yellow when exposed to heat. Yellowing is typically seen on the broken edges, and it is not possible to eliminate it in subsequent finishing processes. For this purpose, it is recommended that the fabrics should be stored in black plastic bags, especially away from chemical substances. Cellulose/Elastane blends should not be stored for more than two months. In this way, the edges of fabric folds are prevented from being "cold fixed". The location of fabric rolls should be changed continuously during storage.[20]
- The raw fabric containing elastane, which needs to be stored for extended time intervals, should be washed with plenty of water and rinsed to prevent colour fading and rotting caused by unsaturated fats.[20]

Although the process steps are chosen according to the desired appearance and the desired performance of the fabric during use, for elastane-containing fabrics, a workflow can be applied in the form of relaxation (releasing internal stresses) → heat setting (thermal fixation) → boiling → bleaching\dyeing\printing → finishing processes. We can list the points to be considered in the finishing processes of all elastane-containing textile materials as follows:

- It is generally recommended to use a pad-batch or fabric beam dyeing system for open-width fabrics to keep the fabric straight and reduce the applied tension. Fabric edges should be glued to prevent curling when winding the fabric onto the fabric beam. In addition, tubular and open-width fabrics can also be processed in jets or winch-becks. If the fabric is treated with a jet, tension-free soft flow or overflow type should be preferred in order to preserve the elastic recovery properties of the fabric. The length of the batch to be taken into the machine should be adjusted so that one cycle takes 2–3 minutes.[20] In addition, it would be more appropriate to use the so-called "pipo" jets, in which the fabric is pulled over a shorter distance and therefore less strained.
- Fabrics containing elastane are generally 25% heavier than equivalent 100% cellulosic fabrics. Therefore, when working with elastane-blended fabrics, it is recommended to increase the liquor ratio by 20–30% in order to prevent the formation of wrinkles.[20]
- To prevent wrinkles that may occur due to sudden cooling of the bath, the bath should be cooled down to 50 °C before it is drained. Also, attention should be paid to slow heating and cooling of the liquor.[20]
- Silicone oils (4–5%) on the elastane must be removed before dyeing. Since polyurethane fibre is a fibre that is damaged by high pH and temperature, the safest way to remove silicone oils is to wash the fabric with a suitable degreasing agent in the dyeing machine before the pre-fixation.[21]

It should not be forgotten that, beyond the fibre type, the properties of the yarn (yarn hairiness, yarn twist, trash content, etc.) and the fabric (density, mass per unit area, knit/weave structure, etc.) will also cause differences in dye uptake. Because when the twist of the yarn or the density, knit/weave structure (single jersey, rib, interlock, etc., in knitted fabrics; plain, twill, satin, etc., in woven fabrics) of the fabric changes, there will be differences in properties such as the tightness/looseness of the structure and the porosity of the fabric, and dye uptake will change.

9.4.1.2.2 Effect of Fibre Pretreatment Processes on Dyeing

Exhaustion and fixation of reactive dyes vary depending on factors such as:

- physical properties of the material to be dyed,
- degree of bleaching and crystallisation and
- activation degree of fibres by mercerisation process.[9]

The pretreatment processes applied before dyeing affect the substantivity of the fibres for the dye. Cotton fabrics are subjected to processes such as desizing (for woven fabrics), alkali treatment and bleaching before dyeing. If they are to be dyed in dark shades, bleaching may not be necessary. However, if they are to be dyed in light and pastel shades, they must be bleached and they are subjected to a mild peroxide bleaching (using less peroxide). When a bleached material is dyed, it appears lighter than an unbleached one. This is because the base is whiter. Apart from this, it is necessary to desize a woven fabric, because the sizing layer will give the fabric a hydrophobic feature, making it difficult for the liquor to be absorbed by the fibres, in other words, for the fibres to get wet and swell.

In addition to the pretreatment processes mentioned above, mercerisation is sometimes required. The mercerisation process is the pretreatment process that most clearly affects the dye uptake ability of cotton materials. The ability of mercerised cotton products to take up dyes is considerably better compared to non-mercerised cotton products.[4] Since the caustic soda solution used in mercerisation is even more concentrated than 20% (27–30 °Bè), it can react with the hydroxyl groups in the cellulose macromolecules either to form sodium alcoholate (sodium cellulose) or to bind to macromolecules by absorptive attraction forces by penetrating into crystallites (intramicellar reaction). Although the second (75%) of these two reactions usually occurs, the first (25%) also occurs in small amounts.[22]

$$\mathrm{Sel(OH)_3 + NaOH \rightarrow Sel(OH)_2ONa + H_2O}\ \text{(Formation of cellulose alcoholate)}$$

$$\mathrm{Sel(OH)_3 + NaOH \rightarrow Sel(OH)_3.NaOH}\ \text{(Binding with absorptive attraction forces)}$$

As a result of these reactions, the cross-section of the fibres swells and the crystal structure changes. As a result of this, the dye uptake rate and amount of cotton fibres, which gain a more open structure, increase. In addition, since mercerised fibres with a smoother surface reflect light more regularly and the path length of light passing through the cross-section of swollen fibres increases, the mercerised fibres dyed with the same amount of dye appear darker than the non-mercerised fibres. When dyeing with reactive dyes, in addition to these, since it is easier for the dye to react with cellulose macromolecules, a larger part of the dye prefers to react with the fibres instead of water, and as a result, the dyeing efficiency increases.[22] It is stated that when dyeing mercerised cotton, dye savings can be achieved by 50–70% in dark shades and up to 30% in light shades, compared to normal cotton. As the stress applied to the material during the mercerisation process decreases, the dye uptake increases more. Therefore, if the main purpose of mercerisation is to increase the dye uptake, caustification, that is, tension-free mercerisation, should be performed.[23]

One of the pretreatments affecting the dyeing of cellulosic fibres with reactive dyes is biopolishing, which is performed for removal of protruding fibres with cellulase enzymes. Cellulase enzymes consist of complex mixtures of endo-gluconases, exo-gluconases, cellobiohydrolases and β-glucosidases. As a result of enzymatic treatment with cellulase, effects such as a strong decrease in the pilling tendency of the material and thus a smoother surface structure on the fabrics, improvement of the handle and covering ability of the fabric and the ability to absorb water into the structure of the fibres are obtained.[24] Moreover, due to the cleaning of the fibre surface during the enzyme treatment, deeper shades can be obtained in dyeing.[25]

9.4.1.2.3 *Effect of Dye Type and Molecular Structure on Dyeing*

By adding reactive groups to dye molecules with high substantivity to cellulose fibres (e.g. to substantive dyes), the uptake of reactive dyes by the fibres will be high, and thus the proportion of the part that reacts with water or remains in the liquor will decrease. However, on the other hand, the dye part, which has lost its ability to form covalent bonds with cellulose by reacting with water (by hydrolysis) will be taken up by the fibres like a normal substantive dye, so it cannot be completely removed from the material during soaping and rinsing at the end of dyeing. The wet fastness of dyeing will be low, since this non-removable part will bleed during washings during use.[4] For this reason, it is undesirable for the substantivity of reactive dyes to be too high. Reactive dyes are generally produced with low-medium substantivity and by adjusting the dyeing conditions (temperature, salt addition, etc.) during dyeing, sufficient substantivity is created and dyes are taken up by the fibres. After the dyeing is finished, the conditions that will reduce the substantivity are selected during the washing processes (salt is removed and hot washings are carried out) and the hydrolysed dye part is removed from the fibres.

Substantivity has no relation with the reactivity of the dye. There are many different substantive dyes with the same type of reactive group, and the ones with good substantivity are recommended by the dye manufacturers for exhaust dyeing (usually in these dyes at the end of their trade names, "E" which is the initial letter of the word "exhaust" or "A" which is the initial letter of the word "all-round" is present), and those with low substantivity are recommended for dyeing by padding method or printing (usually in these dyes at the end of their trade names "P" which is the initial letter of the words "Pad" or "Print").[4] It should be noted here that reactive dyes are also divided into groups with low, medium and high substantivity. However, even a reactive dye with a high substantivity has a very low substantivity compared to direct dyes. Here, high substantivity means that the reactive dye in question has a relatively higher substantivity among other reactive dyes.

In reactive dyes, as the molecular weight of the dye increases, its substantivity increases. In reactive dyes with the same reactive group, the substantivity varies depending on the chromophore group. On the other hand, reactive dye with a certain chromophore group can transform into ester, vinyl or hydroxy forms during dyeing, and as a result, its substantivity changes. For example, substantivity of sulphatoethylsulphone dyes decreases in the following order: vinyl form ≫ hydroxy form > ester form. As can be seen from Table 9.2, while the uptake of the ester form of the dye is 20%, the hydroxy form is 38% and the vinyl

Table 9.2 Substantivity of various forms of reactive dyes to fibres.

Dye chromophore	X	Substantivity (dye uptake in the presence of 50 g/L Na_2SO_4)
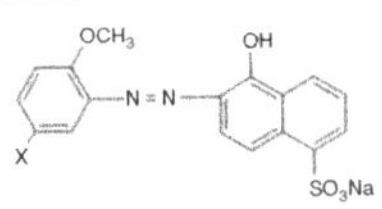	$-SO_2CH_2CH_2OSO_3Na$	20%
	$-SO_2CH{=}CH_2$	80%
	$-SO_2CH_2CH_2OH$	38%
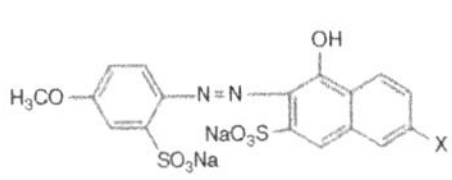	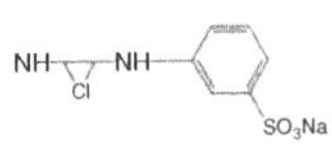	75%
	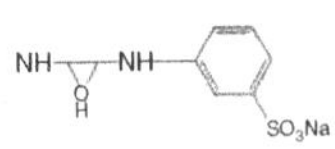	76%
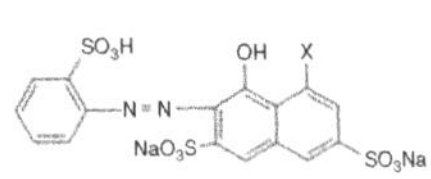	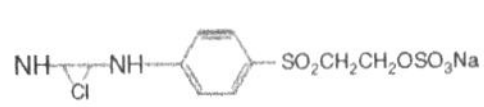	43%
	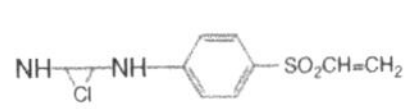	82%
	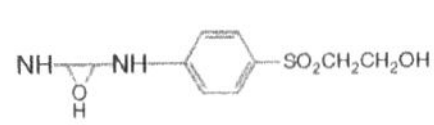	63%

Ref. [9]/Sumitomo Technical Documents.

Table 9.3 Various forms of reactive dyes based on sulphatoethylsulphone, monochlorotriazine and sulphatoethylsulphone/monochlorotriazine.

	Sulphatoethylsulphone	Monochlorotriazine	Sulphatoethylsulphone/ monochlorotriazine
Before alkali addition	$D{-}SO_2CH_2OSO_3H$ **Ester form**	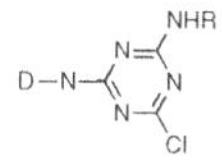	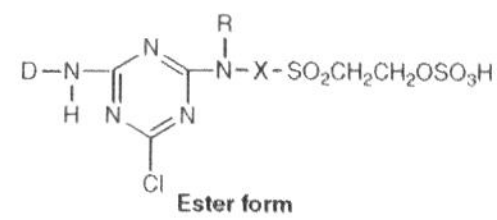 Ester form
After alkali addition	$D{-}SO_2CH{=}CH_2$ **Vinyl form**		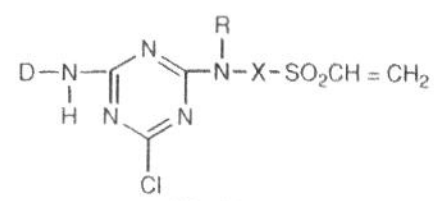 Vinyl form
After hydrolysis	$D{-}SO_2CH_2CH_2{-}OH$ **Hydroxy form**	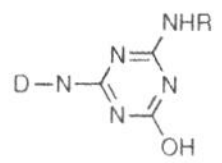	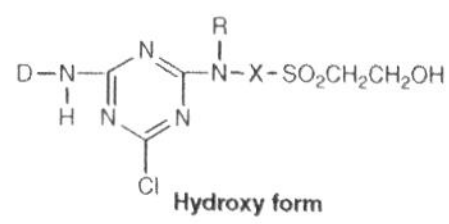 Hydroxy form

Ref. [9]/Sumitomo Technical Documents.

form is 80%. Due to the high substantivity of the vinyl compound, more dye can react with the fibres in the fixation step in exhaust dyeing. On the other hand, due to the low substantivity of the hydroxy form, hydrolysed dye formed at the end of dyeing can be easily removed in the washing process.[9] Table 9.3 summarises various forms of sulphatoethylsulphone and MCT-based monofunctional reactive dyes and sulphatoethylsulphone/monochlorotriazine-based bifunctional reactive dyes.

9.4.1.2.4 Effect of Liqour Ratio and Dye Concentration on Dyeing

Generally, the substantivities of reactive dyes are quite low and are significantly affected by the liquor ratio. As the liquor ratio decreases and the dye concentration in the liquor increases, substantivity increases.[4,9] In addition, dye fixation efficiency also increases.[8] For this reason, it is necessary to work with the shortest possible liquor ratio in order to obtain high efficiency dyeing. On the other hand, when the liquor ratio is decreased, it becomes difficult to obtain uniform dyeing as the migration decreases. Therefore, the liquor ratio should be determined correctly by taking into account the balance between dyeing efficiency and levelness.[4,9] In a study carried out using 30 g/L salt, if the liquor ratio is reduced from 1:20 to 1:3.5, it was observed that the substantivity of reactive dyes increased at proportions ranging from 34 to 53%. The least increase was observed in monofluorotriazine-based dyes, and the highest increase was observed in vinylsulphone/vinylsulphone-based dyes. The increase in dye fixation efficiency with the decreasing of the liquor ratio was greater than the increase in substantivity. For this reason, it can be said that decreasing the liquor ratio in dyeing with reactive dyes will provide the advantages of reducing water, energy, dye, salt and alkali consumption.[8] Since the liquor ratio affects the colour yield obtained in dyeing, it is important to keep the liquor ratio under control in order to obtain repeatable results in dyeing. When the liquor ratio changes, dyeing conditions (such as the amount of salt) must be adjusted to ensure that the dye absorption remains the same.[9]

9.4.1.2.5 Effect of Salt on Dyeing

In dyeing cellulosic fibres with reactive dyes according to the conventional exhaust method, high amounts of salt are needed to increase dye uptake, due to the low substantivity of reactive dyes.[26] Required salt (sodium sulphate or sodium chloride) amount in reactive dyeing is determined depending on the following factors:

- type of cellulose fibre
- the pretreatment processes of the fibre,
- the type of dye used,
- the colour depth and
- dyeing conditions (liquor ratio, temperature, etc.).

The amount and the type of salt used closely affect the result of dyeing. Optimal salt amounts are obtained from the tables or graphics in the dye catalogues of the dye manufacturers.[2]

The amount of salt to be used in dyeing primarily depends on the fibre type. Since the substantivity of regenerated cellulose fibres for reactive dyes is higher than that of cotton, less salt is needed for dyeing these fibres compared to cotton. Similarly, the need for salt is lower in dyeing mercerised cotton. How the amount of salt should be changed depending on the fibre type and the pretreatment process of the fibre is given in tables and/or graphics in the catalogues of dye companies.

In Figures 9.20 and 9.21, optimal salt usage amounts are given according to temperature, liquor ratio and colour depth.[2] Accordingly, the amount of salt needed increases as the colour depth, liquor ratio and dyeing temperature increase. More salt is needed for darker shades to make it possible for fibres to absorb more dye. On the other hand, as the liquor ratio or dyeing temperature increases, substantivity will decrease, and more salt will have to be added to the liquor to compensate for this.

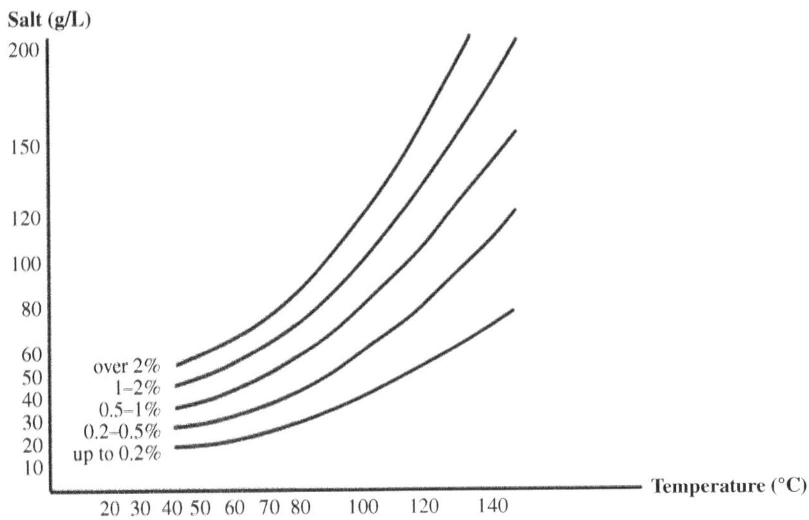

Figure 9.20 Relationship between temperature and salt concentration. Ref. [2].

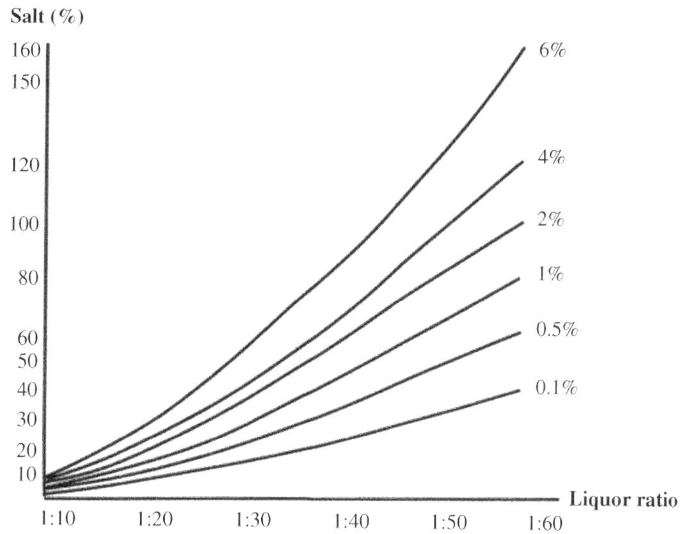

Figure 9.21 The relationship between liquor ratio and salt concentration at different colour depths (%). Ref. [2].

The dye exhaustion and accordingly the dye fixation degree increase as the salt concentration increases. In fact, salt does not directly influence fixation in reactive dyeing but has the function of increasing dye uptake. However, since a certain proportion of the dyes taken up by the fibres in the first step of dyeing is fixed to the fibres in the next step, the higher the amount taken up in the first step, the higher the amount of dye fixed in the second step. However, adding more salt than necessary in dark shades or dyeing at short liquor ratios reduces the solubility of the dye in the dyebath and increases the aggregation significantly, thereby reducing the dye exhaustion. In the case of using more salt than necessary, the reduction in dye exhaustion is greater when using sodium chloride salt compared to using sodium sulphate. Furthermore, the use of more salt than necessary increases the aggregation and reduces the migration ability of the dye, making it difficult to obtain a uniform dyeing. On the other hand, it should not be forgotten that using insufficient salt will result in poor fixation degree and therefore low colour yield due to low dye exhaustion.[9]

Three types of salts are commonly used in dyeing processes with reactive dyes. These are:

- Sodium chloride (common salt) (NaCl),
- Calcined sodium sulphate (Glauber's salt) (Na_2SO_4) and
- Crystalline sodium sulphate ($Na_2SO_4.10H_2O$).

In practice, the first two of these salts do not differ in terms of the amount used, and when the third one is used, it should be used approximately twice the amount of the other two. Because there are 10 moles of water in the structure of crystalline sodium sulphate, this constitutes about 56% of the molar mass of the salt.[27] Therefore, it is necessary to use slightly more than twice the amount of crystalline sodium sulphate compared to calcined sodium sulphate.

There are no systematic data regarding which salt type should be preferred, and companies use both NaCl and Na_2SO_4. The following factors should be considered as salt type selection criteria:

- its price
- its effect on uniform dyeing

- its effect on dyeing efficiency and dye uptake rate
- its effect on the solubility of the dye
- its risk of corrosion (especially during dyeing with hot-dyeing reactive dyes).

In terms of price, sodium chloride is cheaper than sodium sulphate. However, when the salt recycling is desired, it is more advantageous to use sodium sulphate salt instead of sodium chloride. Because the salt contained in the dyeing liquor, which has been rendered colourless by chemical and mechanical purification, is concentrated by the membrane technique and it is easier to recycle sodium sulphate by filtration with membranes.[27] Another important issue regarding the salt type in dyeing is the solubility of the dye in the salt-containing liquor. The solubility of some dyes decreases significantly in liquors containing sodium chloride. In this case, it is recommended to use sodium sulphate salt.[25] Moreover, in the literature comparing these two salts in terms of corrosion properties, it is stated that sodium sulphate is less corrosive for dyeing machines than sodium chloride.[28] Today, it can be said that sodium chloride salt is generally used in dyehouses since it is cheaper and suitable for dosing. Since the solubility of sodium sulphate salt varies with temperature, the salt solidifies in water and this creates problems in dosing. However, it is worth noting that sodium sulphate salt is used in cases where it is difficult to achieve uniform dyeing, especially in turquoise and bright green shades.

Although cellulose fibres are normally non-ionic in nature, these fibres acquire a slightly anionic character as a result of processes such as bleaching before dyeing. Since both pretreatment and dyeing processes of cellulose fibres are carried out in alkaline medium, the fibres have a negative charge. Since most of the dyes used in dyeing these fibres are anionic in nature, the dye is also (−) charged. During dyeing, dye and fibre tend to repel each other because they have the same charge. In this case, this charge must be reduced or completely eliminated in order for the dye to be absorbed by the fibre. As it is known, during dyeing of cellulose-based fibres with anionic dyes, the dye molecule is absorbed by the fibre, not the dye ion. Therefore, the less the dye molecule is dissociated, that is, the lower the negative valence, the equilibrium given below is shifted to the fibre side.

$$\text{Dye in liquor} \rightleftharpoons \text{Dye in the fibre}$$

Therefore, it is generally known that salts increase the uptake of dyes in dyeing cotton fibres with anionic dyes (direct, vat, sulphur, reactive, etc.). The two main reasons for this are that salt cations reduce the electrostatic repulsion between the dye and the cellulose fibre by reducing the negative zeta potential of the cellulose fibre and increase the aggregation of the dye.[29]

An important issue in dyeing with ionic dyes is the Donnan membrane equilibrium. Such ion equilibria occur in the presence of a barrier (membrane) that is permeable to solvent and small ions but impermeable to large ions. For example, if we take Na^+ and Cl^- as small ions and the dye anion (DSO_3^-) as large ions, and we assume that the fibre with dye anions that do not diffuse into the fibre on its surface, plays the role of the semi-permeable membrane mentioned above, while the dye anion cannot diffuse freely into the fibres, small Na^+ and Cl^- ions can be taken up by the fibre until their chemical potentials in the fibre and in the solution are equal. However, the fibre and solution must also be maintained electrically neutral. Due to the dye anions on the fibre surface, there is an excess of (−) charge here, and this can only be neutralised with a higher Na^+ concentration on the fibre surface. As a result, it is seen that Na^+ ions cannot be evenly distributed between the solution and fibre despite the freedom of movement, and the concentration of Na^+ ions is higher in the region where the dye anions are collected.[4] The positively charged sodium ion goes to the fibre-water boundary surface, effectively lowering the negative zeta potential of the fibres, thus removing the barrier that prevents the dye-fibre interaction.[27]

Another effect of the increase in the Na^+ cation concentration in the liquor with the addition of salt is that the anionic dyes (direct, reactive, etc.) used in cotton dyeing shift the dissociation equilibrium (i) to the left. According to Le Chatelier's principle, when the Na^+ cation concentration in the medium increases, the dissociation equilibrium of the dyes into ions shifts to the left and aggregation increases, which is another effect of salt that enhances dye uptake in cotton dyeing.[27]

Although adding salt to the liquor increases the dye uptake in dyeing cotton fibres with anionic dyes, dyeing efficiency and dye uptake rate may differ depending on the salt type. When we compare sodium chloride and sodium sulphate salts, sodium chloride leads to a more rapid dye uptake and is not recommended for risky dyeings (such as turquoise colours). The reason for this can be explained chemically as follows:

$$DSO_3Na \rightleftharpoons DSO_3^- + Na^+ \quad \text{(i)}$$

$$NaCl \longrightarrow Cl^- + Na^+ \quad \text{(ii)}$$

When sodium chloride and sodium sulphate are used equally in dyeing, the amount of sodium cation coming from sodium chloride to the liquor is higher than that from sodium sulphate. By calculation, it can be determined that the amount of Na^+ cation contained in 1 gram of NaCl is equivalent to the amount contained in 1.22 grams of Na_2SO_4. Therefore, when these two salts are used in equal amounts, the amount of sodium cations in the liquor is not equal and the uptake is more rapid in sodium chloride, which gives more sodium to the liquor. However, when compared in terms of colour yield, sodium sulphate leads to slightly darker depths.[27] This situation cannot be explained by classical theories that consider only ionic interactions between the dye and fibre used to explain the effects of salts on the dyeing of cotton fibres with anionic dyes (direct, reactive, etc.). It is thought that this difference in both salt types is due to chloride and sulphate anions, not sodium cations. If this difference was due to sodium cations, it would be expected that dyeing with sodium chloride would be darker. Because when using equal amounts of sodium chloride and sodium sulphate, the amount of sodium cation coming from sodium chloride is approximately 1.22 times higher.[30] In order to explain the difference between the effects of these two salt types on dye uptake, it is necessary to consider the effects of salts on physical interactions as well as ionic interactions between dye and fibre. This situation can be explained using the lyotropic series. The order of the anions and cations in the lyotropic series is as follows:

$$SO_4^{=} > H_2PO_4^- > Cl^- > Br^- > NO_3^-$$

$$Li^+ > Na^+ > K^+ > Rb^+ > Cs^+$$

Chloride is a neutral ion in the lyotropic series; anions before chlorine are called cosmotropes, and anions after chloride are called chaotropes. Cosmotropic ions hold water molecules more effectively and increase the uptake of dye by increasing the physical interaction between dye and fibre. Since the hydration capacity of sodium sulphate is higher than that of sodium chloride, it seems as if the amount of water in the liquor decreases and therefore the dye concentration per unit volume increases, and as a result, the uptake is increased.[31] Therefore, sodium sulphate increases dye uptake by both decreasing the ionic repulsion between the dye and the fibre and increasing the physical interactions between the dye and the fibre.

Many researchers point out that the salt in textile dyeing wastewater can cause significant problems.[32] It is stated that the salt concentration in the wastewater in cotton dyeing made by the exhaust method is generally around 2,000–3,000 ppm.[33] Many countries have set emission limit values as 2,000 ppm and below.[34] From these values, the effect of salt on the wastewater load is clearly seen. For this reason, it is of great importance to develop a process that will enable cotton to be dyed with reactive dyes with little or no salt without causing loss of yield. Since the reason why a high amount of salt is needed in dyeing cellulose fibres with reactive dyes is the low affinity of the fibres for these dyes, it is necessary to increase the affinity of either dyes or cellulose fibres in order to dye with lower salt amounts or without salt. Based on this idea, special reactive dyes with higher substantivity were produced. Many dye manufacturers have developed new dye types and application methods that require 2/3 of the normal salt. Most of these dyes are polyfunctional dyes and can exhibit high fixation. Thus, additional gain is obtained by reducing the amount of unfixed dye in the wastewater. Reactive dyes that require less salt (low salt) have better solubility because less salt is needed for their uptake by the fibres. Therefore, they remain dissolved even in high concentration solutions in dyeing machines that work at low liquor ratios. The fact that the salt consumption in the dyeing of cellulosic fibre by the exhaust method has been reduced by 1/3 compared to that required in conventional reactive dyes, positively affects the wastewater salinity and wastewater treatment processes. However, reactive dyes that require a low amount of salt are dyes with high affinity and are therefore more difficult to remove by washing. However, dyes which have low affinity in hydrolysed (unfixed) form and therefore easier washing-off properties are increasingly being brought onto the market.[26]

Since the cellulosic fibres carry a negative charge, it would thus seem entirely logical to synthesise reactive dyes, which carry a positive charge. The dye was prepared as shown in Figure 9.22.[11]

The second option to reduce or eliminate the amount of salt in reactive dyeing is to increase the affinity of cellulose fibres for reactive dyes. Cotton fibres have a slightly negative charge in the aqueous medium. With ionic modification, these charges can be converted to a cationic state, and the affinity of the obtained cationic cotton for anionic dyes increases compared to untreated cotton.[35] The possibility of making the anionic character of cotton cationic (converting the negative charge into positive) by treating it with cationic substances has been known for many years, and studies on this subject are still ongoing. Changing the ionic structure of cotton increases the affinity of anionic dyes for cotton and helps to overcome various problems (high salt requirement in reactive dyeing, low washing fastness of direct dyes, etc.) encountered during the dyeing of cotton with conventional dyes. The cationisation process increases the dyeability of cotton with already used dyes, while making the fibres dyeable with dye groups such as acid and

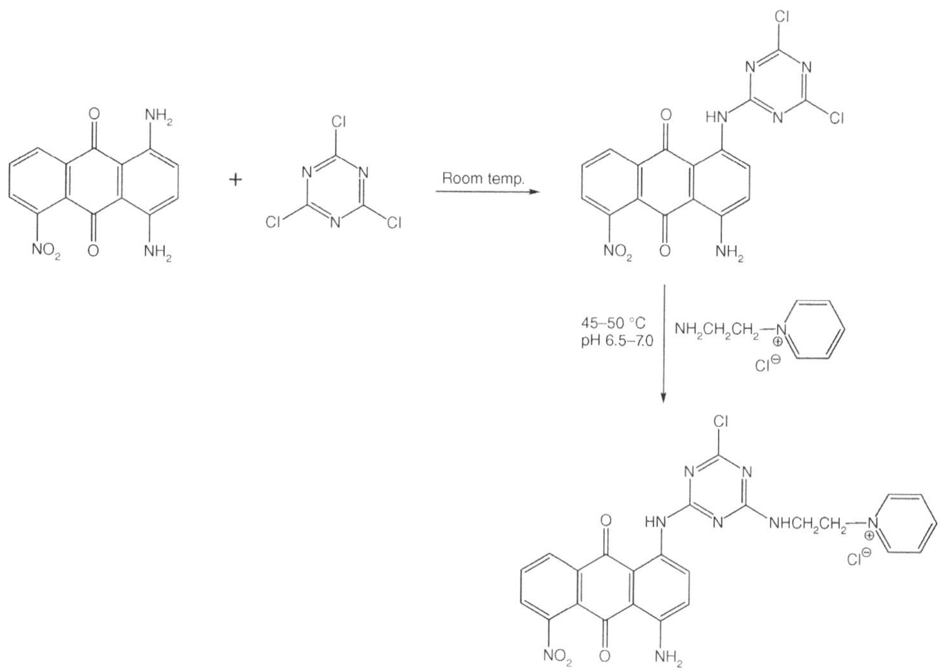

Figure 9.22 Synthesis of a cationic reactive dye for cellulosic fibres. Ref. [11]/John Wiley & Sons.

metal complexes. Quaternary cationic substances or amines with various reactive groups (epoxy, active halogen, ethoxylate or amino) that can react with cotton are used in most of the studies to improve the substantivity of cotton for anionic dyes.[36] Cellulose can be cationised by reacting with amine groups or quaternary ammonium groups thanks to its hydroxyl groups. Cationic cotton is produced by the etherification reaction of cotton, especially with quaternary ammonium (3-chloro-2-hydroxypropyl trimethyl ammonium chloride) or tertiary ammonium components.[37,38] Cationisation of cotton can be done either discontinuously by the exhaust method or continuously by the padding method.

In Figure 9.23, the bonding mechanisms of two different commercial products to cotton fibres: one based on polyethylene polyamine compound, which can be bonded to fibres by electrostatic attraction forces, and the other based on a polyaminochlorohydrin quaternary ammonium compound, which can covalently bond with fibres, are shown. It is possible to dye cotton fibres with reactive dyes without using salt after the cationisation process performed in this way; however the affinity of the fibres for dyes increases as a result of the cationisation of the fibre surface, and dye penetration is weakened as the dye molecules on the surface prefer to stay here. As a natural consequence of this, the rubbing fastness of fabrics dyed without using salt with reactive dyes after the cationisation process is lower than that obtained in conventional dyeing. In addition, it is expected that the light fastness values of the dyed material after pretreatment with the cationising agent will decrease due to the amine groups in the structure of the cationising agent. Because the decomposition of cationic groups with the effect of light and the formation of free radicals and other

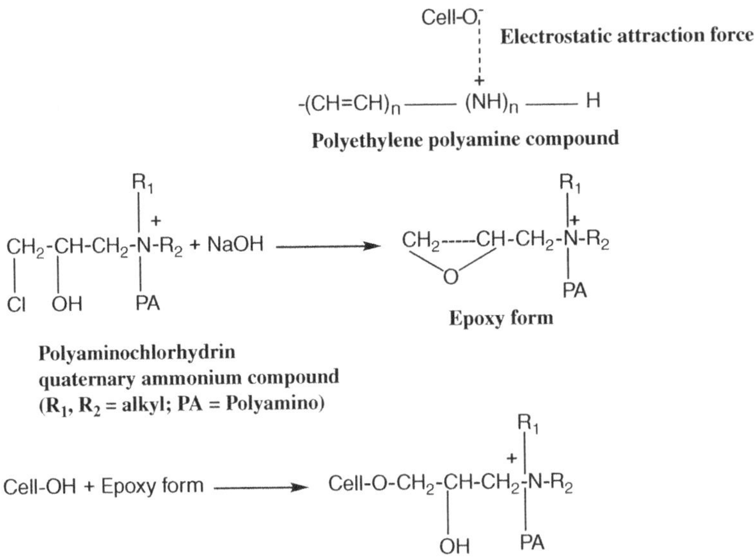

Figure 9.23 Binding mechanisms of cationising agents based on polyethylene polyamine and polyaminochlorhydrin quaternary ammonium compound to cotton fibres. Ref. [40]/Elsevier.

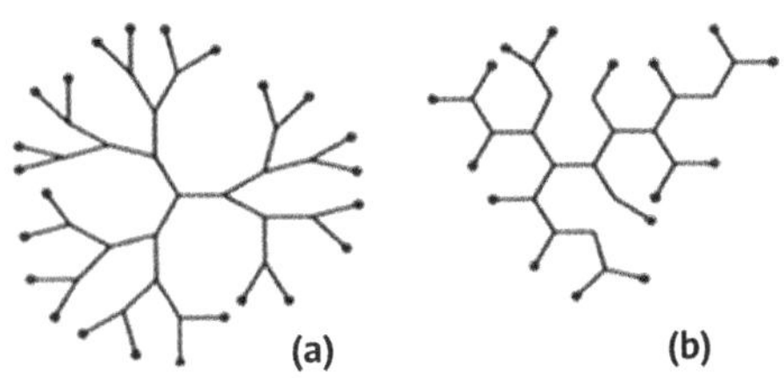

Figure 9.24 General structure of (a) dendrimer and (b) hyperbranched polymers. Ref. [43]/with permission of Elsevier.

chemicals accelerates the degradation and fading of dyes.[39] It should also be noted that cationising agents are polyelectrolyte in nature and significantly increase the wastewater load. Therefore, more ecological cationisation alternatives are being developed.

One of the subjects studied in order to make cellulosic materials dyeable with reactive dyes without the use of salt is the chemical modification of fibres with dendrimers. The term dendrimer was first proposed by Tomalia in 1985.[41] Simultaneously, the Newkome group independently synthesised macromolecules with similar structures. They defined these macromolecules as "aborols", which means tree, based on the Latin word "arbol".[42] Figure 9.24 shows two types of highly branched polymers, regularly branched dendrimers and "hyperbranched polymers" in which branching occurs randomly.[43]

Dendrimers are chemically formed by combining three different components in a special way. These are a multifunctional centre, inner layers and a polyvalent surface.[41] There are dendrimers with a wide variety of end groups. After pretreatment of cellulose fibres with dendrimers with amine end groups, it is possible to reduce or eliminate the use of salt in reactive dyeing, since cationic groups are added to the fibres.[32,44,45] Oxidative pretreatment is recommended to facilitate the binding of dendrimer molecules to cellulose fibres.

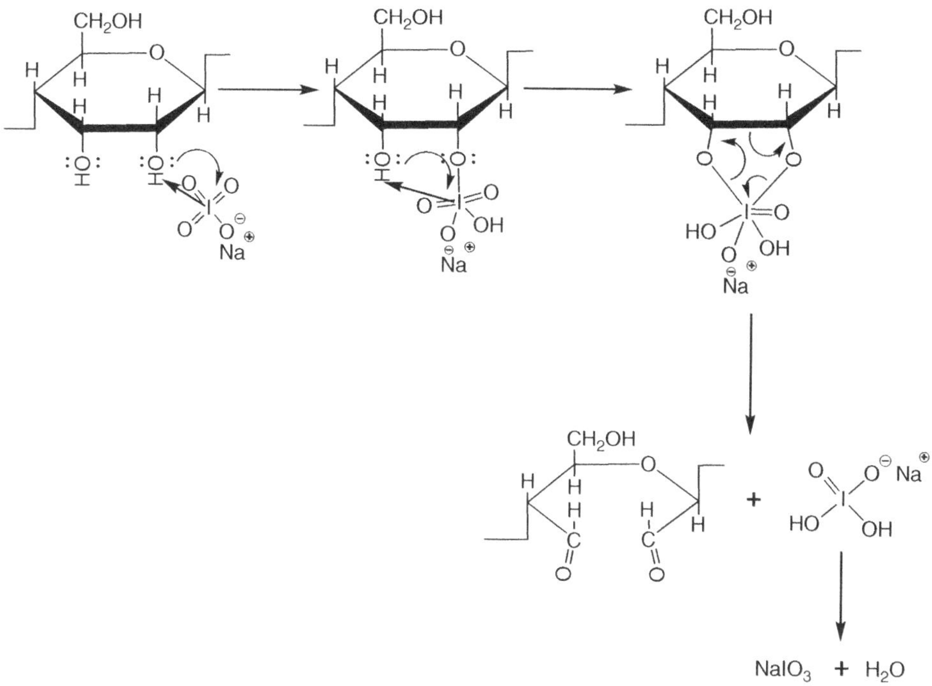

Figure 9.25 Mechanism of oxidation of secondary alcohol groups in the structure of cotton fibres by m-periodate. Ref. [32].

As shown in Figure 9.25, when cellulose fibres are treated with m-periodate, m-periodate selectively reacts with secondary alcohol groups in the fibres and oxidises them to aldehydes under severe conditions.[46]

Dendrimer molecules with amine end groups react and bond more easily with aldehyde groups formed by oxidation, not with alcohol groups in cellulose fibres[32] (as shown in Figure 2.4). Thus, the fibres are cationised and the need for salt in reactive dyeing is reduced.

With the reuse of waste becoming a priority in recent years, instead of being left to decay, shellfish are valorised by chemical or biological methods and new products are obtained. Chitin and its main derivative, chitosan, are the leading products obtained in this way.[47] Chitin and chitosan, a natural and non-toxic biopolymer, are mainly obtained from crab and shrimp shells. In addition, insect shells are also very rich in terms of chitin.[48]

Chitosan can be used as a more ecological alternative for the cationisation of cellulosic fibres. It has been demonstrated in many studies in the literature that oxidative pretreatment of cellulosic fibres before chemical modification with chitosan containing amine groups will increase the amount of chitosan transferred to the fibres. Because the amine groups present in chitosan molecules can be bonded to the fibres by reacting more easily with the aldehyde and carboxylic acid groups formed as a result of their oxidation, not with the primary and secondary alcohol groups in the cotton fibres.[49] Thus, the fibres are cationised and the need for salt in reactive dyeing is reduced. The reaction mechanism is given in Figure 9.26.

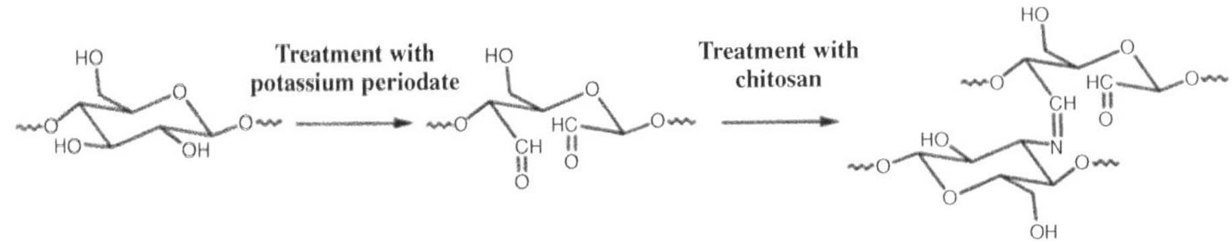

Figure 9.26 Mechanism of binding of chitosan to cotton after oxidative pretreatment. Ref. [50]/Elsevier.

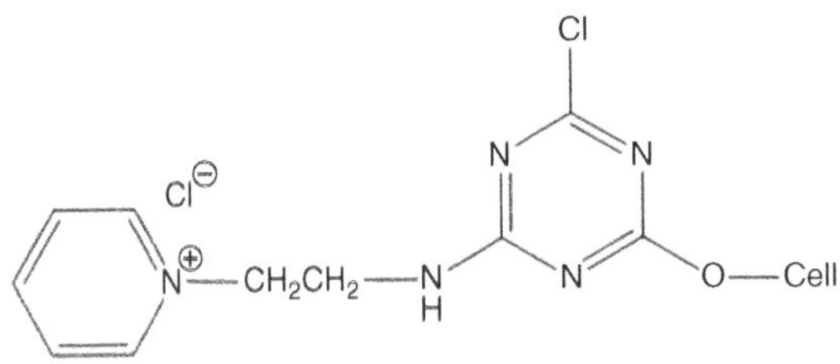

Figure 9.27 DCPEAT active cotton. Ref. [11]/John Wiley & Sons.

Another option to improve permanence in a salt-free dyeing system is to create reactive fibres instead of reactive dyes. For this purpose, 2,4-dichloro-6-(2-pyridino-ethyl-amino)-s-triazine (DCPEAT) was studied. If this compound is applied and fixed to cellulosic substrates under well-defined conditions, the following modified substrate will be obtained (Figure 9.27).[11]

9.4.1.2.6 *Effect of Alkali on Dyeing*

Alkali is required in the dye liquor for reaction between dye and fibre to proceed.[51] Therefore, dyeing of cellulosic fibres with reactive dyes is done in an alkali medium. In this medium, the nucleophilic character of cellulose (the form that has a high tendency to react with the reactive group of the dye) increases. In addition, alkali in the medium neutralises the hydrogen chloride released during the reaction of dyes containing heterocyclic rings with chlorine substituted as reactive group, as shown in Figure 9.13, and prevents them from damaging the fibres. In reactive dyes based on β-substituted ethane derivatives, it is possible to form the vinyl group, which is the main reactive group, by eliminating the substituent, as shown in Figure 9.15, in an alkali medium more quickly.[4]

Although an alkali medium is mostly used for dyeing cellulosic fibres with reactive dyes, there are also reactive dyes that can dye cellulosic fibres in different mediums. While dyes containing phosphonate groups are applied to the material in an acidic medium, nicotinic acid derivatives dye the material in a neutral medium.[5] These dyes were originally developed for one-bath, one-step dyeing of PES/CO blends with a combination of disperse/reactive dyes. Reactive dyes suitable for dyeing in a neutral medium are mononicotinate triazine (MNT)-based. These dyes are obtained by replacing the chlorine atom attached to the triazine ring in monochlorotriazine (MCT)-based reactive dye with nicotinic acid. Figure 9.28 shows the chemical structure of MNT-based reactive dyes.[52]

As can be seen from Figure 9.29, there is no need for alkali use in the dyebath, since weak nicotinic acid is released as a result of the reaction of MNT-based dyes with cellulose. An effective fixation at pH 7 can be achieved in dyeing with MNT-based dyes.[52]

According to the exhaust method, the appropriate pH of the bath is around 11 depending on the dyeing temperature. Similar to the amount of salt, the amount of alkali should be adjusted depending on the colour depth.[9] In addition, the amount of alkali to be used depends on the reactivity of the dye, and stronger alkali should be used in dyes with low reactivity, and less and weaker alkali should be used in dyes with high reactivity.[4]

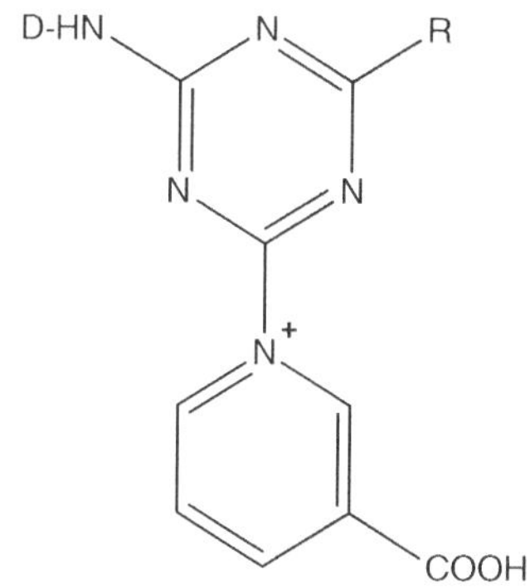

Figure 9.28 Chemical structure of mononicotinate triazine-based reactive dyes. Ref. [52]/John Wiley & Sons.

The amount of alkali to be used varies depending on the dyeing temperature, colour depth and reactivity of the dye, as well as the liquor ratio. When the liquor ratio is changed, instructions on adjusting the amount of alkali are provided in the dye catalogues. When the liquor ratio is increased, the alkali concentration that needs to be added to the liquor must be slightly reduced. In Table 9.4, an example of the change of alkali concentration to be used in dyeing is given depending on the liquor ratio.

For example, for 2% dyeing, the value to be used at 1:10 liquor ratio is 13 g/L soda ash; if the liquor ratio is to be 1:5, this value should be increased by 20% and should be 15.6 g/L. If it will be 1:20, this value should be reduced by 20% and it should be 10.4 g/L. This situation, which seems to be a contradiction at first glance, can be better understood if the amount of alkali (not its concentration) added to the dyebath is calculated based on the liquor ratio. When 13 g/L of soda ash is used in 2% dyeing to dye 10 kg of fabric in a 1:10 liquor ratio, the amount of alkali to be used will be 1.3 kg, since the liquor volume will be 100 L.

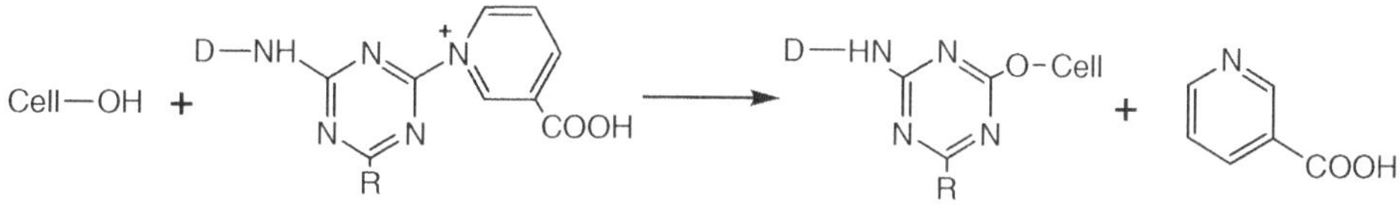

Figure 9.29 Reaction of mononicotinate triazine-based reactive dyes with cellulose. Ref. [52]/John Wiley & Sons.

Table 9.4 Variation of alkali concentration to be used depending on the liquor ratio.

Liquor ratio	Alkali concentration
1:5	+ 20%
1:10	The standard value given for that colour depth in the catalogue
1:15	– 10%
1:20	– 20%

Adapted from Ref. [53].

On the other hand, if the same concentration of alkali is added when the liquor ratio is increased to 1:20, the amount of alkali to be used will be 2.6 kg, since the liquor volume will be 200 L this time. However, since the alkali concentration is reduced by 20%, that is, 10.4 g/L soda ash is used, the amount of soda ash to be added to the liquor is 2.08 kg for a 1:20 liquor ratio. As mentioned before, alkali has two functions in reactive dyeing: to ensure that the pH of the liquor is at an appropriate level, and to neutralise the acid that occurs in the reaction of the reactive dye with the fibre (in heterocyclic ring-based dyes) or to ensure the formation of the active form of the dye (in vinylsulphone-based dyes). If the alkali concentration required for the first of these is "a" g/L, it will still need to be "a" g/L when the liquor ratio is increased. Therefore, the amount of alkali required for the first function of alkali increases as the liquor ratio increases. However, the amount of acid that is released in the reaction of a certain amount of reactive dye with the fibre and needs to be neutralised (not its concentration) or the amount of dye that needs to be converted into active form is constant even if the liquor ratio changes. Then, if the alkali concentration is kept constant when the ratio of the liquor is increased, the amount of alkali added to the liquor will increase, so unnecessary alkali is added more than the amount that will be consumed in the reaction of neutralisation or the conversion of the dye into the active form. On the other hand, when the liquor ratio is increased, the alkali concentration is slightly reduced; thus, unnecessary alkali usage is prevented. However, the amount of acid (not its concentration) (which is needed to be neutralised) that is released in the reaction of a certain amount of reactive dye with the fibre or the amount of dye that needs to be converted to the active form is constant, even if the liquor ratio changes. For this reason, if the alkali concentration is kept constant when the liquor ratio is increased, the amount of alkali added to the liquor will increase, so unnecessary alkali is added beyond the amount that will be consumed in the reaction of neutralisation or the conversion of the dye into the active form. On the other hand, when the liquor ratio is increased, the alkali concentration is slightly reduced; thus, unnecessary alkali usage is prevented.

As mentioned earlier, a 1-digit increase in pH results in an approximately 10-fold increase in reaction rate. The type and therefore the strength of the alkali used affect the temperature at which the reaction can occur. While some dyes react with cellulose fibres only at high temperatures in the presence of medium-strength alkali, they can react even at room temperature when there are high amounts of strong alkali in the medium.[4]

In addition to the rate of reaction of reactive dyes with fibre, the rate of reaction with water, that is, hydrolysis, also depends on pH values. Although reactive dye solutions are generally resistant to hydrolysis for a long time on their own, this resistance disappears after the addition of alkali. Therefore, alkali addition should be made as late as possible in dyeing with reactive dyes.[4]

Addition of alkali according to the chosen dyeing method can be applied in different periods of dyeing, such as:

- after the dye uptake reaches an equilibrium (**two-step method**),
- starting the dyeing with a weak alkali (or a small portion such as 1/3 of alkali) and then adding a strong alkali (or a large portion such as 2/3 of alkali) (**method of adding some alkali from the beginning**) and
- at the beginning of dyeing (**all-in method**).[2]

The addition of alkali to the liquor can be done directly, or it can now be done in the form of dosing using automatic dosing equipment. According to the dyeing method applied in reactive dyeing, mixtures such as caustic soda, soda ash, sodium bicarbonate, trisodium phosphate or soda ash/bicarbonate, soda ash/caustic soda are used as alkali.[4] In Figure 9.30, an example of the variation of the relative fixation degree depending on the alkali type and concentration is given in a dyeing made using different alkalis.

When Figure 9.30 is examined, it is noteworthy that the optimum concentrations to be used vary depending on the type and power of alkali used in dyeing, and that significant reductions in fixation efficiency may occur if alkali is used above the optimum level, especially in dyeings made with caustic soda. It can be said that a combination of 1–20 g/L sodium carbonate or 5 g/L sodium carbonate and 1–2 mL/L caustic (40 °Bè) is generally used in reactive dyeing. In the past, 1–15 g/L trisdoium phosphate could also be used.[9]

As well as soda ash alone, soda ash/caustic soda combination can also be used alone for dyeing cotton fabrics with both cold-dyeing and hot-dyeing dyes. Especially when working with reactive dyes with low substantivity, it is recommended to add weak alkali (sodium bicarbonate or soda ash) to the liquor first, and then a stronger alkali (soda ash or caustic soda) to the liquor, in cases where the pH value suddenly changes, causing excessive

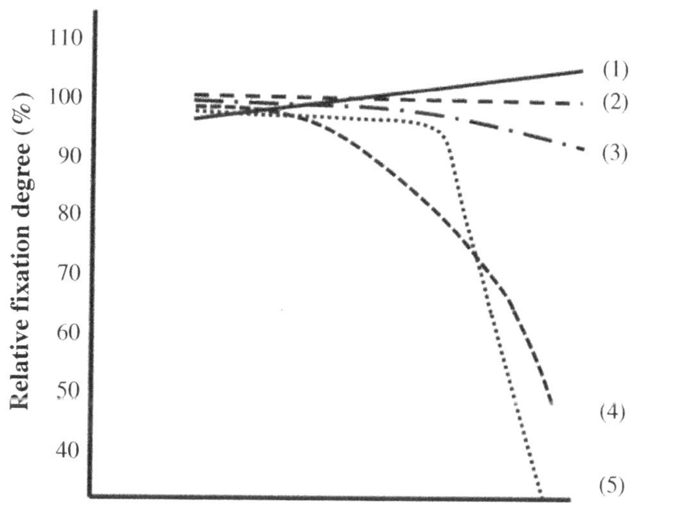

(1) Sodium carbonate (g/L)		10	15	20	25	———
(2) Trisodium phosphate (Crystal) (g/L)		5	10	15	20	– – –
(3)	Sodium carbonate (g/L)	3	5	10	15	—·—
	Caustic soda (40 °Bé) (mL/L)	1	2	3	4	
(4) Caustic soda (40 °Bé) (mL/L)		1.5	3	5	10	-----
(5)	Sodium bicarbonate (g/L)	2	2	5	5	....
	Caustic soda (40 °Bé) (mL/L)	2	4	6	2	

Figure 9.30 Variation of relative fixation degree depending on alkali type and concentration in 1% dyeing with a reactive dye. Ref. [9]/Sumitomo Technical Documents.

dye bleeding.[4] Sodium bicarbonate is often used in the method of adding some alkali from the beginning. Since sodium carbonate can turn into sodium bicarbonate with the effect of acidic gases in the atmosphere, sodium carbonate that has been kept for a long time may contain a certain percentage of sodium bicarbonate.[9] Replacing some of the weak alkali with strong alkali (e.g. instead of 13 g/L soda ash, 5 g/L soda ash +0.7 mL/L 50% caustic soda) requires more careful work, but it is advantageous in terms of reducing alkali consumption. Since the use of caustic soda alone is a more risky method, it can be used in dark dyeings with some dyes (e.g. black) and in systems where alkali dosage can be carried out in a controlled manner. When using caustic soda, care should be taken to add caustic soda gradually (dosing progressively).[4] Due to the strong alkalinity of caustic soda, errors in weighing may cause deviations in colour shade. Therefore, the use of caustic soda alone is not normally recommended. However, if it is used together with sodium carbonate or sodium bicarbonate, buffering effect is obtained and the problems caused by the use of caustic soda alone are not observed. However, more attention should be paid to the amount of alkali when using a soda ash/caustic soda combination compared to using soda ash or trisodium phosphate alone. In particular, when soda ash and caustic soda are used together, soda ash should be added first, and then caustic soda or a pre-prepared mixture of caustic soda and soda ash should be added to the dyebath. When using soda ash and sodium bicarbonate together, care should be taken not to use more bicarbonate than necessary. The pH decreases significantly as the sodium bicarbonate ratio in the mixture increases. As the pH decreases, the fixation of reactive dyes naturally decreases.[9] While this decrease is very evident in dyeing with some dyes, it may be less pronounced in dyeing with others.

If table salt obtained from sea water is used as salt, a significant amount of calcium and magnesium ions from the salt will pass into the dyeing water. If trisodium phosphate is added to the liquor to provide alkaline medium, calcium and magnesium phosphate precipitate and it is difficult to remove them from the product. If alkaline medium is achieved with sodium hydroxide, calcium and magnesium hydroxide precipitate, but these can be removed with a good rinse. Therefore, when using sodium chloride obtained from sea water, it may be recommended to use caustic soda (sodium hydroxide) as an alkali.[4] However, today, there are no prescriptions recommending the use of trisodium phosphate in new catalogues, and this situation is no longer a problem.

9.4.1.2.7 Effect of Water Quality on Dyeing

One of the sources of practical problems in textile processes is metal ions that can come from water, cellulosic fibre and machinery. In addition, salt and other auxiliaries used in different processes can also give metal ions to the medium. Although the operating waters are softened and hardness is decreased to 5 °dH (German hardness), what needs to be taken into account are the fibre itself and the ions that may come from the salts.[54] The concentration of polluting ions that can be found in the dyeing water must be within the following limits.[55]

- Hardness < 5 °dH
- $Fe^{++} < 0.1$ ppm
- $Mn^{++} < 0.02$ ppm

- $Cu^{++} < 0.005$ ppm
- $NO_3^- < 50$ ppm
- $NO_2^- < 5$ ppm.

The absence of hardening metal ions in the water to be used in textile finishing processes is one of the main features sought. The reason for the hardness in the water is the alkaline earth metal salts dissolved in the water. The hardness formed by calcium and magnesium, and rarely strontium and barium bicarbonates and carbonates dissolved in water is called temporary hardness, on the other hand the hardness formed by neutral salts of calcium and magnesium excluding bicarbonate and carbonates is called permanent hardness, and the sum of these two hardnesses gives total hardness. Since the concentrations of Ca^{+2} and Mg^{+2} ions, which are generally dissolved in water, are much higher than other hardening cations, it is accepted that the hardness in water is practically caused by Ca^{+2} and Mg^{+2} salts. During the removal of these salts by softening processes, other ions present in very small amounts in water are also removed.[30]

Although hard water is not harmful to human health, these waters cause significant problems in cleaning and industrial use. Ca^{+2} and Mg^{+2} salts precipitate in heating pipes and steam boilers and prevent heat conduction. It can cause problems by forming sediment and deposit on the machine, roller and fabric surfaces.[56] The most important problem caused by hard water in dyeing processes is the change in the structure of the dye. In this case, the colours may completely deviate, and the solubility or diffusion properties of the dyes may be weakened. As a result, uneven dyeing occurs and fastness values may decrease. In addition, metal ions that give hardness can cause the washing agents to decompose into water-insoluble Ca^{+2} and Mg^{+2} salts and to precipitate on the fabric.[30] Since Ca and Mg soaps do not participate in the washing process, some of the soap amount is lost depending on the hardness of the water. In addition, the precipitates formed cause an uneven dyeing and may adversely affect the rubbing and washing fastness of the material.[57] Since Ca and Mg ions have a strong adsorbing capacity on the fibres, they cannot be removed even during post-processing such as soaping and may cause a "white powder" appearance on the material after drying and finishing.[9]

Apart from hardening metal ions, water containing heavy metal ions such as iron, copper and manganese can also damage the dyeing by forming unwanted precipitates with dyes in textile dyeing.[30] Especially iron and copper ions have a significant effect on the colour shade. For example, the dye molecule given in Figure 9.31 can bind with the metal ion in two different ways, and in both cases, a darkening occurs in the colour.[9]

For this reason, when the values specified at the beginning of the section are exceeded, a sequestering agent should be used to eliminate the negative effects of these ions. However, sequestering agents increase the dyeing efficiency by binding heavy metal ions and hardness ions, on the one hand, but decrease the dyeing efficiency by demetallising the dye molecule, on the other hand, especially in dyeing performed with metal-containing dyes. Therefore, it is not recommended to use sequestering agents for dyeing with all reactive dyes. Choosing the right sequestering agent is of great importance for a dyehouse to achieve success.[30]

In reactive dyeing, the pH of the water is also of great importance in obtaining uniform and reproducible dyeing. In reactive dyeing, the initial pH is targeted to be around 6.5–7,

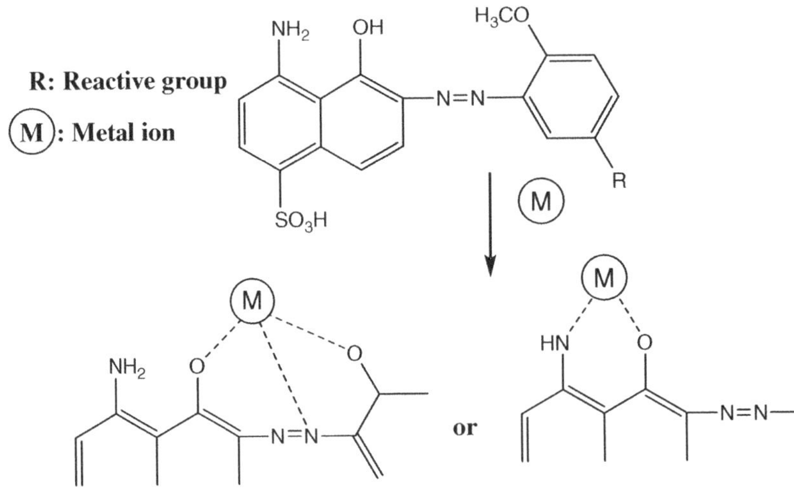

Figure 9.31 Binding of metal ions with reactive dye molecule. Ref. [9]/Sumitomo Technical Documents.

while the presence of bicarbonate in the water causes the pH to rise. If there is excessive bicarbonate in the water used, two main drawbacks arise. The first of these is the pre-fixation of the dye before sufficient migration has occurred. Especially if the reactivity of the dye is high, the risk of pre-fixation increases. This negatively affects levelness and reproducibility. In addition, in the dyeing process using soda ash (sodium carbonate), the bicarbonate in the water forms a buffer with sodium carbonate (the pH value of the sodium bicarbonate/sodium carbonate buffer is 8.82), in which case the pH values required for adequate fixation (around 11) cannot be reached in the dyeing process. This is because the buffer system formed prevents the pH from rising. This negatively affects fixation and reduces the colour yield. For this reason, it is useful to test the alkalinity of the water to be used in dyeing.[2]

9.4.1.2.8 *Effect of Auxiliary Chemicals on Dyeing*

Penetrating agents, lubricants, sequestering agents and protective auxiliaries against reductive effects can be added to the liquor during reactive dyeing. These chemicals should be added to the liquor before the addition of alkali and in the specified order. Each chemical must be dissolved in water before being added to the dye liquor.[9] The structure and functions of these chemicals are briefly explained below.

- ***Penetrating agent:*** In cases where it is difficult for the dye liquor to penetrate the material, such as fabrics made of tightly twisted yarn or tightly prepared wounds, it is necessary to add a non-ionic or non-ionic/anionic type penetration facilitator to the liquor. Generally, non-ionic surfactants have a better penetration facilitating effect. However, it should be noted that the use of high concentrations reduces the dyeing speed and dye uptake. Therefore, the decrease in fixation efficiency and foam formation should be considered when choosing a penetrating agent. Generally, the usage amount varies between 0.5 and 2 g/L.[9]

- ***Lubricant:*** Abrasion marks may occur, especially in the dyeing of heavyweight knitted materials. In such cases, it is beneficial to add a lubricant to the liquor to prevent crease formation and uneven dyeing caused by friction on the fabric. However, since some lubricants have the disadvantage of causing dye aggregation and forming excessive foam, these properties should be tested before use.[9]
- ***Sequestering agent:*** Sequestering agents that prevent the precipitation of metal ions by forming a complex with these ions can be called by different names such as complexing agents, chelating agents, dispersing agents or levelling products according to their effects. The term chelate refers to the tendency to bind the ion. The chelating ligands form water-soluble complexes with metal ions and remove the metal ion from solution.[57] The term sequestering agent encompasses all of the "complexing and dispersing" properties. Sequestering agents prevent the precipitation of these ions by forming stable and water-soluble complexes with metal ions, and they also exhibit a dispersion effect by suspending calcium and magnesium as carbonates in solutions containing carbonate ions.[56] Depending on the chemical structure of the sequestering agent, the water hardness decreases to a certain extent. Therefore, depending on the type of sequestering agent, the amount required also changes.[30] Today, it is possible to classify sequestering agents in five main groups:

  1. Phosphates
  2. Amino polycarboxylic acids and their derivatives
  3. Phosphonates
  4. Acrylates
  5. Polyhydroxy carboxylic acids.

The most widely used of these sequestering agents today are acrylates and phosphonates. Polyhydroxy carboxylic acids are also sequestering agents that have found new uses.[54,56] Phosphates are no longer preferred because they create environmental problems (such as abnormal vegetation formation due to fertiliser effect). Strong sequestering agents such as EDTA or NTA, which are in the amino polycarboxylic acid and derivatives class, are not only environmentally unfriendly,[58] but also cause colour change when used with metal-containing reactive dyes, as they extract metal in the structure of the dye molecule (Figure 9.32). Therefore, the use of such sequestering agents is not recommended today.[9]

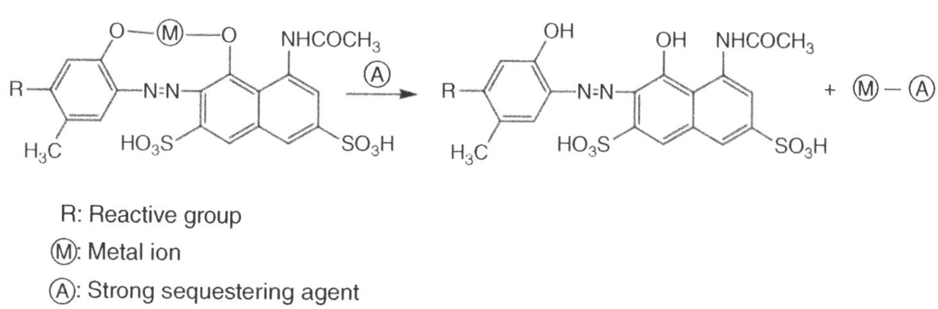

Figure 9.32 Separation of metal from metal-containing reactive dyes by strong sequestering agents. Ref. [9]/Sumitomo Technical Documents.

Phosphonate- and acrylate-based sequestering agents have high ion-binding capacity and dispersing power. Furthermore, acrylates block precipitation caused by water hardness. However, phosphonate- and acrylate-based sequestrants have poor biodegradability. In addition, phosphonates can cause colour changes in dye baths by forming complexes with metal ions in the dye structure.[54,56,58] One of the most interesting properties of phosphonates is the threshold effect. This can simply be defined as stabilising supersaturated solutions. Polyphosphonates prevent precipitation. By adsorbing on the crystals formed by cations in solution, they prevent them from growing and precipitating.[56,59] Polyhydroxy carboxylic acid-based sequestering agents have all the positive properties of acrylate and phosphonate-based sequestering agents. In addition, their biodegradability is very good (BOD_{14}: 65%).[54]

- ***Protective chemical against reductive effects:*** It is beneficial to add a protective auxiliary to the liquor against reductive effects in order to prevent colour change problems that may occur as a result of deterioration of reactive dyes by reductive effects during dyeing. Aldehyde groups formed in cotton or viscose fibres damaged by chemical processes can cause problems by causing reduction of reactive dyes. To prevent this, 1–3 g/L sodium m-nitrobenzene sulphonate is used.[9]

9.4.1.2.9 Effect of Dyeing Temperature and Duration on Dyeing

As the liquor temperature rises, the reaction rate between the dye and the fibres increases.[4] The occurrence of the reaction and the determination of the reaction time vary greatly according to the temperature, and the reaction rate increases 2–3 times, depending on the chemical structure of the dye and dyeing conditions, with every 10 °C temperature increase.[30] But on the other hand, with increasing temperature, the rate of reaction of the dye with water (hydrolysis) also increases and the increase here is greater, as shown in Figure 9.19 before. In methods working at high temperatures and with strong alkalis, some of the dye attached to the fibres may also detach and transform into a form which does not have affinity for the fibres.[4]

As explained before, the reaction temperatures of various reactive groups used in different brands of reactive dyes with cellulose macromolecules are very different from each other. Some of them can react at cold temperatures, while others can only react at cold temperatures. Since the optimum temperature for dyeing is not only dependent on the type of reactive group but also on the structure of the chromophore group (since depending on the chromophore, the molecular size and thus the ability to obtain uniform dyeing will vary), various dyes with the same reactive group also differ among themselves. For this reason, the most suitable dyeing temperature for each dye is indicated in the catalogues.[4] For example, although turquoise-coloured reactive dyes, which have a phthalocyanine-based chromophore, are included in the cold-dyeing dye group containing vinylsulphone groups as a reactive group, the dyeing process is carried out at higher temperatures (such as 85 °C) in order to achieve uniform dyeing with them.

Figure 9.33 shows an example of the relationship between the dyeing temperature (between 40 and 80 °C) and the colour yield obtained in dyeing with dyes having various reactive groups. When the figure is examined, it is seen that the sulphatoethylsulphone/MCT type heterobifunctional reactive dye has the least dependence on temperature among various dye types. One of the most important advantages of combining two different types of reactive groups is that the change in dyeing temperature between 50 and 80 °C causes very

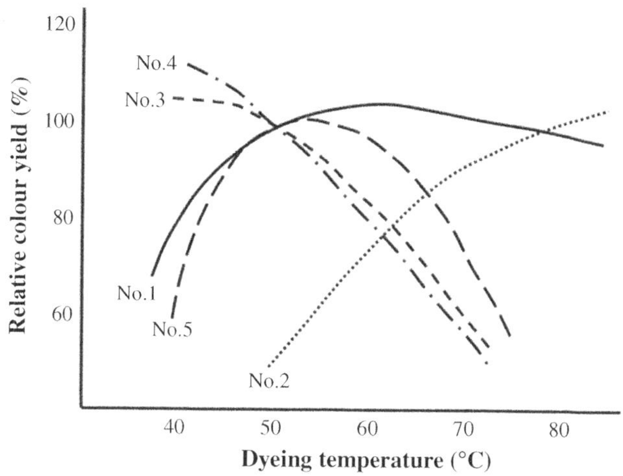

Figure 9.33 The relationship between dyeing temperature and obtained colour yield in dyeing with dyes with various reactive groups. Ref. [9]/Sumitomo Technical Documents. (No. 1: Sulphatoethylsulphone/monochlorotriazine, No. 2: Monochlorotriazine, No. 3: Monofluorotriazine, No. 4: Difluoromonochloropyrimidine, No. 5: Sulphatoethylsulphone)

little colour change. Because in this temperature range, the reactivity of these two different reactive groups balances each other.[9]

In terms of energy savings, dyeing temperatures should be as low as possible. However, in the dyeing of cellulosic fibres, swelling does not occur in the fibres at temperatures of 40 °C and below, and diffusion is insufficient and dyeing evenness is poor. Levelness increases with increasing temperature, but in this case the energy consumption also increases.[9] Dyeing temperature varies between 40 and 60 °C for dyes with high reactivity, and 80–95 °C for dyes with low reactivity.[8] Various temperature-time diagrams can be used for dyeing with reactive dyes. As shown in Figure 9.34, there are basically three temperature-time models: starting the dyeing at low temperature and gradually increasing the liquor temperature (temperature-raising method), dyeing at a constant temperature (isotherm dyeing) and starting the dyeing at high temperature and decreasing the temperature.[4] Although the first two of these are more common, in some cases, the temperature-time diagram, which is called the "migration method", can be used by starting the dyeing at high temperature and cooling the liquor.

Example dyeing graphs for dyeing at increasing temperature and at constant temperature (isotherm) are given in Figure 9.35. Constant temperature dyeing is more common with high reactivity dyes.[4]

Starting the dyeing at a high temperature and continuing at this temperature or in the cooling liquor is used to prevent excessive swelling of the fibres and thus to avoid the difficulty of obtaining well-penetrated dyeings, especially in the dyeing of products made of regenerated cellulose fibres and mercerised cotton with a tight structure. In this case, while the advantage of high temperature in increasing diffusion is utilised at the beginning, the drawback of decreasing substantivity with increasing temperature is partially eliminated since the temperature is lowered over time. This method, which is called the migration method, is also applied for colours such as turquoise and green with which it is difficult to obtain uniform dyeings.[4] In this method, in the first step of dyeing, the dye is allowed to

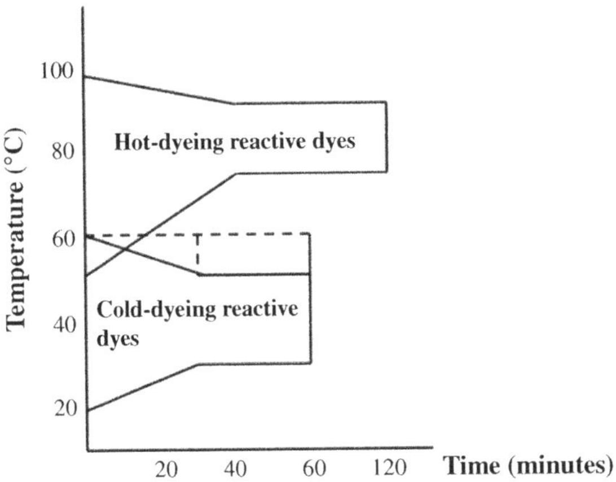

Figure 9.34 Temperature-time program models for cold- and hot-dyeing reactive dyes. Ref. [4].

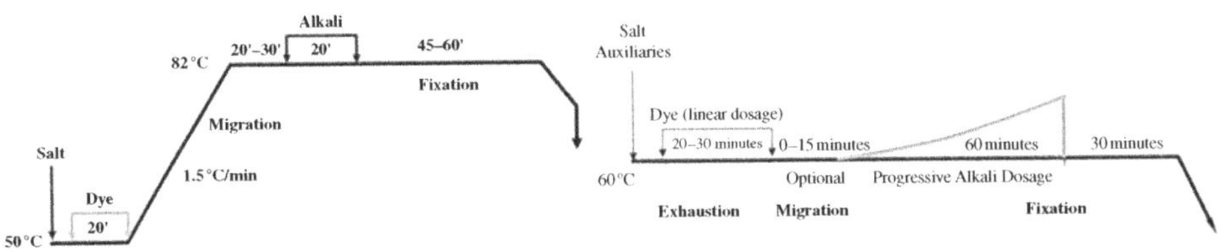

Figure 9.35 Sample graph of dyeing with reactive dyes according to the temperature raising (on the left) and constant temperature methods (on the right). Ref. [3]/Benar Kimya.

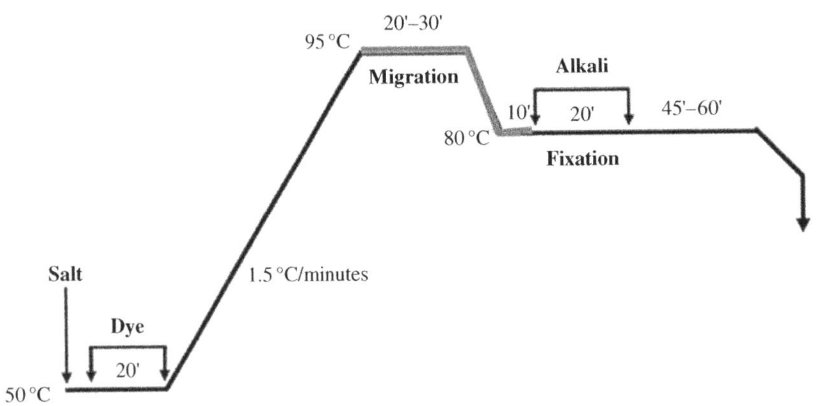

Figure 9.36 Example graph of dyeing with reactive dyes by migration method. Ref. [3]/Benar Kimya.

migrate at a temperature above the dyeing temperature (15–30 minutes). Then, the dyeing temperature is reduced and the second step of dyeing is continued, similar to the standard method.[2] Since in this method, from the moment of dyeing, there is a temperature that can cause the dye to react quickly with the fibres, special attention should be paid to the fact that the liquor is not alkaline at the beginning, and then, not all of the necessary alkali should be added, but first a small amount of weak alkali and then the remainder of the alkali should be added or it should be added to the liquor by alkali dosing.[4] In Figure 9.36, an example graphic is given for dyeing by **migration method**, which is applied in the form of starting the dyeing at a low temperature first, increasing to high temperature to benefit from the migration before adding alkali, and then lowering to the normal dyeing temperature, adding alkali and continuing the dyeing.

9.4.1.2.10 *The Effect of the Characteristics of the Dyeing Machine Used*

The type of dyeing machine used has an important effect on dyeing efficiency and levelness. While loose fibres and yarns are dyed in apparatuses, fabrics can be dyed in rope form in jets or in open-width form in jiggers or fabric beam dyeing machines. The more intensive and fast the contact between the textile product and the dyeing liquor in the dyeing machine, the higher the critical speed (dye uptake speed at which dyeing starts to become uneven) and the easier it becomes to achieve uniform dyeing. Therefore, the critical speed increases as the movement of the textile product accelerates in the machines and as the number of liquor circulations rises in the apparatuses. For example, in a jet dyeing machine where the contact with the liquor is intensive and both the product and the liquor are moving, the critical rate of dyeing in the jet dyeing machine is much higher when the fabric pass speed is the same as that of a winch, compared to the critical rate of dyeing in the winch with the same recipe. In this way, uniform dyeing can be achieved more easily.[4]

9.4.1.3 Parameters Affecting Uniformity and Reproducibility

The factors listed in Section 9.4.1.2 not only affect dye uptake and fixation and washing processes but also significantly affect the levelness and reproducibility of dyeing. If all these factors cannot be kept constant between batches or when transferring laboratory and sample dyeing recipes to production, problems will be encountered in colour reproducibility. Due to the differences in dyeing conditions, about 5–10% reprocessing may occur, although it varies from enterprise to enterprise. As a result of these re-dyes, both the cost of dyeing increases and scheduling becomes difficult as the production capacity decreases.[60] A study was conducted on the effect of faulty dyeing on the process costs during the dyeing of cotton knitted materials with reactive dyes in 300 kg batches in a jet dyeing machine at a 1:8 liquor ratio. Accordingly, when the cost of right-first-time dyeing, that is, dyeing without taking a sample, is assumed to be 100 units, the cost of dyeing that requires sampling is 105, the cost of dyeing that requires shade correction via dye addition to the same bath is 117, the cost of dyeing that requires shade correction via dye addition to a new bath is 134, the cost of dyeing that requires shade correction after drying is 166, and the cost of dyeing that requires dyeing into black after drying is 217.[61] Each shade correction and redyeing not only increases the cost of dyeing but also leads to reduced production and missed deadlines.[60]

In recent years, ecological, economic and timely delivery criteria have gained prominence in the dyeing process in line with both the quality requirements and consumer expectations. It has become the primary duty of textile dyers to deliver the dyed goods in the desired quality and in the desired time in order to get a share from the increasingly narrowing market due to the rapid change in fashion, smaller order sizes and the complexity of the product structure. Right-first-time dyeing and reproducibility are very important and intensifying competition conditions force every company to act fast. Achieving and even increasing the desired market share under intense competition conditions made it necessary to minimise the failures in the dyeing process and, if possible, to achieve the desired colour right-first-time. There are two mechanisms for uniform dyeing in dyeing made by the exhaust method:[60]

- Migration (post-levelling of dye)
- Controlled sorption (level exhaustion of dye from the beginning).

when the dyeing process is carried out for a sufficient time, a dynamic equilibrium occurs between "the dye remaining in the liquor ↔ the dye taken up by the fibres". After the equilibrium is established, dye exchange continues from the liquor to the fibres and from the fibres to the liquor. Therefore, in some dyes, even if the dye uptake is fast, levelling is possible as a result of migration. This situation, that is, levelling, is also seen in reactive dyes. However, this feature is valid as long as reactive dyes do not form covalent bonds with fibre macromolecules. Since the dye uptake rate is high at the beginning of dyeing in dyes with high substantivity, it is likely that the dyeing will be uneven. However, during the period before alkali is added to the liquor, the dye is only taken up by the fibres, and since covalent bonding has not yet occurred, some degree of displacement (migration) of the dye takes place. It can be seen that by decreasing or increasing the parameters in Table 9.5, the properties of the dye such as substantivity, reaction rate and, accordingly, fixation efficiency and migration can be changed.[60]

As can be seen from Table 9.5, increasing the liquor ratio and temperature or decreasing the salt decreases the substantivity and fixation yield; on the other hand, it facilitates migration and thus the uniform dyeing. Conversely, while the increase in pH increases the reaction rate and fixation yield, it decreases the migration ability of dye molecules due to covalent bonding. The increase in temperature also increases the reaction rate, but as explained before (Figure 9.19), when the temperature increases, the reaction rate of the dye with water (hydrolysis) increases to a greater extent than the reaction rate with the fibre.

In order to partially correct unlevelness by migration, it is necessary to extend the dyeing time. Since the extension of the dyeing time negatively affects both the cost and the physical properties of the fabric, strategies based on controlled adsorption have gained

Table 9.5 Effect of various parameters on fixation efficiency and migration during reactive dyeing (+ increase, − decrease).

Parameter	Effect	Result
+ pH	+Rate of reaction	± Fixation yield
		− Migration
+ Temperature	+Rate of reaction	− Fixation yield
	− Substantivity	− Fixation yield
		+ Migration
+ Salt	+ Substantivity	+ Fixation yield
		− Migration
+ Liquor ratio	− Substantivity	− Fixation yield
		+ Migration

Adapted from Ref. [60].

importance in today's dyeing strategy. The basis of this method is to ensure that the dye is taken up by into the fibres uniformly from the beginning of the dyeing and to finish the dyeing without the need for a levelling step afterwards. Controlled dyeing is required to obtain right-first-time dyeing. With the controlled dyeing carried out in optimal conditions where the dye is taken up uniformly, both cost-effectiveness and preservation of product quality will be achieved.[60]

To achieve reproducible dyeing, the laboratory, sample and production stages must have an error-free and systematic workflow. This is possible with standard fabric, standard dyeing processes and standard conditions. Deviations that may occur in any of these will adversely affect reproducibility. Fluctuations in dyeing conditions are frequently encountered as a result of the high workload within the enterprise and the efforts to meet deadlines. This often leads to inadequate control of production steps or a lack of attention to quality, specifically in the weighing and measurement of chemicals and auxiliary materials by employees. An example of typical deviations that may occur in dyeing processes can be seen in Table 9.6.[60]

When the dyeing performance of an enterprise that uses the exhaust method and whose acceptance limit for the total colour difference (ΔE) is 1.2 is examined, it is estimated that approximately 0.3 units of the colour difference are due to the dye, and 0.9 units are due to the changes in the material to be dyed and dyeing processes. As can be understood from this, the dye standard has a significant effect on the reproducibility of dyeings. Considering the demands regarding reactive dyes from the user's point of view, the most important issues to be considered in the optimal dyeing process and selection of the optimal dye are cost-effectiveness, high efficiency, high fastness, reproducibility, uniform dyeing and ecological criteria. In order for the dyeings to be of optimal reliability and reproducibility, first of all, the dyes must be of a certain standard and the best balance must be found between the substantivity, diffusion, fixation and washing properties of the dyes used in the combination.[60] Even reactive dyes with the same reactive group may differ from each other in terms of exhaustion and fixation rates. Dyes with close exhaustion and fixation rates have good combinability. Figure 9.37 shows the relationships between the half-dyeing time and the dye affinity values of trichromia dyes of various types of reactive dye groups. The closer the positions of the dyes of trichromia are to each other, the better the combinability of the dyes.[9]

In order for the dyeings to be optimally reliable and reproducible, the second condition is that the normal fluctuations that may occur during dyeing should not affect the

Table 9.6 Deviations in dyeing conditions.

Parameter	Deviation
Salt concentration	10%
Alkali concentration	10%
Liquor ratio	1:2
Fixation temperature	5 °C
Fixation time	15 min
Under or over loading of the fabric to be dyed in the machine	10%

Adapted from Ref. [60].

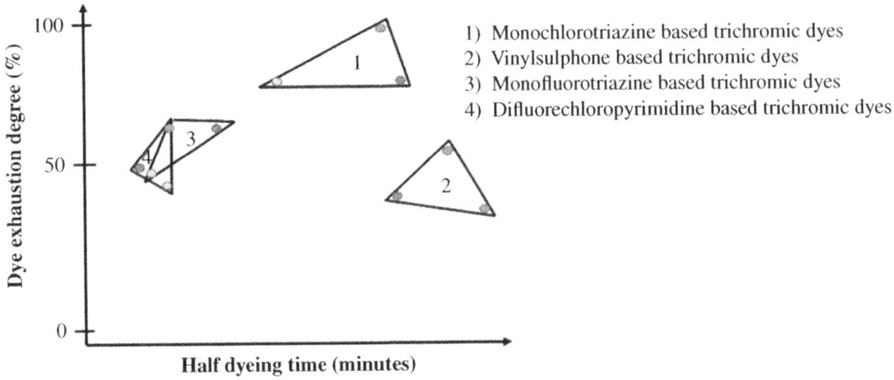

Figure 9.37 Examples of the relationships between the half-dyeing time (min) and the dye affinity values (%) of trichromic dyes of various types of reactive dye groups. Ref. [9]/Sumitomo Technical Documents.

reproducibility. If dyes that are sensitive to fluctuations mentioned in Table 9.6 are used, it will be more difficult to obtain reproducible dyeing. For this reason, it is necessary to use dyes that are not affected or are only slightly affected by small fluctuations in dyeing parameters. In addition, it is beneficial to carry out operations such as weighing, dissolving, filtering and discharging the dye carefully and always in the same way.[60]

A large part of the total tolerance of an enterprise is due to small deviations in parameters such as material quality and weight, liquor ratio, liquor temperature and dyeing pH. As it is known, in practice, the parameters in the dyeing conditions are not always constant, and there may be slight fluctuations in the transition from laboratory to sample dyeing and production, or between batches. For this reason, the differences that may occur in the dyeing properties of yarn and fabric, starting from fibre, may lead to errors in dyeing. Therefore, through the necessary controls that can be made at the beginning, the errors that may occur in the pretreatment and therefore in the dyeing will be reduced.[60] Below, information is given separately about the other factors affecting the reproducibility in reactive dyeing, apart from the dye type.

- **Type of fibre and properties of yarn/fabric:** It is known that the dyeability of regenerated cellulosic fibres and mercerised cotton is better than that of normal cotton, and that even dyeing properties of cotton types may differ from region to region. Therefore, the differences that may occur in the fibres to be used in dyeing and the yarn, woven or knitted fabrics made from these fibres (bright-dull fibre, mature-immature fibre, texture density, mass per unit area, yarn hairiness, yarn twist, trash content, heavy metal and alkaline earth metals in the fabric, etc.) may cause errors in dyeing. While it is necessary to use the same lot of fabric in the laboratory dyeing, sample dyeing and mass dyeing, this is sometimes not possible in practice. Slight differences between the fabric supplied by the customer for sample dyeing and the fabric to be used in the actual production may adversely affect reproducibility, and nuance adjusting will be required to match the colour. For this reason, it is important to make the necessary controls beforehand so that the dye uptake behaviour of the fabric to be dyed is not different.[60]

- **The pretreatment processes of the fibre:** The majority of the faults in the finishing processes are due to the pretreatment steps. For this reason, optimal pretreatment processes should be selected by making necessary controls at every stage of production. The pretreatment process to be applied before dyeing should be suitable to balance or reduce the small differences that may be present in the fabric. Otherwise, due to uneven hydrophilicity and damage to the fibres from time to time, the uniform dyeing process will become significantly more difficult. Bleaching agent, tenside, size, reducing agent, acid and alkali wastes should not be left on the fabric to be dyed, and it should be ensured that the fabric has uniform and sufficient hydrophilicity. The colour difference that may come from the pretreatment should be eliminated by taking the fabrics to be dyed in the laboratory and in the sample, as much as possible, from the fabrics prepared for dyeing in the mass dyeing.[60] However, since this is not possible in practice, care should be taken to ensure that the pretreatment degrees of the fabric used in the laboratory and the fabric to be dyed in the factory are as similar as possible.
- **Liquor ratio:** With the increasing liquor ratio, the substantivity of the dye and dyeing efficiency decrease, while an increase in migration is observed. On the other hand, the increase in the liquor ratio supports the reaction of the dye with water, and thus the hydrolysed dye fraction increases. For this reason, it should be checked whether one is working at the same liquor ratios by showing the necessary sensitivity to the preparation of the liquor, both in sample dyeing and mass dyeings.[60] In dyehouses, there may sometimes be differences in the liquor ratio between the machines used in the laboratory and production, or between lots, which negatively affects the reproducibility of dyeing.[9] Sometimes standard conditions cannot be met in practice due to the under or over loading of the fabric in the machine. On the other hand, it should be taken into account that the conditions in the sample dyeing machine (such as higher liquor ratio and manual operation) are often not the same as those used in the mass dyeings (such as shorter liquor ratio and full automation). While working at 1:4–1:8 liquor ratios in production machines, problems may arise in transferring sample dyeing results to production when necessarily higher liquor ratios are used in sample dyeing machines.[60]
- **Type and amount of salt:** The type of salt used in the mass (large-scale) production and in the laboratory must be the same. Whether sodium chloride or sodium sulphate is used in dyeing, the salt quality must be very good and always at the same standard. In addition, care should be taken to ensure that the salt concentration used in a particular recipe is the same. However, sometimes there may be fluctuations in the salt concentration of about $\pm$ 10% as a result of both the weighing error and the errors caused by the salt being in an unsuitable environment due to its crystallisation and hygroscopic properties.[60] For this reason, due attention should be paid to salt weighing or dosing (if liquid salt is used) and to the storage conditions of the salt.
- **Type and amount of alkali:** Even small deviations in pH value during dyeing will adversely affect the levelness and reproducibility of dyeing.[2] For this reason, it is necessary that the type and amount of alkali used in the mass production and the laboratory and between the various lots should be the same. In addition, the necessary care should be given to the alkali dosing.
- **Water quality:** First of all, the properties of the operating water should be checked, and care should be taken to ensure a certain standard. In addition, it should be noted

that the water used in the mass dyeing and the water used in the laboratory dyeing should be the same. The pH of the process water should be 7, and its hardness should not exceed 5 °dH. In addition, it should not contain Ca^{++} and Mg^{++} ions, and other foreign substances it contains should not exceed the limit values. One of the most common foreign substances in water is sodium bicarbonate, and the difference in the amount of bicarbonate negatively affects the reproducibility of the dyeing. In this case, either the addition of a complexing agent or an alkali system that is less sensitive to sodium bicarbonate (caustic instead of soda ash or a soda ash/caustic soda combination) should be used.[60]

- **Dyeing temperature:** As the dyeing temperature rises, there are differences in the reaction rate, substantivity and yield of the dye. Therefore, it will be difficult to obtain reproducible dyeing as a result of small differences in dyeing temperature. Optimal dyeing temperatures of the dyes chosen, especially in combination dyeings, should be the same or close to each other. In addition, dyes must not be affected by a temperature fluctuation of ± 5 °C. In other words, the exhaustion and fixation properties of dyes should not depend much on temperature. In this case, it will be easier to obtain reproducible dyeing.[60]
- **Dyeing time**: In reactive dyeing, once the dye fixation step is complete, it is generally not beneficial to attempt exhausting the remaining dye in the liquor by extending the dyeing time. Because the dye is covalently bonded to the fibres, it should be taken into account that there will be no possibility of levelling afterwards and that after fixation step, most of the remaining dye to be taken will be hydrolysed dye, that is, the part that needs to be removed by washing will increase. Otherwise, either the washing load will increase or there will be a decrease in fastness values. However, in practice, a fluctuation of ±5–10 minutes can often occur in the fixation time. While it is important that the dye used is not affected by such minor changes, showing the necessary sensitivity to dyeing times will also facilitate reproducible dyeing.[60]
- **Characteristics of the dyeing machine used:** With the overflow dyeing machines taking the place of the winch-becks in the dyeing of knitted fabrics, the opportunity to work at shorter liquor ratios has arisen. Since the movement of both the fabric and the liquor is provided, the contact between the fabric and the liquor increases, and the problem of unevenness and reproducibility is greatly reduced with modern dyeing machines with automatic control programs and dosage systems. However, even if dyeing is carried out under the same conditions, under or over loading the machine, the rope length $\left[\textbf{rope length}(\textbf{m}) = \left(\textbf{fabric weight}\left(\textbf{g}/\textbf{m}^2\right)\right)/\textbf{linear weight}(\textbf{g}/\textbf{m})\right]$ and fabric speed $[\textbf{fabric velocity}(\textbf{m/min.}) = (\textbf{rope length}) * (\textbf{cycles per minute})]$ differences may cause colour differences between lots. For this reason, care should be taken to ensure that the number of turnovers per minute between batches is always the same in order to obtain reproducible results in dyeing. Once the amount of fabric to be dyed and the linear weight are known, the rope length can be easily calculated. When there is a change in the rope length of the material to be dyed, for example, when the rope length becomes shorter, the fabric speed should be reduced to ensure that the number of cycles per minute is the same.[60]

Although laboratory and sample dyeing conditions are tried to be adapted to production conditions as much as possible, the differences that may occur in the liquor circulation, fabric movement and rope length cause the contact of the fabric with the liquor to be different. For example, due to the fact that the length of the rope is much shorter in sample dyeing than in production, the number of cycles per minute is often different between sample and production. This causes colour differences. As a result, it should be taken into account that even small differences in the number of cycles per minute can negatively affect the reproducibility, and the sensitivity shown for the parameters such as dye and chemical weighing, liquor ratio, temperature, etc., should also be shown at points such as the operating speed of the machine.[60]

Today, through the use of optimal dyeing conditions and dyeing machines with automatic temperature-time control and salt-alkali dosage systems, the problems such as:

- difficulty in obtaining uniform dyeing,
- low reproducibility of dyeings,
- problems occurring in transferring laboratory recipes to production,
- difficult colour matching in colours such as olive, beige, earth colour and
- unnecessarily high dyeing costs.

can be largely eliminated.[8] However, it is useful to remind once again that if the necessary sensitivity is not shown to the above-mentioned parameters, colour differences will occur between batches, between sample production and mass production, and between laboratory and mass production, and reproducibility will decrease.

9.4.1.4 Dissolution of Dyes

Generally, dyes are made into paste with a small amount of warm water (40–50 °C) and then hot water (80–90 °C) is added with stirring. When preparing a high concentration dye solution, urea is also added to dissolve the dyes. Since the aqueous solutions of reactive dyes are not durable at high temperatures for a long time, after the dyes are completely dissolved, the dye liquor is made up to the desired volume with cold water and thus cooled to room temperature. A sequestering agent can also be added during dissolving, but in this case, attention should be paid to pH. Generally, aqueous solutions of reactive dyes have a weakly acidic (5–6.5) pH. At these pH values, reactive dyes are quite stable. However, if the pH of the dye solution becomes alkaline due to the water or auxiliary chemicals used, the dyes tend to hydrolyse. Therefore, care should be taken to ensure that the pH is not alkaline. Table 9.7 summarises an example of the hydrolysis of reactive dyes depending on the alkalinity and temperature of the liquor.[9]

9.4.1.5 Dyeing Methods According to Salt and Alkali Addition

If a certain product is to be dyed with a certain dye in a certain machine, most of the factors listed in the parameters affecting reactive dyeing are fixed. Therefore, the most important dyeing variables in exhaust dyeing are:[4]

- Salt (type, amount, method of addition),
- Alkali (type, amount, method of addition) and
- Temperature-time diagram.

Table 9.7 Alkali and temperature stability of a reactive dye.

Temp.	NaOH	$Na_3PO_4.12H_2O$	Na_2CO_3	$NaHCO_3$	No alkali
25 °C	X–Δ	Δ–O	O	O	O
40 °C	X	Δ–O	O	O	O
50 °C	X	X–Δ	Δ–O	O	O
60 °C	X	X–Δ	Δ–O	O	O
95 °C	X	X	X	X–Δ	O
130 °C	X	X	X	X	O

(O: Almost no hydrolysis, Δ: Significant hydrolysis, X: Almost complete hydrolysis)

Ref. [9]/Sumitomo Technical Documents.

Since the difference between the reaction rates of cold- and hot-dyeing reactive dyes can vary by a factor of 1000, it becomes clear that when working with reactive dyes with low reactivity, the temperature and pH should be kept higher compared to dyes with high reactivity.[4] Whether it is cold-dyeing or hot-dyeing reactive dyes, 5–6 different dyeing methods can be applied for each dye class. Instead of explaining these methods separately in each dye class, reactive dyeing methods according to the method of salt and alkali addition will be explained by dividing them into four categories:[4]

- two-step method,
- method of adding some alkali from the beginning,
- all-in method and
- programmed alkali addition method.

a. Two-step method

This classical method, which used to be the most common application method for dyeing all reactive dyes by the exhaust method, today finds more applications in cold-dyeing dyes. The basis of the method is to add the dye and salt in the liquor first, and after the dye exhaustion reaches to equilibrium (usually 30–45 minutes after dyeing starts), to add alkali to the liquor and to continue dyeing for 30–120 minutes.[4]

In this method, the dye is only taken up by the fibres during the period before the alkali is added, and there is no fixation. In other words, the dye is first taken up by fibres and then it is fixed to the fibres, which is initiated by the addition of alkali. So, temperature is not important in the initial phase. That is, optionally, dyeing can be done at increasing (Figure 9.35), constant (isotherm) (Figure 9.35) or decreasing (Figure 9.36) temperatures.[4]

In Figure 9.38, examples of dye exhaustion and fixation curves are given in dyeings made according to the two-step method with two different reactive dyes with high and low substantivity. When these dye exhaustion curves are examined, it is seen that the dye is taken up quickly in the first 5–10 minutes and after a while, the dye uptake stops at 20–75% in cottons and 30–85% in regenerated cellulose and mercerised cottons. However, since there is no covalent bonding of the dye molecules to the fibres, an effective migration (post-levelling) can occur at this time. After the addition of alkali to the liquor, most of the dye taken up

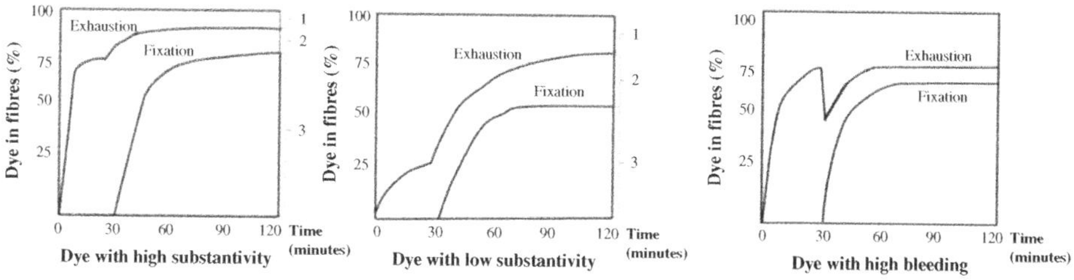

Figure 9.38 Examples of dye exhaustion fixation curves in dyeing according to the two-step method (1: Hydrolysed dye remaining in the liquor, 2: Hydrolysed dye in the fibres, 3: The dye fixed to the fibres). Ref. [4].

by the fibres begins to fix immediately. Therefore, since the fixed dye removes itself from the equilibrium, some more dye is also taken up by the fibres and fixed. However, since the rate of reaction of the dye in the liquor with water will increase with the addition of alkali, even if the dye part taken after this point is high, a significant part of it cannot react with the fibres and exists as "hydrolysed dye (hydrolysate)" that must be removed by washing.[4]

Since the initial uptake rate of the highly substantive dye seen in Figure 9.38 is very high and the subsequent levelling (migration) ability is relatively low, the risk of this dyeing being uneven is high. Because after the addition of alkali to the liquor, a large part of the dye taken up by the fibres begins to be fixed to the fibres in their immediate location. Although it is not a definite rule, it can be said that 90–95% of the dye taken by the fibres without adding alkali under normal conditions and approximately 50% of the dye in the liquor can be covalently bonded to the fibres, while the rest is hydrolysed. If an example is given, if the dye uptake is 70% before adding alkali, the amount of dye that will be fixed after alkali addition is approximately 78% ($70 \times 90/100 + 30 \times 50/100$). If the amount of dye taken before adding alkali is 30%, fixation will be approximately 62% ($30 \times 90/100 + 70 \times 50/100$).[8]

The effect of this step on the levelness of the dyeing is important, as with low substantive dyes, a rapid dye exhaustion and fixation can occur immediately after the alkali addition, as can be seen from Figure 9.38. In cases where the dye uptake is high (when working with dyes with high substantivity) before alkali is applied, the levelness of the entire dyeing is determined by this step, since most of the fixed dye (2/3 or more) will be taken up in this step. Especially in this type of dye, if all of the salt to be added to the liquor is added from the beginning, the dye exhaustion will take 5–10 minutes. Since it will be completed very quickly, the risk of uneven dye exhaustion will be high. Some of the measures that can be taken to reduce this danger are:[8]

- **Reducing the amount of salt:** Generally, 30–100 g/L salt should be used in reactive dyeing, depending on the colour depth. Sometimes the amount of salt is deliberately reduced. In this case, although the dyeing yield decreases a little, the savings in salt consumption and the ease of obtaining a uniform dyeing partially compensate for this drawback.[4]
- **Adding salt in portions:** One of the most effective measures to prevent the dye from being taken up suddenly at the beginning (reducing the slope of the dye uptake curve) is to add the salt in portions instead of adding all the salt in the first place. In this case,

the addition of increasing portions (1/10, 2/10, 7/10) instead of equal portions provides much better results.[4]

- **Adding dye to the liquor with a dosing pump:** Especially in cases where the dye uptake occurs faster than the time required for the dye added to the liquor to be homogeneously dispersed in the liquor, uneven exhaustion is inevitable. The dye is not added into the liquor all at once, but added slowly over a period of 15–30 minutes, reducing this risk as well as reducing the initial very high dye uptake rates. Dosing the dye manually is not a reliable solution. Ideally, at the beginning of dyeing using dosing pumps, the dye is added to the liquor within 10–30 minutes with a linear program and then the alkali dosage is started.[62]
- **Extending the dye exhaustion step (delaying the addition of alkali):** If the dye is taken up too quickly at the beginning of dyeing, the addition of alkali can be delayed. Thus, the initiation of fixation is delayed, and the unfixed dye part has the opportunity to migrate.[4]

b. Method of adding some alkali from the beginning

One of the most important problems encountered in the two-step dyeing method is that the affinity of some dyes for fibres decreases significantly in alkali medium compared to neutral medium. In other words, as can be seen in Figure 9.38, some of the dye previously taken up by the fibres bleeds into the liquor after the addition of alkali.[4]

Generally, since the pH value of the liquor and the substrate do not change equally at every point during the addition of alkali, there is a high risk that this dye bleeding, reuptake and fixation will occur unevenly. Therefore, when alkali is added while working according to the two-step method, alkali is added in portions, not all at once, in cases where the dye bleeding is high. In this way, the risk of sudden changes in the pH value, and thus the risk of bleeding large amounts of dye is reduced to some extent. In this case, a weak alkali (sodium bicarbonate or soda ash) is usually added to the liquor first, followed by a stronger alkali (soda ash or caustic soda). For example, the same result can be obtained by using 2 g/L soda ash and 0.5–2 mL/L caustic soda (38 °Bè) instead of 5–20 g/L soda ash in cold-dyeing dyes. The subsequent addition of alkali in portions does not increase the dye yield but reduces alkali consumption and promotes proper fixation. Since 2 g/L soda ash is added to the liquor from the beginning, dye fixation starts as soon as the appropriate temperature is reached. Therefore, dyeing at constant temperature (isotherm) is not possible in this method. For cold-dyeing dyes, dyeing is started at low temperatures and after 30 minutes, temperature is raised to the dyeing temperature (40–50 °C) within 30 minutes. On the other hand, for hot-dyeing dyes, the dyeing starts at low temperature and the liquor temperature is gradually increased (within 45–60 minutes) to the dyeing temperature (80–95 °C). The remainder of the alkali (soda ash or caustic soda) is usually added 30 minutes after the dyeing temperature is reached, and in portions if desired.[4] In Figure 9.39, a sample dyeing graph is given according to the method of adding some alkali from the beginning.

Figure 9.39 shows examples of dye exhaustion and fixation curves in dyeing according to the method of adding some alkali from the beginning. As can be seen from the figure, both dye uptake (especially if salt addition is made in portions) and dye fixation curves have a rather gradual character when this method is used. So, levelness of dye exhaustion is good. Since the fixation of the dye is adjusted by both temperature and the addition of alkali,

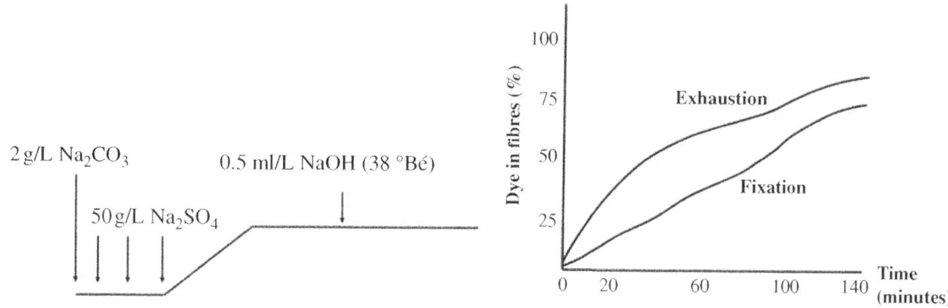

Figure 9.39 Sample dyeing graph and exhaustion/fixation curves according to the method of adding some alkali from the beginning. Ref. [4].

good results can be obtained in this method even if the temperature cannot be adjusted precisely (with an automatic control program).[4]

c. All-in method

This method, which became widespread at the end of the 1970s, gives the best results, especially when working with hot-dyeing dyes and dyeing apparatus or machines that can be precisely adjusted for temperature and time (with automatic control program). The advantage of the method is that all the salt and alkali to be added into the liquor are added from the beginning, and thus no intervention is required during dyeing. At the same time, the reproducibility of dyeing is facilitated.[8] However, this dyeing method is not used much nowadays, especially since cold-dyeing reactive dyes are used, although it can be preferred in laboratory dyeing because it is generally more practical.

In high reactivity (cold dyeing) dyes, when the all-in method is applied, it is generally preferred to use weaker or less alkali in order to prevent the fixation from starting effectively and immediately. However, in order to provide an adequate fixation level, the dyeing temperature must be kept high (e.g. 50–80 °C instead of 40 °C). This eliminates the advantage of these dyes in terms of energy. On the other hand, there is a danger of decreasing the dyeing efficiency, as increasing the temperature will significantly reduce the substantivity of many dyes.[8]

Although there is alkali in the medium in low reactivity (hot dyeing) dyes, fixation does not start below 50 °C. Therefore, if the dyeing starts at 25 °C and the temperature is increased slowly (approximately 2 °C/minutes), as in the sample dyeing graph given in Figure 9.40, the dye is taken up within the first 20–45 minutes after the dye is added to the liquor, but it is not fixed as in the example given in Figure 9.40. Since there is alkali in the medium, swelling of the fibres is also good, and the dye that has not been fixed yet has the opportunity to post-levelling. After the temperature exceeds 50 °C, the fixation of the dye begins with an increasing speed. In the meantime, since the dye diffusion will increase and accelerate, uniform dyeing can be achieved even in tight-structured and heavy weight products[8]

In reactive dyes, in order for the dye to react with the hydroxyl groups in the cellulose macromolecules (usually with the primary alcohol groups at the sixth C atoms), it is not enough to come close to the macromolecules to a certain degree, it must come into direct contact and collide. In other words, in these dyes, the penetration of dye molecules into the

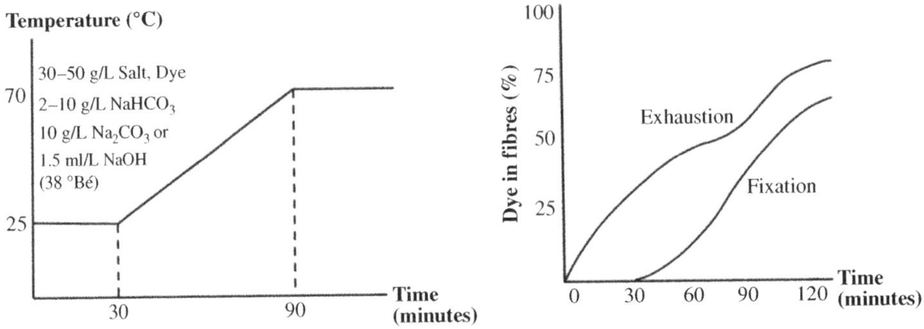

Figure 9.40 Sample dyeing graph and exhaustion/fixation curves according to the all-in method. Ref. [4].

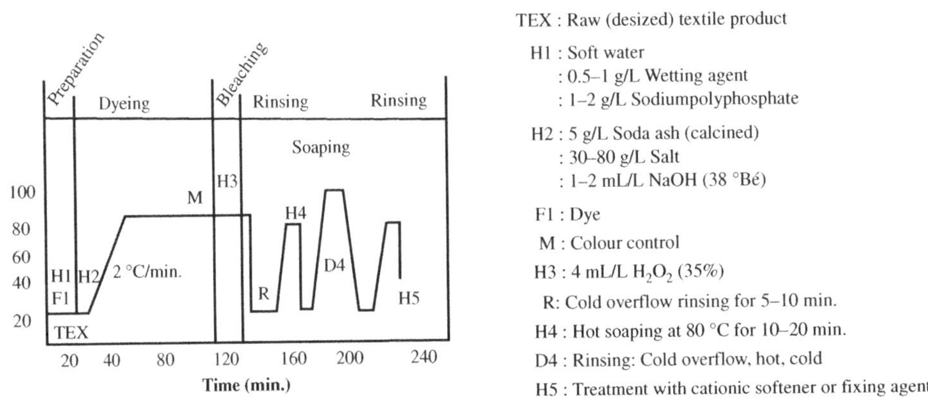

Figure 9.41 Graph related to the dyeing and hydrogen peroxide bleaching of raw cotton products with hot-dyeing dyes using all-in method. Ref. [4].

fibres is of particular importance. For this reason, the effect of pretreatment processes is evident in mercerised cottons or regenerated cellulosic fibres where the swelling of the fibres and therefore the pores is better, and the irregularities in pretreatment and drying cause irregularities in dyeing. However, when dyeing with hot-dyeing dyes according to the all-in method, if suitable complexing agents are added to the liquor, a pretreatment (extraction) effect will also be achieved during dyeing with the effect of alkali and high temperature, and therefore the fact that the pretreatment is not done well will not cause as many problems as in other dyeing methods.[8] However, since sizing agents, excluding polyacrylates, can also react with reactive dyes, good desizing in woven fabrics is mandatory for all kinds of reactive dyes and dyeing methods.[4] It is also possible to dye raw (only desized) cotton fabrics with hot group reactive dyes. As in the example given in Figure 9.41, if 3–4 mL/L perhydrol (35% H_2O_2) is added to the dyeing liquor 20–30 minutes before the end of the dyeing at a temperature of 80–95 °C, very good results can be obtained in terms of removal of the impurities and brightening of the colours.[8] As a result of this study, which is very advantageous in terms of process integration and energy consumption, the washing fastness of dyeings increases after peroxide bleaching, especially in dyeing in dark shades.[4] However, since the number of reactive dyes with very good hot peroxide fastness values is limited, this method could not find a wide application area.[8]

d. Programmed alkali addition method

When the methods described so far in dyeing with reactive dyes are examined, it is understood that at least one of the following variables, but in many cases both,

- salt,
- alkali,
- dye and
- temperature.

must be precisely controlled for the adjustment of the uptake and fixation rates of the dyes and the reproducibility of the dyeings. In the two-step method, temperature control is not important. However, in order for the dye to be taken up properly from the beginning, the dye or salt, and sometimes even alkali, must be added into the liquor in portions. As long as this is done manually, good reproducibility and colour matching are difficult. In the method of adding some alkali from the beginning, both temperature and alkali control are in question. The dyeing levelness provided is good. However, when both the temperature and the addition of alkali are controlled manually, reproducibility and colour matching problems may occur in this method as well. In the all-in method, since the entire process is under temperature control, automatic temperature-time control equipment is essential for a proper dyeing. While working with dyeing machines with this equipment, uniform and reproducible dyeing can be obtained. However, the fact that this method is suitable for hot-dyeing dyes and that the temperature must be kept higher than normal in order to be used in cold-dyeing dyes causes the energy consumption of this method to be higher. Therefore, the all-in method is less cost-effective.[8]

In the above-mentioned classical dyeing methods, both salt and alkali additions by hand and fluctuations in heating cause low reproducibility of dyeing. Especially when working on old-style machines sufficient dyeing levelness and reproducibility cannot be achieved.[63] Reduction or elimination of faulty dyeing can be achieved with the appropriate machine and method. By reducing the faulty dyeing process, energy, water, chemical and time savings are achieved; in other words, the unit cost of dyeing is reduced. The latest development in the exhaust method in this field is the dosage dyeing system, and dye and/or chemical substance (salt and alkali) additions can be made automatically according to the desired program. Therefore, dyeing processes can be done isotherm (constant temperature).[8,63] In Figure 9.42, an example dyeing graph is given for dyeing according to the programmed alkali addition method.

With the programmed alkali addition method, it is possible to make uniform and reproducible dyeings, and the rate of faulty dyeing is reduced from 5 to 10% to less than 1%.[63] With the application of this method, the temperature fluctuations in conventional systems are eliminated, the dyeing time is shortened as dead time losses and the number of colour nuance adjustments are reduced, and even the dye fixation efficiency can be increased, since the possibility of hydrolysis reaction is reduced thanks to the appropriate pH adjustment.[8,63]

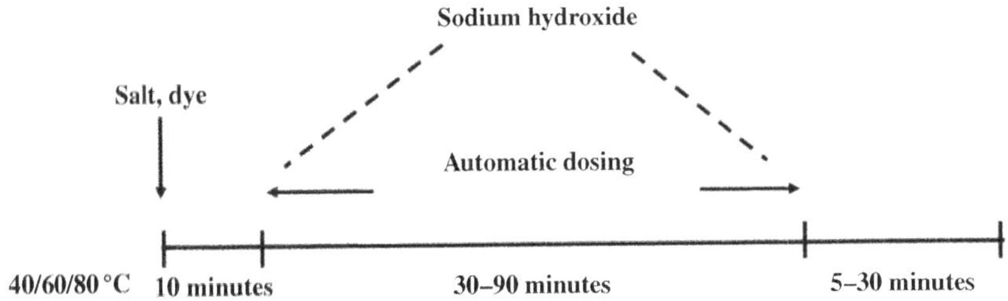

Figure 9.42 Programmed alkali addition method. Ref. [2].

In order to obtain uniform dyeing with reactive dyes, both the conditions of proper exhaustion of the dye and proper fixation must be met. Since there is no fixation before the alkali is added, and there is only the matter of taking the dye, if the dye is added to the liquor with a linear program within 10–30 minutes, especially in dyes with high substantivity, the dye can be taken up quite well from the beginning. Since the dye molecules are not yet fixed, some migration may occur at this time.[8]

Figure 9.43 gives examples of relative dye fixation curves (calculated dye fixation values based on time, based on 100% assuming the actual final dye fixation yield). When these curves are examined, there is a high risk of uneven dyeing as most of the dye is fixed very quickly in curve a. In curve b, level dyeing is easier to occur, since even the highest dye fixation rate is not very high. However, since the dead times at the beginning and end of dyeing are very long, the total dyeing time is very long. Since the highest dye fixation rate in the c curve is the same as in the b curve, uniform dyeing is the same. However, since there are no dead times at the beginning and end of dyeing, the total dyeing time is much shorter. In reactive dyes, there is a zone in which dye fixation occurs more slowly at the beginning and at the end, due to the fact that a migration phase is required at the beginning and the concentration of the dye that can be fixed (not hydrolysed) in the liquor decreases towards the end of the dyeing. Therefore, it is not possible to provide the fixation completely similar to the c curve.[8] Therefore, in the programmed alkali addition method, the fixation curve in general is similar to the curve indicated by d.

When Figure 9.43 is examined, it is seen that the relative fixation curves of reactive dyes are similar to the S form. It is possible to examine these fixation curves in three sections:

- ***Migration phase:*** In the initial part of the dyeing curve, the fixation curve has a feature that rises very slowly. The time elapsed until the reactive dye fixation is 10% after starting the alkali dosage is called the "migration phase". In this period, which varies between 10 and 20 minutes, the dyeing becomes uniform.[8,63]
- ***The main fixation phase:*** The part that gets steeper after the first part forms the main fixation area. Here, 25–65% of the relative dye fixation takes place, and the fixation rate (relative fixation degree) in this region can be calculated as follows:

$$r_s = \tan\alpha = \frac{65 - 25}{t_2 - t_1}$$

t_1 = time required for 25% fixation
t_2 = time required for 65% fixation

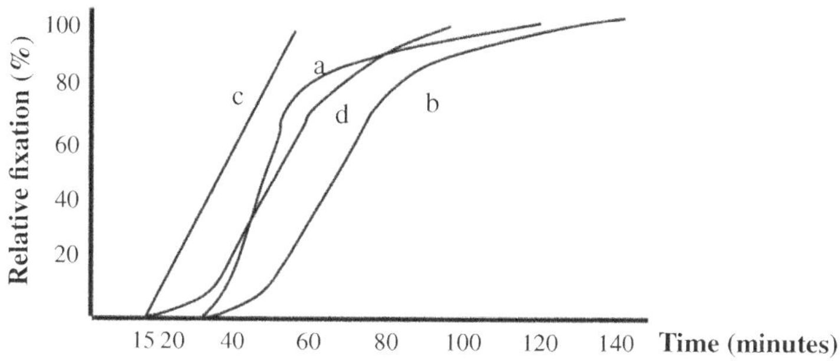

Figure 9.43 Examples of relative dye fixation curves. Ref. [8]/with permission of Tekstil & Teknik.

According to this formula, the relative fixation rates of the b, c and d curves given in Figure 9.43 and thus the smooth dyeing conditions are the same.[8,63]

- ***Completion phase of fixation:*** With the completion of alkali dosing, the fixation is not considered to be totally completed. For this reason, it is necessary to continue for a certain period of time in order for the relative dye fixation to reach 95%.[63] After the alkali dosage is over, the time required for the relative dye fixation to reach 95% and which is defined as the **"completion time of the fixation"** (for the alkali dosage period to be 60 minutes and for 50% progression) varies between 5 and 40 minutes. In combination dyeings, the longest one of the dyes in the combination is considered as the completion time of the fixation for the combination dyeing. To shorten the completion time of the fixation measures listed below can be taken:
  - increase the dyeing temperature by 10–20 °C after the alkali dosage is completed,
  - double progression in alkali dosage so that the second progression is high or
  - using slightly more total alkali than usual and applying higher progression.[8]

With automatic dosing equipment, dosing can be done for all intermediate stages between 0 and 100%. Although linear, degressive and progressive dosing is possible, linear dosing is especially recommended for dyes with high reactivity, thus ensuring that the dye is properly taken up by the fibre before the addition of alkali. Degressive dosing is not used for reactive dyeing, but mostly progressive (increasing amounts) dosing.[63] As it is known, an increase in the pH value by one step causes the fixation rate of reactive dyes to increase approximately 10 times. Therefore, jumps in pH value cause sudden and large increases in the fixation rate, leading to uneven dyeing. Ideally, the pH value should always rise at the same rate in the form of an oblique line. However, since the pH value is the negative logarithm of the concentration of H^+ ions, the amount of alkali needed to increase the pH of a neutral liquor by one unit (i.e. raising the pH of a liquor from 7 to 8) is much less compared to that required to raise the pH value from 10 to 11. Therefore, alkali addition must be made in progressively increasing portions so that the pH value always rises in an oblique line at the same rate.[8]

Figure 9.44 shows the effect of the degree of progression applied in alkali dosage on the dye fixation curve. At low degrees of progression (closer to linear dosing), the initial pH increases are rapid, so the fixation starts quickly and the risk of uneven dyeing increases because the migration phase is very short and the relative fixation rate is high (i.e. rapid

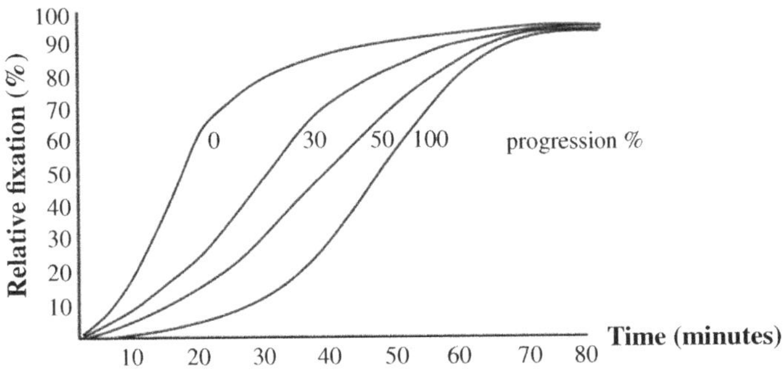

Figure 9.44 The effect of the degree of progression applied in alkali dosage on the dye fixation curve. Ref. [8]/with permission of Tekstil & Teknik.

fixation occurs). On the other hand, at very high degrees of progression, the initial pH increases are very slow, and the fixation starts slowly, and when the alkali dosing time is halfway through, only 15–20% of the total dye fixation is realised. In the second half of the alkali dosing period, as the amount of alkali added increases, the rate of fixation also increases. Therefore, the relative fixation rate is higher than that achieved with progressions of around 50%.[8] Although dosing time varies between 20 and 120 minutes, usually 60 minutes is sufficient. For goods that are difficult to dye, it is recommended to extend it up to 90 minutes.[63] Except for special cases and normally constituting 90% of the total dyeing, working with 60 minutes dosing time and 30, 50 and 70% progression respectively in dark shades, medium shades and light shades (also for mercerised cotton and regenerated cellulose products) provides good results.[8]

Normally the reduction of dosing time (from 60 to 45 minutes) is achieved by applying double progression. Thus, the total dyeing time is also shortened.[63] For example, in dyeing performed at 50% progression and 60 minutes. dosing time, there is no risk of uneven dyeing since the migration phase is 20 minutes and the relative fixation rate is around 1.2%/minutes. When dosing time is reduced to 45 minutes with the same degree of progression, there is a high risk of uneven dyeing as the migration phase decreases by 15 minutes and the relative fixation rate approaches 4%/minutes. However, if the progression program of the first curve (50% progression and 60 minutes dosing time) in the first 25 minutes and a 90% progression and 45 minutes dosing time program in the last 20 minutes is applied, the migration phase will be 20 minutes and the relative fixation rate will be around 2.2%/minutes. Thus, it is possible to shorten the alkali dosing time by 15 minutes without increasing the risk of uneven dyeing by applying double progression.[8]

The recommended double progression versus single progression values are given in Table 9.8.

9.4.2 Dyeing by Impregnation Method with Reactive Dyes

As mentioned before, the share of reactive dyes used in dyeing by the impregnation method is around 26%. Approximately 15% of this is pad-batch, and the remaining 11% is made up of other methods.[2] Since good results can be obtained in dyeing based on impregnation

Table 9.8 Recommended double progression versus single progression values. Ref. [63].

Single progression	Double progression
30%	1st Progression 25%
	2nd Progression 70%
50%	1st Progression 50%
	2nd Progression 90%
70%	1st Progression 80%
	2nd Progression 100%

with reactive dyes, a wide variety of semi-continuous and continuous dyeing methods are available. These methods are divided into two groups as follows:

a. **Semi-continuous methods:** Pad-batch, pad-jig, pad-roll.
b. **Continuous methods:** Pad-steam, pad-dry, pad-thermofixing, pad-wet fixation.

In addition, dyeing by the impregnation method can also be classified as follows:

a. **One-bath methods:** One-bath pad-batch, one-bath pad-roll, one-bath pad-steam, one-bath wet pad-steam, pad-dry, pad-thermofixing.
b. **Two-bath methods:** Two-bath pad-batch, two-bath pad-steam, two-bath pad-wet fixation, pad-jig.[4] Among them, the most commonly used methods are:
   - pad-batch,
   - pad-steam and
   - pad-thermofixing.

Since the application of the dye to the fabric is done on the basis of impregnation in dyeings made according to these methods, the substantivity of the dye used should be as low as possible and the renewal of the impregnation liquor should be quick (using low-volume padding vessels) to obtain dyeings with the same darkness (tailing-free). In dyes with low substantivity, it is easier to remove the part of the dye that is not fixed to the fibres. This also makes it possible to easily remove the hydrolysate in the washings made in the continuous washing machines after dyeing.[4]

Anionic products should be used as wetting agents in dyeing, and the selected wetting agent should not cause the formation of foam and should be resistant to alkalis and electrolytes.[53,64]

According to the resistance of the dye used against hydrolysis, they are dissolved with cold, warm or hot water. Since the solubility of some dyes is low during dyeing, it may be necessary to add urea to the liquor. Since urea is a hydrotropic substance, it increases the solubility of the dye. In addition, it is known that urea has the effects of swelling the cellulose fibres, lowering the aggregation degree of dye molecules and increasing the diffusion of dye. Urea should never be dissolved in a hot bath together with dye. First, the dye must be dissolved, then the urea must be added to the liquor. As the urea absorbs heat as it dissolves, the dyebath will cool slightly. After the urea is dissolved, the volume of the liquor should be made up to the desired level by adding water.[4]

9.4.2.1 Dyeing by Semi-continuous Methods

9.4.2.1.1 Dyeing by Pad-batch (Cold Pad-batch) Method

Pad batch is the most economical and eco-friendly process among the reactive dyeing processes being used.[65] The energy and water consumptions are at the lowest and salt addition is eliminated, thus rendering it more eco-friendly and the dye fixation is also high.[66] Dyeing made by the pad-batch method provides an economical dyeing of both cotton and regenerated cellulose fibres. This situation can be clearly seen from the cost comparison for the dyeings made by exhaustion and pad-batch method given in Figure 9.45.[67]

The main advantages of this method can be listed as:

✓ Low investment costs,
✓ High efficiency,
✓ Suitable for both long-length and short-length lots,
✓ High colour yield and fixation degrees,
✓ Well-penetrating dyeing,
✓ Low chemical, water and energy requirement,
✓ Possibility of dyeing without chloride and sulphate salts,
✓ Short cleaning and preparation times and
✓ Low dyeing cost.

The main disadvantages are as follows:

- Hydrolysis of the dye,
- Risk of tailing (head-end difference),
- Listing (Edge-centre-edge difference) and
- The colour to be obtained can be seen only after a long time.

I) One Bath Pad-batch Method

According to the one-bath pad-batch method, the fabric is impregnated with a dye and alkali-containing liquor at room temperature batched, and kept for 6–24 hours depending on the dye class and colour depth.[2] After dwelling, post-treatments such as rinsing and soaping can be done discontinuously in jiggers, jets or perforated fabric beam dyeing apparatus, but they are also generally carried out continuously in open-width washing

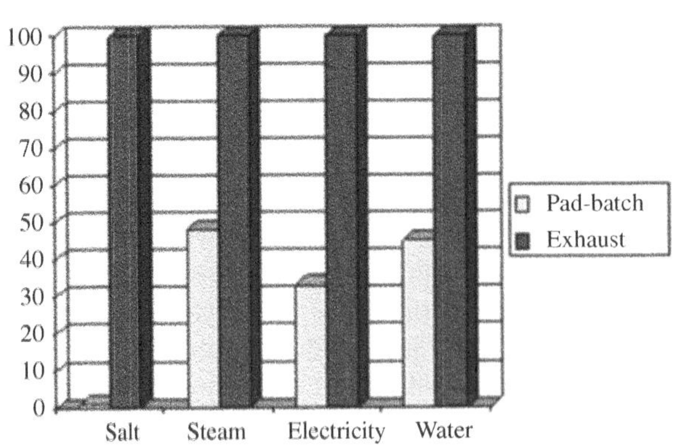

Figure 9.45 Cost comparison for exhaust and pad-batch dyeings. Ref. [67].

Figure 9.46 Process steps in dyeing by one-bath pad-batch method. Ref. [68]/Ciba Technical Documents.

machines.[4] Figure 9.46 shows the process steps in dyeing according to the one-bath pad-batch method.[68]

This method, which requires the least equipment and energy consumption, has been very popular in dyeing with reactive dyes. The most important difficulty of pad-batch dyeing is the low stability of reactive dyes in alkali liquors. It is possible to summarise the recommendations of various dye factories in two groups to overcome this drawback:[4]

1. Dye solution (wetting agent and urea if used) and alkali solution are prepared separately and mixed with each other in desired proportions (4:1 ratio) while flowing into the padding mangle with the help of an automatic dosing pump. For this, an automatic dosing pump must be present in the enterprise. And also, the padding vessel should be as small as possible and the temperature should be below 25 °C in order to reduce the risk of hydrolysis. It is also beneficial to have the pick-up value as low as possible (60–70% for cotton, 80–100% for regenerated cellulose).[4] If the dye solution is prepared hot (60 °C), it must be cooled before being combined with alkali in the dosing pump. Otherwise, the risk of hydrolysis will increase.[4,53,64] In pad-batch dyeing, the accuracy of dosing is very important for colour reproducibility, process optimisation, minimising the use of dyes, auxiliaries and wastewater load.
2. In addition to the alkalis used, salts such as water glass and sodium bicarbonate are added to the liquor to buffer the basicity strength and increase the stability of the liquor (60–120 minutes at 20–25 °C, 30–120 minutes at 30–35 °C). If caustic soda is used as alkali, water glass is used, and if soda ash is used, sodium bicarbonate is used. Amount of water glass/caustic soda to be used varies depending on following issues:
   - the temperature of the impregnation liquor,
   - the colour depth and
   - the density of water glass (There are sodium silicates with different ratios of Na_2O:SiO_2 in the market under the name of water glass. It is generally recommended that the ratio of Na_2O:SiO_2 should not be more than 1:3.3).

Separate lists are given for liquors at temperatures of 20–25 °C or 30–35 °C in the catalogues. Whenever possible, operations should be carried out at 20–25 °C. However, especially in hot summer months, operating water can reach 30–35 °C.[4] For enterprises located in tropical climates where the temperature of the impregnation bath rises up to 30–40 °C, it is recommended to increase the amount of silicate used.[53, 64]

In pad-batch dyeing,

- Soda ash alone
- Soda ash/bicarbonate combination

- Soda ash/caustic soda combination
- Caustic soda/silicate combination.

can be used depending on the dye. As the alkali type used becomes weaker, the hydrolysis resistance of the dye liquor increases, but the fixation rate decreases. This means longer dwelling times are required. On the other hand, when caustic soda and calcined soda ash are used as alkalis, the rate of fixation is high, so the dwelling time is short.[4]

It is stated that the addition of salt to the liquor also increases the rate of fixation. Usually, salt is added to the alkali solution. It should be noted that in dyes with limited solubility, the dye may precipitate in the liquor after mixing the dye solution and the alkali solution containing salt in the dosing pump. Because the addition of salt partially reduces the solubility of the dye. In such cases, the amount of salt is reduced or the use of salt is completely omitted. If necessary, urea is added to the dye solution. In this case, dye and urea are prepared in a separate container, alkali and salt are prepared in a separate container, and they are mixed with each other just before starting the impregnation.[4]

Since alkali liquors prepared with low-reactivity (reacting rapidly only at high temperatures) dyes are usually stable for one hour or longer in the cold (at 20 °C), they can also be used without a special dosing pump. If a catalyst is added to the impregnation liquor while working with these dyes, the reaction rates increase. Thus, dwelling times can be shortened. On the other hand, since the stability of the alkaline dye liquor will decrease, it is necessary to work with a dosing pump.[4]

In silicate methods, the stability of the liquor is better than caustic soda alone or a soda ash/caustic soda combination. However, silicate causes some problems listed below:

- It increases the waste load.
- Excess silicate, especially in dark colours, complicates the washing process.
- Silicate residues can accumulate on the pad rollers and cause 10% deviations in the pick-up values in dyeing. Therefore, padding rollers need to be ground.

Photographs of fibre cross-sections showing the variation of diffusion/penetration properties depending on the dye and alkali type used in dyeing can be seen from Figure 9.47.

Care should be taken to ensure that the edges of the fabric impregnated with the alkaline dye solution in the padder overlap while winding onto the batch. In addition, after the winding is completed, it should be covered and sealed with polyethylene foil in an airtight manner. Otherwise, since the carbon dioxide in the air will neutralise some of the caustic soda on the fabric, especially on the edges, the fixation in these areas will not be complete and the colour shade will remain lighter. Even in very long-length lots, if the batching process is prolonged (in cases where the winding process takes longer than 45 minutes), and therefore, if the contact with the air increases, the dwelling time is kept longer than specified in the catalogues; despite the decrease in the amount of caustic soda, it can be ensured that the fixation is complete. The batches must be rotated to prevent the liquor from draining under the influence of gravity during dwelling. The dwelling time varies depending on the dye and is at least six hours. Keeping the material longer than the holding time given in the catalogues corresponding to the temperature does not provide any additional benefit.[4] In addition to the impregnation step, the temperature is also very important in the fixation step. It is important for the levelness and reproducibility of the dyeing that the temperature of the environment where the batches are kept is constant and homogeneous.[53,64]

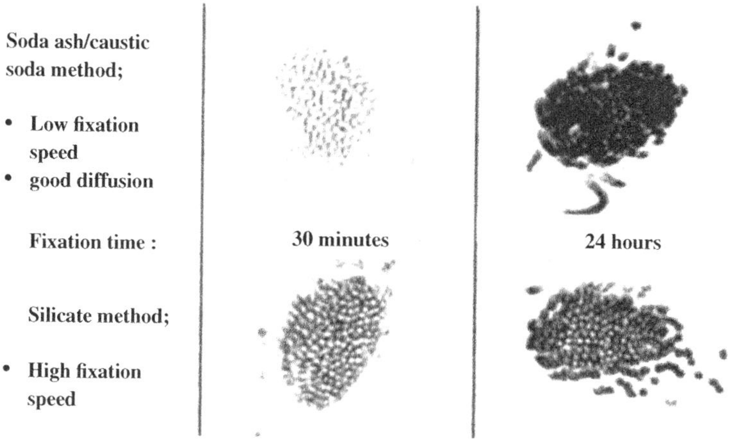

Figure 9.47 Photographs of cross-sections of the fibre showing the variation of diffusion/penetration properties depending on the dye and alkali type used in dyeing. Ref. [64]/Dystar.

II) Two Bath Pad-batch Method

Although the solutions of reactive dyes alone are highly resistant to hydrolysis, this resistance decreases in some cases when an alkali is added to the liquor. In such cases, instead of impregnating the fabric with dye and alkali (single bath), first impregnating with dye and after drying impregnating with an alkali solution (including salt to prevent dye bleeding) is preferred. However, this kind of two-bath work is much more elaborate and costly because it requires intermediate drying compared to the one-bath method. On the other hand, the obtained dyeings do not always give better results than the one-bath method. Because, during the intermediate drying after the application of the dye liquor, some of the unfixed dye will migrate through the fibres towards the surface, making it difficult to obtain dyeing that has penetrated well into the fibre. This also causes the dyed fabric to appear darker in colour. In addition, if the intermediate drying is done carelessly and the dryer does not work properly, surface migration occurs and as a result, unevenness may occur on the two edges, centre-to-edge or two sides of the fabric.[4]

If a perfect dyeing appearance cannot be obtained compared to the one-bath method in dyeing the product made of cellulose fibres, the two-bath pad-batch method can be applied. In addition, this method is also important in dyeing polyester/cellulose blends with a disperse/reactive dye combination. The process steps for dyeing by the two-bath pad-batch method are explained below.[4]

a. ***Impregnation with neutral dye solution:*** In addition to dye, wetting and antimigration agents can be added to the dye impregnation liquor. The impregnation temperature is usually 20–25 °C. In cases where it is difficult for the dye to penetrate into the fibres, the temperature can be increased up to 80 °C. After impregnation, the pick-up value should be 65–80%.[4]
b. ***Intermediate drying:*** Intermediate drying is done in hot-flue. In order to reduce migration, it is recommended to start intermediate drying at 80–90 °C and increase the temperature gradually.[4]
c. ***Impregnation with alkali and salt:*** Impregnation with chemicals is done in a padder with a low-volume vessel. Care should be taken that the temperature of the liquor does

not exceed 15 °C. The amount of salt in the liquor must be high so that the dye on the fabric does not bleed. After impregnation, the pick-up value should be 70–90%.[4]

d. ***Cold batching:*** After the batches are well covered with polyethylene foil, they are kept at room temperature by rotating them. The dwelling time is 2–8 hours.[4]

9.4.2.1.2 *Dyeing by Pad-roll Method*

The difference of the pad-roll method from the one-bath pad-batch method is that the fabric impregnated with an alkali liquor containing dye is left to wait in hot thermo-holding rooms. This method, which does not always provide an advantage compared to pad-batch and requires special pad-roll facilities (padder – IR channel – thermo-batching chambers), has also lost its importance today. However, since the diffusion of the dye will be better at high temperatures, it is advantageous to use the pad-roll method for some products (heavy viscose, lining fabrics, etc.) where the diffusion of dye is difficult. When working according to this method, the risk of moire is low.[4]

Impregnation with alkali liquor containing dye is done at 20–25 °C. Soda ash, sodium bicarbonate or soda ash/caustic soda mixture can be used as alkali, depending on the dye. When necessary, urea is added to the liquor to increase the solubility of the dye, and salt is added to reduce migration. It is recommended to add 5–20 g/L common salt or calcined sodium sulphate, especially when working with dyes with low substantivity. Some reactive dyes are sensitive to impurities and sizing residues on the product to be dyed and can be reduced. When working with such dyes, it is beneficial to add 2–6 g/L special organic oxidation agent to the impregnation liquor. The impregnated fabric is heated by passing through the IR (infrared) channel and enters the thermo-holding chamber. The dye is fixed in the thermo-holding chamber by keeping it at 50–60 °C for cottons and at 60–70 °C for regenerated celluloses for 2–8 hours.[4]

9.4.2.1.3 *Dyeing by Pad-jig Method*

In general, dyeings made by the pad-jig method do not provide an advantage compared to the ones made by the pad-batch method; on the other hand, they require more machinery, energy and labour force. For this reason, they have lost their importance today. The fabric is impregnated with a neutral dye solution containing a suitable wetting agent. Although the impregnation temperature is usually 20–25 °C, it can be increased, where necessary, up to 70–80 °C for better penetration of the dye into the product. Then, the fabric is squeezed to obtain a pick-up value of 60–75% for cotton and 75–90% for regenerated cellulose and treated with a liquor containing caustic soda in the jigger. However, a high amount of salt is added to the liquor so that the dye on it does not bleed into the jigger liquor. It is stated that if an intermediate drying is done in a hot-flue before the fabric enters the jigger, the obtained dye yield increases by 30%. But on the other hand, the cost of dyeing increases.[4]

9.4.2.2 Dyeing by Continuous Methods

9.4.2.2.1 *Dyeing by Pad-steam Method*

I) Single Bath Pad-steam Method

In the one-bath pad-steam method, dyeing efficiency is high, dye penetration into the fibres is good and the whole appearance of dyeing is excellent. However, this method has disadvantages such as the low stability of the alkaline dye liquors, the need for attention for intermediate drying, and the high need for machinery and energy. This method is mostly used for dyeing the ground in long-length cottons and prints. In the one-bath pad-steam method, after the fabric is impregnated with an alkaline dye solution and dried, it is printed with discharge paste, followed by the fixation of the dye in the areas outside the printed areas during steaming. Working according to this method consists of the following steps:[4]

a. Impregnation with alkaline dye liquor
b. Intermediate drying
c. Steaming
d. Aftertreatment.

These steps are described separately below.

a. ***Impregnation with alkali dye liquor:*** If necessary, urea and a suitable anionic wetting agent are added to the dye solution. In order to protect the dye against reducing effects, up to 10 g/L of organic oxidising agent should be used. Sodium bicarbonate and soda ash are generally recommended as alkali. Although the stability of the liquor increases when sodium bicarbonate is used, longer steaming times are required. After the alkali is dissolved in a separate place, it is mixed with the liquor containing dye and auxiliaries just before starting the dyeing. The impregnation temperature is generally around 20–25 °C. It can reach 50 °C in some dyes.[4]

b. ***Intermediate drying:*** Intermediate drying is generally done in hot-flue. In addition to the measures to prevent migration (reducing the pick-up value, working at a low drying temperature, IR pre-drying, adding anti-migration agent or salt to the impregnation bath), another issue to be considered is that the drying temperature should not exceed 90–120 °C and over-drying does not occur. Otherwise, some of the dye starts to get fixed during drying; this shows negative effects in the subsequent steaming. The dried material is sensitive to acid vapours and even air carbon dioxide. For this reason, if it is not going to be steamed immediately, it is recommended to wrap it in rolls and cover it with polyethylene foil. In fact, since they are also sensitive to sunlight, it would be appropriate to cover the polyethylene foil with a black cloth.[4]

c. ***Steaming:*** Steaming can be done in any continuous steamer. It is necessary to steam for 6–3 minutes at 102–105 °C with saturated steam that does not contain acid residues. When working in HT steamers (at 120–130 °C), the time is reduced to 2–1 minutes. In addition, if a catalyst is added to the impregnation liquor, the steaming time is shortened.[4]

d. ***Aftertreatment:*** After setting by steaming, the fabric is aftertreated in an open-width washing machine.

II. Single Bath Wet (Without Intermediate Drying) Pad-steam Method

Compared to the first method, this mode of operation, which is more advantageous due to the elimination of intermediate drying, is especially suitable for use in heavyweight products where intermediate drying is not economically feasible and in cases where intermediate drying increases the risk of dye migration.[53] For example, since processes involving intermediate drying lead to poor dyeing in viscose, corduroy and terry goods, it is more appropriate to use the pad-steam method without intermediate drying for these

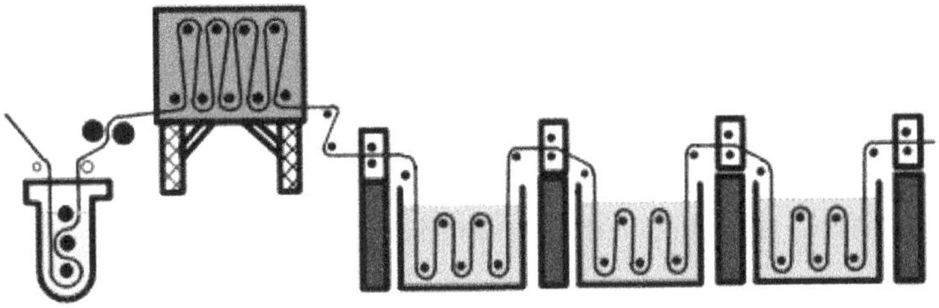

Figure 9.48 Process steps in dyeing by one-bath wet pad-steam method. Ref. [68]/Ciba Technical Documents.

products, where the dye penetration into the fibres is better.[4,53] Moreover, this method is especially suitable for fabrics with good dye absorbency. It is recommended for dyeing mercerised cotton and regenerated cellulose, as it provides a good dyeing efficiency in dyeing.[4] Figure 9.48 shows the process steps in dyeing by the one-bath wet pad-steam method.[68]

Various working methods are recommended by various dye manufacturers for their dyes. Soda ash/sodium bicarbonate mixture, caustic soda alone or caustic soda/water glass mixture is recommended as alkali. Since the stability of dye liquor containing caustic soda alone is very low, the dye and caustic soda solution must be prepared separately and mixed while being fed into the liquor by means of special dosing pumps. If a caustic soda/water glass mixture is used, the durability of the liquor increases as the water glass acts as a buffer, as explained earlier. In this case, it is possible to work safely without a special dosing pump by mixing the dye and alkali solutions just before starting the dyeing. It is known that the liquor temperature should be as low as possible (20 °C) and the padding vessel should be small. Depending on the colour depth, salt is added to the liquor as well as alkali. Steaming is done in a normal continuous steamer at 103–105 °C for 30–60 seconds. In regenerated celluloses, it is beneficial to dwell for an hour or more before steaming the fabric, as this increases the diffusion of the dye into the fibres.[4]

III. Two-bath Pad-steam (Pad-dry-pad-steam and Pad-pad-steam) Methods

It is the classical method of continuous dyeing of woven fabrics. It is especially used for dyeing cotton, long-length fabrics.[53] The advantage over the one-bath method is that the operating conditions can be adjusted within wide limits. Since the product to be dyed is impregnated separately with the neutral dye solution and the alkali, the type and concentration of the alkali can be adjusted as desired, and accordingly, the required steaming time can be shortened or extended.[4]

The recommended procedure in this method is as follows:

a. Impregnation with neutral dye liquor
b. Intermediate drying
c. Impregnation with liquor containing alkali and salt
d. Steaming
e. Aftertreatments[4]

Figure 9.49 shows the process steps for dyeing by the two-bath pad-steam (pad-dry-pad-steam) method.[68]

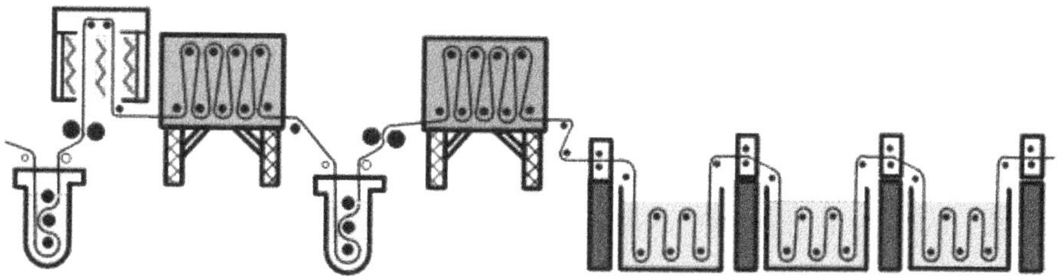

Figure 9.49 Process steps in dyeing by two-bath pad-steam method. Ref. [68]/Ciba Technical Documents.

In some cases, it is also possible to work without intermediate drying. However, in this case, the amount of dye that will bleed from the fabric into the alkali liquor is high. In studies without intermediate drying, a 30% lighter shade is obtained with the same amount of dye compared to intermediate drying. If intermediate drying is not done, there should be at least a 20% difference between the wet pick-up in the first and second padder.[4]

Since the stability in the neutral dye liquor is high, the impregnation temperature can be increased to 50 °C in cases where it is necessary to increase the penetration of the dyes into the fibres, and even up to 80 °C for some dyes, but the impregnation is usually done at 20–30 °C. Auxiliaries such as wetting agent, anti-migration agent, and organic oxidising agent can also be added to the liquor. Although it is recommended to perform intermediate drying in a hot-flue, other dryers can be used, provided that various conditions are observed. 200–300 g/L salt is added to the alkali liquor to prevent the dye on the fabric from bleeding. After the impregnation, steaming is started with or without intermediate drying. Steaming is carried out at 100–105 °C for 20–100 seconds depending on the type of dye, the amount of caustic soda used and the dye concentration.[4]

9.4.2.2.2 Dyeing by Pad-dry Method

This method consists of impregnation and drying of the product with an alkali liquor containing dye. Since the drying step is kept longer than a normal drying process, the fixation of the dye is also achieved. Instead it is a method that can be recommended for dyeing in light or medium shades with dyes of high reactivity. It is recommended to use soda ash as the alkali, and sodium bicarbonate for some dyes. The impregnation temperature is 20–25 °C and the pick-up value should be as low as possible. When working according to this method, it is recommended to add urea to the liquor even though the solubility of the dye in water is high. The addition of urea to the liquor increases the dyeing efficiency and facilitates dye penetration into the fibre.[4]

Adding an alkali-resistant alginate thickener to the liquor reduces the risk of migration during drying for products that are difficult to dye (such as fabrics with an embossed surface). Appropriate anti-migrating agents or the addition of 10 g/L salt also has the same effect. It is also beneficial to add a suitable organic oxidising agent (2–10 g/L) to the liquor.[4]

After impregnation of the regenerated cellulose with the dye solution, allowing the product to dwell for 1 hour or more before drying increases dye diffusion and thus dyeing efficiency. Drying conditions are as follows: 6–8 minutes at 100 °C, 4–5 minutes at 110 °C,

3–4 minutes at 120 °C, 2–3 minutes at 130 °C or 1.5–2 minutes at 140 °C. If a catalyst is added to the liquor, the drying time is shortened.[4] Today, this method is not used.

9.4.2.2.3 *Dyeing by Pad-thermofixation Method*

This method is a salt-free one-bath continuous dyeing process that is effective in covering dead cotton. However, this method has disadvantages such as low fastness properties, contamination of the waste air and contamination of the machine with urea.[53] The pad-dry method cannot always be used, especially for dyes with low reactivity. In this case, after the fabric impregnated with dye liquor is dried at lower temperatures, fixation of the dye (thermofixation) is achieved at high temperature is provided in a third step.[4]

As in the pad-dry method, urea must be added to the liquor in the pad-thermofixation method. At high temperatures, some urea decomposes to release ammonia gas, and some urea sublimes and condenses again in the cold parts of the dryer, causing fouling of the dryer and drip stains. In order to eliminate this drawback, some companies recommend using borax instead of urea during the high-temperature fixation for their dyes. In the thermofixing method, soda ash, sodium bicarbonate or a mixture of both is used as the alkali. Although intermediate drying yields the best results in a hot-flue dryer, other nozzle (air spray) dryers or cylinder dryers can also be used. The fixation process can be carried out in a curing machine, stenter, hot-flue or cylinder dryer. The fixation temperatures recommended by various dye manufacturers vary between 120 and 210 °C. At high temperatures, the heat setting time can be shortened considerably. The heat-fixing time varies between 1 and 6 minutes, depending on the heat-fixing temperature, colour depth and the type of alkali. The addition of a catalyst to the liquor shortens the required heat-fixing time.[4]

9.4.2.2.4 *Dyeing by Wet Fixation Method*

Wet fixation is an interesting method that does not require much machinery. It can be recommended for dyeing cotton fabrics with highly reactive dyes (which can easily react with cellulose macromolecules). The principle of this method is based on passing the fabric impregnated with a neutral dye solution through a hot (95 °C) liquor containing a high concentration of salt (to prevent the dye on the fabric from bleeding into the fixing liquor) and alkali after intermediate drying. Under these conditions, the fixation of the dye is completed within a few seconds. It is also beneficial to add a suitable wetting agent or anti-migration agent to the dyeing liquor, if necessary. The impregnation temperature with the dye can be 20–30 °C, or it can be increased up to 50–60 °C to ensure better penetration of the dye. In regenerated celluloses, the appearance of dyeing is not good due to intermediate drying. However, this inconvenience decreases if the fabric dwelled for 1 hour before intermediate drying.[4]

9.4.2.2.5 *Dyeing by Econtrol Method*

In Econtrol dyeing method (see Figure 9.50), the use of urea, soda ash, water glass, caustic soda, etc., which are required in other methods, is not necessary.[2] This significantly reduces the waste load and prescription costs.[69] It is sufficient to use 1–2 g/L wetting agent and 10 g/L sodium bicarbonate in addition to the dye in the liquor. After impregnation with Econtrol drying (containing 25% steam), the dye is fixed within 2 minutes at 120 °C.[2]

Figure 9.50 Econtrol dyeing unit. Ref. [69]/Monforts-Dystar.

The Econtrol process provides controlled colouring of cotton, viscose, Lyocell fibres and their blends. It can be used in a wide range of materials, from low-weight fabrics (100 g/m^2) to heavyweight fabrics such as corduroy (500 g/m^2). With the right dye selection, high fixation, excellent washing properties and high fastness levels can be achieved.[69]

9.5 Washing Processes After Dyeing with Reactive Dyes

The most important problem encountered during dyeing with reactive dyes is that reactive groups in dye molecules can react with —OH groups in water molecules as well as with —OH groups in cellulose macromolecules. Since dyes reacting with water lose their reactivity, dye molecules hydrolysed in this way cannot be attached to the fibres by covalent bonds, and this increases the unfixed dye part, the washing load and the pollution load of wastewater.[8] The amount of hydrolysed dye can be minimised by the use of controlled dyeing processes, especially bifunctional reactive dyes, which allow sufficient dye uptake and fixation. However, despite this, the number of dyes whose fixation efficiency can exceed 90% is very low, and it is inevitable that some dye on the fabric will remain attached to the fibre surface at the end of dyeing due to secondary attraction forces. In order to achieve the desired fastness properties (especially washing and rubbing), this part of the unfixed dye must be effectively removed. Therefore, washing is an important factor in obtaining the best results in reactive dyeing.[70]

Chemical, water and energy consumption in the washing step after reactive dyeing is significant, and half of the total energy consumption and one-third of the total waste load (chemical oxygen demand) in reactive dyeing come from the washing step. In addition, total water consumption has a close relationship with washing processes. In recent years, with the increasing cost of energy and water, as well as the increase in environmental awareness, the washing process has become an important issue that needs to be emphasised.[71] The issue of how to do the best washing economically and while protecting the environment has been questioned. In the review of washing processes, the main effects expected without sacrificing product quality can be listed as follows:

- It is necessary to reduce water consumption. For this, either the number of washing steps should be reduced or the liquor ratios should be shortened. In both cases, it is necessary to ensure the intensive interaction of the product and water.

- It is necessary to reduce the amount of chemicals used in washing, and even to completely eliminate the use of chemicals. Because the use of chemicals both brings an additional cost and increases the amount of organic pollution in the wastewater.
- It is necessary to reduce the costs arising from energy and labour and to save time.[72]

9.5.1 Steps of the Washing Process

Removal of unfixed reactive dye from the fabric takes place in three steps:

- Dilution of dyes and chemicals in the liquor and on the cellulose surface (initial displacement phase)
- Diffusion of the dye hydrolysate that has penetrated well into the fibre but not fixed (diffusion phase)
- Dilution and removal of the dye hydrolysate diffused onto the fibre surface (second displacement phase).[70]

The first phase of washing ensures that most of the hydrolysate in the waste liquor is removed and the electrolyte concentration is reduced to 1–2 g/L to reduce the substantivity of the dye hydrolysate. It is not necessary to raise the liquor temperature in this phase.[73] Because the removal of excess salt and alkali is by dilution. Therefore, it does not matter if the dilution water is hot.[70] Although increased temperature increases the desorption of the dye hydrolysate from the fibre, its use is only economical when warm or hot water can be supplied without additional heating costs. In case of working with alkali sensitive vinylsulphone-based dyes, neutralisation must be done before hot rinsing. Otherwise, the bonds made by the dye part that is covalently attached to the fibres may be cleaved and dyeing may be disrupted.[73]

In the second phase of the washing, the dye hydrolysate should diffuse out of the fibre. For this purpose, higher temperatures (80–98 °C) and longer times are required than the first phase, depending on the dye type and colour depth.[73] In this phase it is beneficial to have the temperature as high as possible. Also, washing agents can be used. Two or three rinses are required to dilute and remove the dye that has passed into the liquor.[70]

The third phase of the washing ensures removal of the dye desorbed into the liquor and termination of the diffusion phase.[70]

To a certain extent, there are points of overlap between these three phases. By separating it into three phases, it is possible to define the washing process in a simpler way.[73]

Washing processes, which are the third step of dyeing, often take a time close to the main dyeing process. The classical washing process of reactive dyeing generally starts with cold rinsing and consists of approximately 5–6 steps such as warm rinsing → hot soaping → warm rinsing → cold rinsing, respectively. In addition, in order to prevent the cleavage of the dye-fibre bond in a hot alkaline medium in vinylsulphone-based dyes, cold-warm pre-rinsing and neutralisation with acid at 40–50 °C are required before starting the hot washing step.[2] A general washing cycle for reactive dyes is shown in Figure 9.51.

The purpose of each washing step mentioned above is summarised below:

- ***Cold rinsing:*** The purpose of this is to remove alkali and salt. If a high amount of alkali remains on the fabric in the hot rinsing step to be made later, especially the

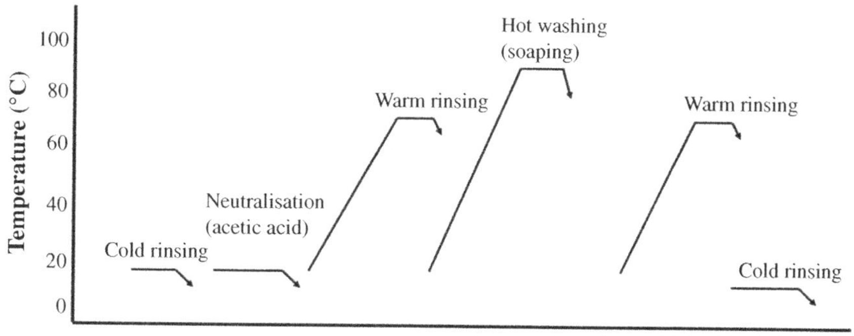

Figure 9.51 A general washing cycle for reactive dyes (the neutralisation process is valid only for vinylsulphone-based dyes). Ref. [70].

vinylsulphone-based reactive dye may be stripped. Therefore, the alkali must be removed from the fabric as well as possible in the first cold rinsing step. In addition, salt in particular should be removed as completely as possible. Because salt residues increase substantivity, making it difficult to remove unfixed dyes.[70]

- ***Neutralisation:*** The neutralisation process is carried out to neutralise alkali residues. In light colours, the neutralisation process can be done simultaneously with cold rinsing. Vinylsulphone-based reactive dyes can be removed from the fibres if there is still alkali residue in the fabric when the hot washing step is started. For this reason, it is recommended that a neutralisation process (with acetic acid) be performed at a pH of 7–8 for vinylsulphone-based reactive dyes.[9,70]
- ***Hot rinsing:*** Increasing temperatures lead to a decrease in the substantivity of the dye, which facilitates the removal of the hydrolysate.[70] In dyeing medium to dark shades, hot rinsing before proceeding to the hot washing increases the washing efficiency.[9]
- ***Hot washing (soaping):*** In addition to raising the temperature, another way to effectively remove the hydrolysate on the fabric is the use of a washing agent. Hot washing is done with 1–3 g/L anionic or non-ionic washing agent at temperatures above 90 °C.[70]
- ***Hot rinsing:*** This process has a significant effect on the wet fastness of dyed materials. Hot rinsing can be repeated more than once, controlling the colour of the waste liquor.[9]
- ***Cold rinsing:*** In this rinsing process, it is aimed to cool both the dyed fabric and the dyeing machine. When softening is to be done, this is done simultaneously with the rinsing step.[9]

9.5.2 Factors Affecting the Washing Process

The main factors affecting the washing processes after reactive dyeing can be listed as:

✓ Substantivity of the dye
✓ Degree of dye fixation
✓ Diffusion rate of the dye

✓ Temperature
✓ Duration
✓ Number of washing steps
✓ Liquor ratio
✓ pH of the washing liquor
✓ Water hardness
✓ Washing agent usage
✓ Enzyme usage
✓ Ultrasonic energy usage
✓ Washing method (drain-fill, overflow).

In addition to these, naturally the type of washing machine has a significant effect on the efficiency of washing processes.

9.5.2.1 Effect of Dye Substantivity

The wash-off properties of the dye depend on the substantivity of the dye molecule. Dyes with low substantivity exhibit good wash-off properties. However, since these dyes have lower exhaustion and degrees of fixation, their efficiency in reacting with fibres is low. For this reason, they are not recommended for exhaust dyeing operations.[9] During washing, affinity forces between fibre and dye hinder the removal of hydrolysed dye. The amount of hydrolysed dye is lower since dyes with high substantivity have a higher tendency to react with fibre rather than water. On the other hand, since the hydrolysed dye portion also has high affinity for the fibre, this unfixed portion is more difficult to remove than that of low-substantivity dyes. Although the hydrolysate can be removed more easily in dyes with low substantivity, in this type of dye, the amount of dye that has been hydrolysed and needs to be removed is higher.[74]

The substantivity of the dyes used in exhaust dyeing is generally high. Since the substantivity depends on the chromophore groups, not the reactive groups, the hydrolysed dye shows substantivity to the fibre to the same extent as the non-hydrolysed dye. This makes it difficult to remove the dye hydrolysate, which causes low wet fastness values. In contrast, the washing process is easier in dyeings performed by the impregnation method, as dyes with low substantivity are used.[70]

Polyfunctional dyes suitable for low salt use, which were developed to reduce the use of salt in dyeing by the exhaust method, are more difficult to wash off, since they are dyes with high affinity.[26]

9.5.2.2 Effect of Fixation Degree of Dye

The degree of fixation of the dye is of great importance in the washing of reactive dyes. It is known that the lower the amount of unfixed dye to be removed during washing, the easier the washing process will be.[70] Vinylsulphone-based dyes undergo chemical structure changes before and after alkali addition and after hydrolysis, and their affinity for cellulosic fibres also changes due to these structural changes. During the dyeing process, the reactive dye in the form of vinylsulphone has high affinity and a high fixation degree, while the affinity of the dye hydrolysate, which is converted to the hydroxyethylsulphone form during the washing process, decreases, thus facilitating the washing process. On the other hand, there is little difference between the affinities of the active and hydrolysed forms of MCT-based reactive dyes. Therefore, there is no direct correlation between the degree of fixation and washability in these dyes.[9]

9.5.2.3 Effect of Dye Diffusion Rate

Another factor affecting the removal of the unfixed dye is the diffusion rate of the dye. The diffusion rate varies depending on the bath temperature. As the bath temperature increases, the kinetic energy, hence the mobility and diffusion rate of the dye molecules increase, and this facilitates the removal of the hydrolysate. In order to obtain an optimal washing result, it is becoming increasingly important to start the washing directly with hot water and to remove as much of the waste liquor as possible in the first washing baths. Another factor affecting the diffusion rate of the dye is the type of cellulose and the pre-treatment processes applied. For example, mercerised cotton shows a higher diffusion rate, and therefore it is easier to remove unfixed dye from mercerised cotton during washing than from unmercerised cotton.[70]

9.5.2.4 Effect of Temperature

The success of the washing is largely dependent on the affinity of the hydrolysate for the fibre. Low-affinity hydrolysate can be removed to a large extent with cold water, while high-affinity hydrolysate can only be removed with hot water. As mentioned before, it can be said that the lower the substantivity of the dye for the fibres and the higher its diffusion, the easier it will be to wash. The most effective way to reduce the substantivity of the dye and accelerate its diffusion during washing is to increase the washing temperature.[2] As can be seen in Figure 9.52, as the washing temperature increases, the removal of the unfixed dye and the resulting alkali perspiration fastness values increase.[9] At increasing temperatures, the pores of the cellulose open and the hydrolysed dye rapidly diffuses into the rinsing liquor. Therefore, direct hot rinsing after the dyeing process prevents the pores of the cellulose from shrinking and prevents the unfixed reactive dye from becoming trapped in the fibre, which occurs when rinsing is started with cold water. Since the bonds formed by heterocyclic ring-based reactive dyes with cellulose are resistant to hot rinsing in the presence of alkali,

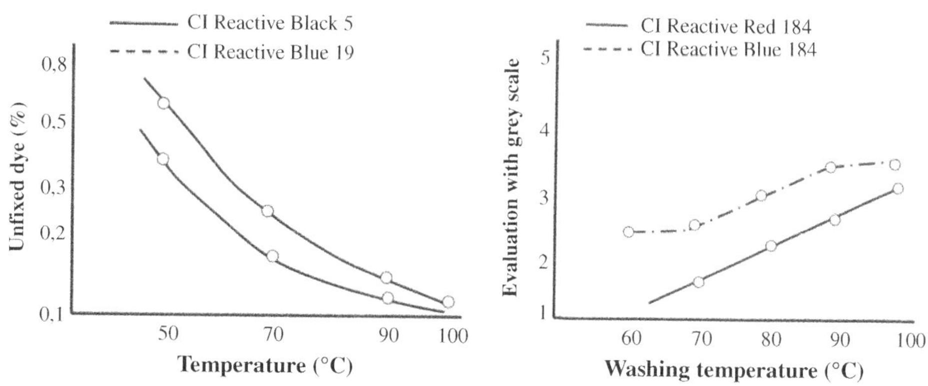

Figure 9.52 The variation of the degree of removal of unfixed dye (a) and the obtained alkali perspiration fastness values (b) depending on the soaping temperature. Ref. [70].

acidic cold rinsing is not required before hot rinsing.[75] However, as mentioned before, since the bonds formed by vinylsulphone-based reactive dyes with cellulose are sensitive to alkali hydrolysis, it is not recommended to start washing with these dyes directly with hot water.[2]

In addition, in a study, it was determined that the temperature had a positive effect on the waste load. It has been stated that the reason for this may be the degradation of organic pollutants at high temperatures.[76] However, despite the mentioned advantages of starting the washing directly with hot water, it is stated that problems may arise as a result of accidental stopping of the machines. In such cases, the very high temperature of the hot rinse water may cause the bonds between the reactive groups of the dye and the hydroxyl groups of cotton or viscose to be irreversibly broken.[26]

9.5.2.5 Effect of Duration

Non-substantive chemicals such as electrolytes rapidly saturate the rinsing liquor when their concentrations are at their highest level (especially in the first steps of rinsing). For this reason, the amount of dye removed in washings over 10 minutes increases only slightly (Figure 9.53). At this point, the bath must be drained and a new liquor must be prepared.[77]

9.5.2.6 Effect of Washing Number

To obtain the best product quality, dyers consider it beneficial to repeatedly wash the product using more hot water. On the other hand, it is very important to keep hot water consumption to a minimum from a cost and efficiency point of view.[70] In a study conducted in the literature, the results obtained from repeated washings show that the washing effect increases as the number of washing steps increases. However, it is stated that the amount of hydrolysate transferred to the washing water after the first washing is much higher than the amount removed at the end of the second and third washings.[78]

9.5.2.7 Effect of Liquor Ratio

Desorption of unfixed dye molecules into washing liquors is governed by principles similar to those of dyeing. Therefore, the relationships used to describe the dyeing equilibrium

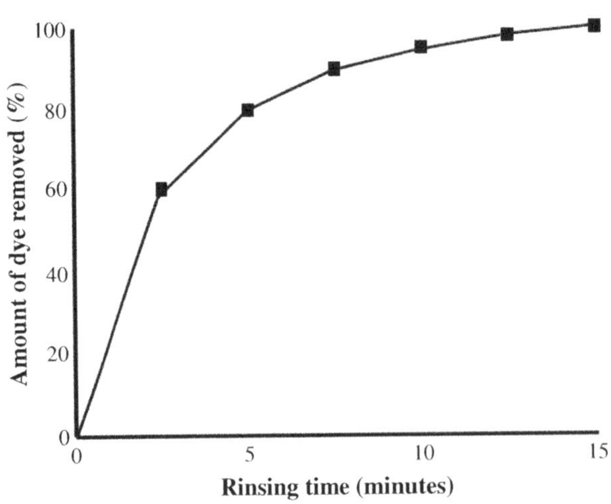

Figure 9.53 Relationship between washing time and amount of dye removed. Adapted from Ref. [77]

and dyeing kinetics can give at least a general idea about the effects in the washing process. Thermodynamically, the transfer of the dye from the solution and its uptake by the fibres is due to the fact that the chemical potential of the dye in the fibres is lower than the chemical potential of the dye in the solution. Accordingly, the uptake of dye by the fibres will stop when the chemical potentials of the dye in the fibres and in the solution are equal to each other, that is, the dynamic equilibrium of "dye remaining in the liquor ↔ dye taken up by the fibres" will occur. This equilibrium is also valid for the equilibrium concentrations of the hydrolysed dye in the fibre and solution at any temperature T, at a certain liquor ratio. High liquor ratios in washing favour dye desorption.[70] In a study reported in the literature, it was determined that the liquor ratio had a significant effect on the washing efficiency, and more dye hydrolysate was removed from the fabric in washings performed at a high liquor ratio compared to washings with at a low liquor ratio. In addition, it has been stated that the positive effect of increasing the liquor ratio on the washing efficiency is much more pronounced than extending the processing time.[78]

9.5.2.8 Effect of pH of Washing Liquor

An important point to be considered during washing processes is the pH at which the reactive dye-fibre bond is stable. Table 9.9 summarises the stability of the reactive dye-fibre covalent bond to various parameters, depending on the type of reactive group.[9] As can be seen from the table, the fibre-dye bond is stable between pH 6 and 7 in reactive dyes that are fixed according to the substitution mechanism (triazine, pyrimidine, etc.). Therefore, these dyes are sensitive to acidic environments unless aftertreatment with cationic fixing agents is applied. The bonds formed by reactive dyes that are fixed according to the addition mechanism with cellulose (vinylsulphone-based) are stable between pH 4 and 5. Therefore, in strongly alkaline washings, wet fastness values can give results ranging from moderate to poor, due to the hydrolysis of the fibre-dye bond at increasing temperatures.[14] For this reason, it is recommended by the dye manufacturers to neutralise these dyes at a pH of around 8 before raising the temperature during rinsing.[70] For this

Table 9.9 Stability of the reactive dye-fibre covalent bond to various parameters, depending on the type of reactive group.

	Acid	Alkali		Heat	
Reactive group	**pH 3–4**	**pH 9–10**	**pH 13–14**	**180 °C**	**Oxidising agent**
Sulphatoethylsulphone	O	O	X	O	O
Monochlorotriazine	Δ	O	Δ	Δ	Δ
Monofluorotriazine	Δ	O	Δ	Δ	Δ
Difluoromonochloropyrimidine	Δ	O	Δ	Δ	X
Trichloropyrimidine	O	O	Δ	O	X
Dichloroquinoxaline	X	O	Δ	X	X

(O: Stable, Δ: Partially stable, X: Significantly unstable)

Ref. [9]/Sumitomo Technical Documents.

purpose, it is recommended to use a volatile acid such as acetic acid. Because, in the case of using non-volatile acids such as phosphoric acid, the acid residue that may remain on the fabric may cause problems in processes such as resin finishing.[9] However, there are studies in the literature stating that the omission of the neutralisation process does not cause a decrease in the quality of the dyed textile material, even with vinylsulphone-based reactive dyes whose neutralisation process is recommended.[70]

In an interesting study in the literature, the effect of using ammonium hydroxide, sodium bicarbonate, potassium hydroxide, sodium hydroxide and buffer systems (pH 8–9–10 –11–12) in the hot washing step on the washing of cotton fabrics dyed with alkali-resistant monochlorotriazine-based reactive dyes was investigated. As a result, it was stated that the use of alkali in the washing process facilitates hydrolysate removal, and the best results are obtained in washings using potassium hydroxide and sodium bicarbonate. In addition, it has been reported that there is no difference in terms of wet fastness values obtained across the various washings treatments.[79]

9.5.2.9 Effect of Hardness Degree of Washing Water

The hardening calcium and magnesium ions in the washing water are also a factor affecting the washing. In fact, reactive dyes have very good solubility and are not sensitive to water hardness. However, studies have shown that when hard water is used in the washing and soaping processes, the removal of hydrolysed dye is much more difficult than when using soft water. Therefore, water hardness also negatively affects wet fastness.[80] The hardening Ca^{++} ions in the water interact with the solubilising groups of the dye, and since the solubility of the dye-metal ion complexes formed in water is very low, the hydrolysate can be removed from the fabric only with great difficulty. In addition, hardness-causing metal ions increase the precipitation of the dye hydrolysate.[81] The effect of hardness-causing metal ions on the removal of hydrolysate differs depending on the chemical structure of the dye. While low substantivity dyes are less affected by calcium ions, those with high substantivity are more negatively affected by calcium ions. On the other hand, experiments with dyes containing different reactive groups showed that the reactive group did not change the sensitivity of the dye to calcium ions in water.[70]

When reactive dyes were first introduced to the market, it was recommended to use detergents only for washing. Later, it was found that the addition of sodium tripolyphosphate to detergents facilitates the removal of dye hydrolysate, and polyphosphate-based products have been developed by certain manufacturers for washing reactive dyes. However, with the developing environmental awareness, it was necessary to find another sequestering agent with equivalent performance that could replace phosphates in washing, taking into account the pollutant loads in wastewater. Considering the ecological, economic and technical aspects, polyacrylic acid and its salts are considered the best chemical class. It is stated in the literature that if soft water below 5 °F hardness is used for washing, there is no need to use a complexing agent.[70]

9.5.2.10 Effect of Washing Agent Use

In the washing processes, surfactants (tensides) are the primary agents due to their good wetting power and detergent effect. Detergency feature is the ability to remove impurities such as oil, wax and dirt and to keep them in the form of an emulsion or suspension in the liquid.[82] All surfactants (detergent, wetting agent, dispersant, emulsifier) have a hydrophobic part of varying length and at least one hydrophilic end group in their structure. The hydrophobic part can be aliphatic, aromatic, or aliphatic and aromatic in character. The predominance of one of the washing, wetting or emulsifying-dispersing properties of tensides compared to the other is closely related to the length of the hydrophobic part and the location and number of the hydrophilic groups in the molecule. Tensides show these effects primarily due to their ability to lower the interfacial tension of fibres and foreign materials. For this reason, surfactants are also called "surface-active agents". Washing agents do not have to have foam-forming ability, but some foam supports the wash. Because the foam creates a capillary effect, allowing the dirt to come out of the fibre more easily.[70]

Today, there are washing agents with very different chemical structures in the market, and generally anionic or sometimes non-ionic washing agents are used for washing after reactive dyeing. Non-ionic washing agents show some cationic properties. A change occurs only on the reactive groups of the hydrolysed reactive dye; the anionic end groups that provide water solubility are not affected by the hydrolysis. Therefore, a non-ionic washing agent having a slightly cationic character can more easily remove anionic hydrolysed dye from the fibre.[70]

There are studies in the literature stating that washing agents do not contribute to the removal of hydrolysed reactive dye from the fabric. In one project, 50 washes were carried out in dyehouses without the use of washing agents, all demonstrating that washing agents could be omitted without adversely affecting product quality. In this way, both chemical consumption and wastewater pollution are reduced.[70]

In practice, the hot washing process is generally carried out at temperatures above 90 °C with 1–3 g/L anionic or non-ionic washing agent. In addition, it is observed that auxiliaries are often used in high doses with the idea that this will provide effective washing. The reason for this "be sure beforehand rather than regret later" mentality is because the garment industry aims to meet the colour shade and fastness demands in small batches and variable specifications. However, it should not be forgotten that the washing chemicals used during washing contribute to organic pollution in the wastewater.[70]

As a result of the search for ecological alternatives to washing agents, the use of cyclodextrin in washing processes after reactive dyeing has begun to be examined. Cyclodextrins (CD) are cyclic compounds that resemble a hollow cone with a truncated top, obtained by enzymatic degradation of starch. The most common cyclodextrins are alpha (α), beta (β), and gamma (γ). They contain 6, 7 and 8 glucopyranose units, respectively. There are also cyclodextrins containing 9, 10, 11, 12 and 13 (δ, ε, ζ, η and θ Cyclodextrin) glucopyranose units. The molecular structure of cyclodextrins is given in Figure 9.54.[83]

The cavity in the centre of the cyclodextrin molecule is lipophilic and provides a lipophilic microenvironment in which, under favourable conditions, "guest" molecules can enter and be included. The complex formed by filling the cavity in the centre of the cyclodextrin molecule (host) by another molecule (guest) is called an inclusion complex.[84]

By utilising the inclusion complex formation properties of cyclodextrins, when cyclodextrin is used in the washing processes of cellulosic fibres after dyeing with reactive dyes, hydrolysed dye that is not covalently bonded to the fibre and cyclodextrin molecules

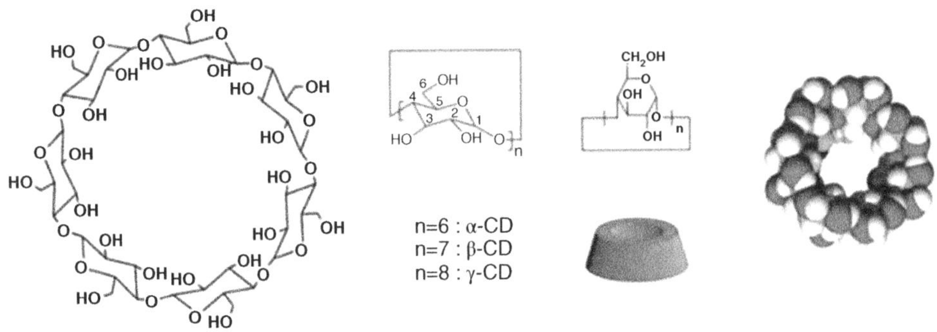

Figure 9.54 Molecular structure of cyclodextrins. Ref. [83]/with permission of Tekstil ve Konfeksiyon.

can form an inclusion complex. In this way, the hydrolysed dye can be removed from the fabric. As a result, cyclodextrin acts as a washing agent. However, it should not be forgotten that cyclodextrins do not form complexes with every dye. The first condition for complex formation is that the molecular size of the dye is suitable for the cavity in the cyclodextrin. In addition, some chemical properties also play an important role in the complex formation. Therefore, cyclodextrins cannot be used as washing agents after dyeing with every reactive dye. First, it should be determined whether there is an interaction between the dye and the cyclodextrin using methods such as chromatographic, thermal, and spectrophotometric, techniques. In a study in the literature, it was revealed that β-cyclodextrin showed a good efficiency in removing CI Reactive Blue 19 and CI Reactive Blue 21 hydrolysates from cotton fabric.[85]

9.5.2.11 Effect of Enzyme Use

In order to remove unfixed dye molecules not only from the fibres but also from the exhaust bath, methods such as applying an enzymatic treatment in washing processes after reactive dyeing have also been developed. Application of the enzymatic treatment is usually done in the fourth or fifth rinsing step.[26]

As can be seen from Table 9.10, enzymatic washing is carried out at 50 °C immediately after the hot washing process above 90 °C. After the machine is filled with clean water (50 °C), buffer is added for pH adjustment, and after pH control (adding acetic acid if necessary), enzyme (0.25 g/L) is added to the liquor and left for 10 minutes. Then the liquor is drained.[26] During the enzymatic washing process, the hydrolysed and unfixed reactive dye in the water is effectively decolourised. Thus, one hot washing step is saved. The main advantage of applying this technique is the savings in water, energy and detergent consumption. Fastnesses either remain at the same level or improve in some cases.[86] Most of the enzymatically reactive dyes can be decolourised. However, as a precaution, it is recommended that they be tested on a laboratory scale first.[26]

9.5.2.12 Effect of Using Ultrasound Energy

As it is known, ultrasound is sound that has a frequency beyond the human hearing limit. Human normal hearing frequencies range from 20 Hz to 20 kHz, while ultrasonic

Table 9.10 Comparison of conventional and enzymatic washing process steps (dyeing by exhaust method).

Conventional washing	Enzymatic washing
5 min rinsing	5 min rinsing
10 min 40 °C	10 min 40 °C (neutralisation)
10 min 40 °C (neutralisation)	10 min 60 °C
10 min 95 °C	10 min 95 °C
10 min 95 °C	10 min 50 °C (enzymatic process)
10 min 50 °C	10 min 30 °C
10 min 30 °C	—

Adapted from Ref. [26].

frequencies range from 20 kHz to 500 MHz. The use of ultrasound for cleaning is already known. The use of ultrasound in washing processes after reactive dyeing has also been investigated in the literature. It has been stated that the best washing fastness values among the washings made with the classical method, ultrasonic probe and ultrasonic bath are obtained in the washings made using an ultrasonic probe.[87]

9.5.2.13 Effect of Washing Method

After the semi- and full-continuous dyeings made by the impregnation method, the washing processes are carried out in continuous open-width or rope washing machines. Washing processes after discontinous dyeing by the exhaust method are generally performed in the same device (jet, overflow, jigger, apparatus, etc.) where the dyeing is done, by discontinuous methods. Basically, there are two main discontinuous rinsing methods, drain-fill and overflow. Information about these methods is given below.

a) Drain-fill rinsing

In the drain-fill method, the rinsing process is done by successive filling, rinsing and draining steps.[26] The efficiency of these processes depends on the following factors:

- Liquor ratio
- The temperature of the rinsing liquor
- The amount of liquor carried by the material (carry-over)
- Cycle time of the fabric rope or rinsing liquor.[70]

The value at which the foreign matter concentration in the textile product decreases after n drain-fill rinses can be calculated according to the following formula:[82]

$$C_{fn} = C_p \times \left(\frac{V_0}{V_0 + V}\right)^n$$

C_{fn}: Final concentration of foreign matter at the end of the n rinse (g/L)
C_p: Concentration of foreign matter in the previous bath (g/L)
V_0: The amount of liquor remaining in the textile product (carry-over) (L)
V: The amount of clean water put into the vessel for washing (L).

It can be seen from the equation that the liquor ratio has a great effect on the dilution, in case of working at low liquor ratios, more rinsing steps are required to achieve the same rinsing effect, but less water is used. For example, after draining a waste dye liquor with a liquor ratio of 1:15 and containing 60 g/L electrolyte, assuming that a material with a 300% carry-over capacity is processed, the electrolyte concentration in the new bath after a fill-and-drain operation (C_{f1}) will be 12 g/L. If the liquor ratio is reduced to 1:7, the C_{f1} value becomes 26 g/L and the C_{f2} value (electrolyte concentration at the end of the second rinse) becomes 11 g/L. That is, in rinsing with lower liquor ratios, more rinsing steps are needed to reach the same point.[88]

The amount of carry-over brought by the material also has a great effect for washing garment pieces in a drum washing machine and for long-term washing of heavy goods or towels in overflow machines. Carry-over in the product can be reduced by working for a longer draining time (more than 10 minutes) after heavily contaminated dyeing and soaping baths, or by sucking the liquor with compressed air or vacuum between washing steps in some machines (e.g. package dyeing, loose fibre dyeing).[70] When the carry-over amount in the product is reduced, the foreign matter concentration can be brought to lower levels after a water rinse, or the number of washes required to achieve a certain washing effect and the total water (hence energy) consumption can be significantly reduced. This decrease is more evident in washings made in a short liquor ratio. According to the formula just given, the number of water rinses (or washings) required to reduce the initial impurity concentration (C_p) to a certain concentration (C_{fn}) can be easily calculated by the following formula:[82]

$$n = \frac{\log\left(C_{fn}/C_p\right)}{\log(V_0/(V_0 + V))}$$

The total water consumption in case of n water rinses can be calculated as follows:[82]

$$V_T = n \times V$$

For example, it is desired that the dye and chemical concentrations remaining in the product as a result of a dyeing process are diluted 100 times as a result of rinsing. Since the amount of liquor remaining in the product (V_0, carry-over) is 300% and the liquor ratio is 1:10, the number of water (n) rinses required for 100 kg of product and in this case the total water consumption (V_T) can be calculated as follows:[82]

$$n = \frac{\log(1/100)}{\log(300/(300 + 700))} = 3,825 \sim 4$$

$$V_T = 4 \times 700 = 2800\ \text{L}$$

The main advantage of drain-fill rinsing is that it is relatively efficient in terms of water consumption. The main disadvantages of classical drain-fill rinsing methods can be listed as:

- longer processing times meaning reduced production efficiency
- thermal shock of the material during the first rinse
- redeposition of superficial contaminants on the material when draining the liquor between rinses.[70]

b) Overflow rinsing

Non-productive filling and draining times in drain-fill rinses can be minimised by overflow rinsing, where clean water is filled into the machine and usually drained through a normal operating level overflow weir. Although overflow rinsing is more advantageous in terms of time, it is a less efficient rinsing method in terms of water consumption since clean water is usually filled into the machine with the valve fully open. Water consumption and waste-water treatment make this method less attractive. To overcome these disadvantages, many dyers use a combination of these classical techniques, first an overflow rinse to remove surface soils and partially dilute soiling for up to 10 minutes, followed by a series of drain-fill rinses.[70]

In overflow rinses, the concentration of dissolved foreign matter in the liquor at the end of successive rinsing can be calculated according to the formula given below:[82]

$$C_t = C_p \times e^{-\frac{V_t}{V}}$$

C_t: Concentration of foreign matter in the rinsing liquor at the end of time t (g/L)
C_p: Concentration of foreign matter in the process liquor at the time of rinsing (g/L)
V_t: the amount of clean water added until the end of the t period (L)
V: Rinsing liquor volume (L)
t: Rinsing time (minutes)

As can be understood from the formula, the liquor ratio is the determining factor that affects the rinsing process.[88] If the example given above is to be dissolved for rinsing by overflowing instead of the drain-fill operation, the total water consumption can be calculated as:[82]

$$V_T = v \times \left(\frac{\log\left(\frac{C_p}{C_t}\right)}{\log e} \right) = 1000 \times \left(\frac{\log 100}{\log e} \right) = 4600\ L$$

As can be seen from this example, the amount of water required to reduce the concentration of foreign matter remaining in the product to a certain level is considerably higher in overflow washing than in drain-fill washing. Therefore, the drain-fill technique is more efficient in terms of water consumption than conventional overflow rinses. Since there are usually hot and warm rinsing steps, reducing water consumption means reducing energy consumption. An advantage of the drain-fill method, compared to the classical overflow method, is that it is possible to store the used concentrated dyeing liquor and rinse water separately. This feature means that these liquors can be reused or at least treated separately to provide heat energy recovery.[26]

9.6 Aftertreatments After Dyeing with Reactive Dyes

Cation-active post-treatment agents, which were mainly developed for direct dyes, are also used for reactive dyed fabrics. It is known that during the dyeing of cellulosic fibres with reactive dyes, some of the dye is hydrolysed by reacting with water. The hydrolysed dye, which cannot be chemically bonded to the fibres and cannot be completely removed from the fabric, affects wet fastness values negatively. If an effective washing process is carried out, aftertreatment may not be necessary. However, especially in loose fibre dyeing and dyeing in dark shades, it is difficult to completely remove the unfixed dye by washing alone. In this case, after soaping and rinsing, aftertreatment with fixing agent is applied to improve wet fastness values. In materials such as woven fabrics, this process can be done in one step with the crease-free finishing process.[9]

Cationic fixing agents (F^+X^-) reduce the solubility of dye molecules in water by blocking the anionic sulpho groups ($D{-}SO_3^-$) in the structure of reactive dye molecules through electrostatic attraction forces (Figure 9.55).[15]

$$D{-}SO_3Na + FX \longrightarrow D{-}SO_3^- \ldots\ldots\ldots\ldots {}^+F + NaX$$

With the electrostatic bonding between anionic dyes and cationic polymers, a water-insoluble polymer salt is formed, which positively affects water, acid and alkali perspiration fastness values. In addition to the electrostatic bond, hydrogen bridges, van der Waals attraction forces and hydrophobic interactions can also occur when the molecules get close enough to each other. It is known that, aftertreatment materials can have other functional groups such as $-OH$, $=NH$, $-NH_2$, $-C{=}N-$ or guanidine groups besides quaternary ammonium groups. Thus, they can form complexes with metal-containing anionic dyes. This bond is very stable and higher washing fastness values can be achieved for reactive dyes containing metal.[15]

Apart from the classical monofunctional fixing agents mentioned above, there are also some special aftertreatment agents with a bifunctional structure that can form covalent bonds with cellulose. High wet fastness values can be obtained with such products.[15]

Cation-active aftertreatment agents used in the past were usually dicyandiamide type fixing agents, which were formaldehyde condensation products. Although these products showed a good fixation effect, caused little change in colour, and had low cost, they had the drawback of lowering light fastness values. Formaldehyde-free aftertreatment materials available on the market today are divided into three groups:[15]

- Polyions
- Polyheterocyclics
- Polyalkylenepolyamines.

The chemical building blocks of all three types of fixing agents are given in Figure 9.56.

Polyions are linear quaternary polyamines and carry the (+) charge on the molecular axis. The effectiveness of these fixing agents varies according to the length of the alkyl group between the charges and the type of alkyl group attached to the nitrogen atom.

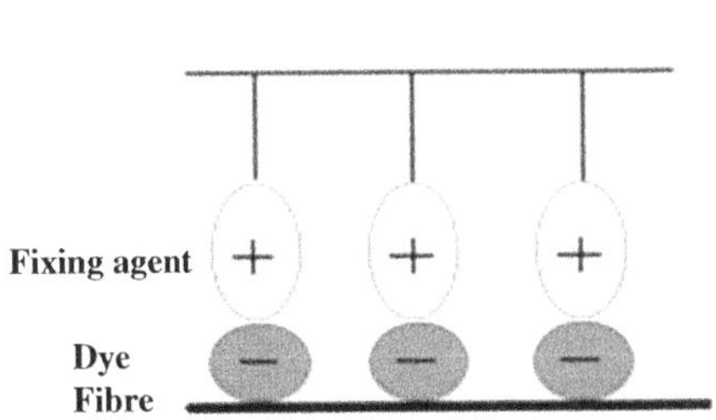

Figure 9.55 Mechanism of dye fixation. Ref. [15].

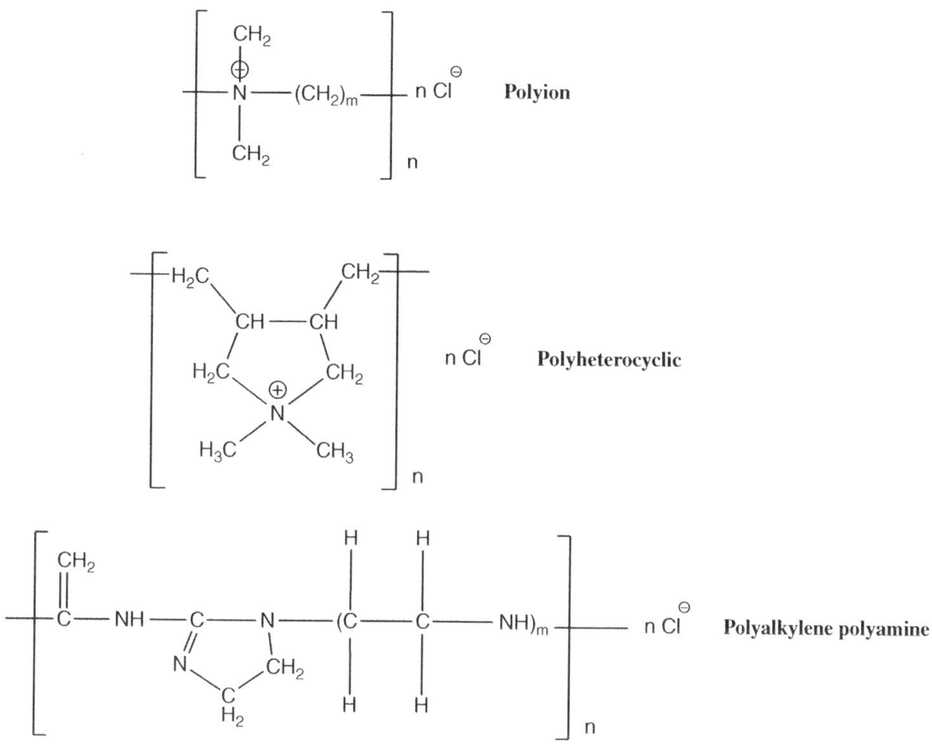

Figure 9.56 Repetitive units of various fixing agent types. Ref. [89].

In polyheterocyclic structures, (+) charges are located next to the molecular chain, and these charges are separated from each other as much as possible by rotating the repetitive structural elements along the molecular axis. Polyalkylenepolyamines contain protonated amines, and they also contain alkylamine groups as well as heterocyclic and other structural elements as the main group.[89]

Fixing agents block the water-soluble groups of dye molecules, thereby ensuring that they are trapped in the fibre, thus improving wet fastness values. In addition, they increase the colour depth by fixing some of the hydrolysed dye to the fibre. However, in addition to incurring an additional cost, fixing agents also have disadvantages such as a change in colour shade, a decrease in light fastness values and hardening of the fabric handle, depending on the type of fixing agents used.[15] It should also be taken into account that some fixing agents can reduce the chlorine fastness of dyed materials. It is thought that amines in the structure of fixing agents have a high absorption capacity for active chlorine in tap water, therefore, active chlorine in materials aftertreated with fixing agents accumulates in the fibres and causes a change in the colour of reactive dyes. On the other hand, some commercial fixing agents can also have the effect of increasing the chlorine fastness of the reactive dyed material. Therefore, the performance of fixing agents should be tested before use.[9]

Aftertreatments by the exhaust method are generally carried out with liquor containing 1–4% or 1–3 g/L fixing agent and acetic acid to adjust pH to 4–5 for 10–20 minutes at 60 °C.

In applications by the impregnation method, after dyeing, 10–50 g/L fixing agent is added to the last vessel of the open-width washing machine and aftertreatment is carried out. When combined with anti-crease finishing, care should be taken to ensure that the fixing agent used is compatible with other auxiliaries to be added to the liquor. If an excessive amount of fixing agent remains on the dyed material, discolouration may occur in metal-complex reactive dyes. Therefore, if aftertreatment with the fixing agent is carried out by the impregnation method, cold rinsing is recommended after the treatment with the fixing agent.[9]

If it is necessary to remove the fixing agent from the fabric for colour correction or stripping, it is required to perform a treatment with 3–6 mL/L acetic acid (48%) or 1–3 mL/L hydrochloric acid or phosphoric acid at 80 °C for 30 minutes.[9]

Another common aftertreatment after reactive dyeing is treatment with softeners. In general, softeners are classified as follows:

- Non-ionic softeners
- Cationic softeners
- Anionic softeners
- Special softeners (amphoteric, pseudocationic, etc.)
- Silicone softeners.

As a result of the condensation of fatty acids with polyamines, non-ionic or cationic softeners are obtained according to the type of amine used and the amount of fatty acid. Amphoteric softeners are obtained by the conversion of suitable fatty acid amine condensates with sodium chloroacetate, and anionic softeners are obtained by sulphation or phosphation of oil compounds in general.[90,91] With the introduction of functional silicone compounds, they started to be used as softeners and even constitute the most important product group. Modern silicone emulsions are macro- and micro-emulsions, also known as milk emulsion, mostly derived from aminofunctional silicone oils. In addition, hydrophilic silicones have been developed to eliminate the silicone stain problem. Macroemulsified silicone softeners remain on the surface of the fibre due to the large size of the particles, thus providing a certain surface softness. In microemulsions, the particle size is smaller than 50 nm, and due to their small particle size, they can penetrate into the fibre and impart excellent inner softness and an effective surface lubricity to the fabric. Hydrophilic organosiloxanes are water-soluble or water-dispersible chemicals, as opposed to aminosiloxanes which are neither water-soluble nor water-dispersible. Therefore, there is no need for a conventional oil/water emulsion in this type of softener, there is no limit on the pH value of the bath, and also the stability of these softeners against shear forces is quite high.[15]

Cationic and silicone-based softeners are generally used as softeners for dyed materials. Non-ionic softeners, on the other hand, are preferred especially for white fabrics as they do not cause yellowing. Commercial cationic softeners available in the market can be classified into six different chemical structures: primary amine salts, tertiary amine salts, quaternary ammonium salts (quartamine), aminoamide salts, imidazoline salts or aminoester salts (esterquart). Although there are certain differences between each class, cationic softeners generally impart good softness. On the other hand, they can cause yellowing, decrease the hydrophilicity of the fabric (the hydrophilicity of quartamine-based cationics is better compared to other cationic softeners), and they can adversely affect the light fastness.

Silicone softeners, on the other hand, provide not only softness, lustre and smoothness to textile products, but also elasticity, sewability and fullness. It is known that coloured textiles treated with silicone appear darker than untreated ones. The point to be considered when working with silicone-based softeners is that the medium should not be alkaline. In an alkaline medium, the silicone emulsion deteriorates and silicone stains may occur on the fabric.[15] Softeners, like fixing agents, can be applied by the exhaust or impregnation method. Aftertreatments by the exhaust method are generally carried out with liquor containing 1–4% softener and acetic acid to adjust pH to 4–5 for 20–30 minutes at 40–50 °C.[9] In applications by the impregnation method, the fabric is impregnated with a liquor containing 20–40 g/L softener and acetic acid at pH 4–5 and pick-up value of 70–80% on a padder and dried at 80 °C for 3–4 minutes.

References

1 Imperial Chemical Industries Limited (ICI). Procion dyestuffs on cellulosic fibres. Dyeing of cellulosic fibres (Chapter 2), UK, 1977.
2 Yurdakul, A., Atav, R. Boya baskı esasları. Bornova, İzmir: Ege Üniversitesi Tekstil ve Konfeksiyon Araştırma Uygulama Merkezi Yayını, 2006.
3 Benar Kimya. Reactive dyeing technical training presentation.
4 Tarakçıoğlu, I. Tekstil Boyacılığı I Teksiri. İzmir, Türkiye: Ege Üniversitesi Mühendislik Fakültesi Çoğaltma Yayınları, 1982.
5 İçlioğlu, H.İ. Pamuklu dokunmuş kumaşların reaktif boyalarla boyanması ve uygulama yöntemlerinin incelenmesi (MSc Thesis). Çukurova Üniversitesi Fen Bilimleri Enstitüsü, Adana, Türkiye, 2006.
6 Başer, İ., İnanıcı, Y. Boya kimyası. İstanbul: Marmara Üniversitesi Matbaası, 1990.
7 Dystar. Reactive dyes technical documents.
8 Tarakçıoğlu, I. Reaktif boyalarla çektirme yöntemine göre boyamalarda bazı yeni gelişmeler. 1990; Tekstil & Teknik, 80–92.
9 Sumitomo Chemical. Technical information Sumifix supra and Sumifix dyes, Exhaust Dyeing.
10 Ciba. Reactive dyeing training presentation, 2003.
11 Lewis, D.M. Developments in the chemistry of reactive dyes and their application processes. Coloration Technology, 2014; 130(6), 382–412.
12 Dystar. *Chemical structure of reactive dyes*, Technical documents.
13 Özgüney, A.T., Özkaya, K., Özerdem, A. Reaktif boyalı örgü kumaşların parça baskıdaki ısıl işlem sonrasında renk değiştirme eğilimlerinin incelenmesi. Tekstil ve Konfeksiyon, 2007; 3, 192–199.
14 Shore, J. Dyeing with reactive dyes. In Cellulosics Dyeing (Ed. J. Shore, Chapter 4, 189–242). Bradford, West Yorkshire, UK: Society of Dyers and Colourists, 1995.
15 Yurdakul, A., Öktem, T., Kumbasar, P., Atav, R., Korkmaz, A., Arabacı, A. Boyama işlemlerinden sonra kullanılan tekstil kimyasallarının ve diğer terbiye işlemlerinin haslık özellikleri üzerine etkileri. Project No: TÜBİTAK-TAM 2002-02, Kasım 2003.

16 Bredereck, K., Strau, F. Reaktif boyalarla boyanmış selüloz esaslı tekstillerin sürtme haslıklarda genel görünüm. Melliand Türkiye, 2001; 4, 123–129.
17 Yurdakul, A., Akçakoca, P. Reaktif boyalarla boyalı örgü kumaşların yaş sürtme haslığını etkileyen parametrelerin incelenmesi. MISAG-156, 2002.
18 Karadağ, Ç. Viskon, modal ve lyocel liflerinin reaktif boyarmaddelerle boyanmalarının optimizasyonu (MSc Thesis), Ege Üniversitesi Fen Bilimleri Enstitüsü, İzmir-Türkiye, 2015.
19 Raheel, M. Konvasiyonel olarak ve solvetten çekilen rayonların yüzey morfolojisi, özgül viskozitesi ve performans özellikleri. VI. Uluslararası İzmir Tekstil ve Hazır Giyim Sempozyumu, 1992, 429–445.
20 Duran, K., Özen, İ. Elastan içeren kumaşların ön terbiyesi. Tekstil ve Konfeksiyon, 1999; 5, 385–390.
21 Elastan Yıkama. http://www.eksoy.com.tr/questiontr.htm /, (accessed: 30 April 2005).
22 Tarakçıoğlu, I. Tekstil terbiyesi ve makinaları. Ege Üniversitesi Emel Akın Meslek Yüksek Okulu, Bornova, İzmir, 1996.
23 Nevell, T.P. Cellulose: structure, properties and behaviour in the dyeing process. In Cellulosics Dyeing (Ed. J. Shore, Chapter 1, 1–73). Bradford, West Yorkshire, UK: Society of Dyers and Colourists, 1995.
24 Duran, K., Ayaz, Ö.Y. Selülazların rejenere selüloz liflerinde kullanımıyla alternatif terbiye prosesleri. Tekstil ve Konfeksiyon, 1999; 5, 390–396.
25 Cavaco-Paulo, A. Mechanism of cellulase action in textile processes. Carbohydrate Polymers, 1998; 37, 273–277.
26 Türkiye Tekstil Terbiye Sanayiciileri Derneği. IPPC Best Available Techniques (BAT) reference document for the textiles industry, 2002.
27 Yurdakul, A., Atav, R., Arabacı, A. Tekstil boyacılığında kullanılan sodyumklorür ve sodyumsülfat tuzlarının reaktif boyamadaki performanslarının karşılaştırılması. TÜBİTAK projesi, 2004.
28 Garrett, D.E. Sodium Sulfate: Handbook of Deposits, Processing, Properties, and Use. Amsterdam, Holland: Elsevier, Academic Press, 2001.
29 Addressing Salt Issues in Textile Dyeing Using an ISQ and ARIZ. http://www.triz-journal.com/archives/2000/01/d/, (accessed: 16 April 2004).
30 Atav, R. Pamuklu kumaşların reaktif boyalarla boyanmasında iyon tutucuların renk verimi üzerine etkileri (MSc Thesis). Tekstil Mühendisliği Anabilim, Fen Bilimleri Enstitüsü, Ege Üniversitesi, İzmir, Türkiye, 2005.
31 Yang, Y. Effect of electrolytes on physical interactions in dyeing. Colourage, 2000, 29–34.
32 Namırtı, O. Nanoteknoloji Ürünü Dendrimerler Kullanılarak Pamuk Liflerinin Reaktif Boyalarla Boyanabilirliğinin Geliştirilmesi (MSc Thesis). Tekirdağ Namık Kemal Üniversitesi, Fen Bilimleri Enstitüsü, Tekirdağ, Türkiye, 2013.
33 Environmental Protection Agency (EPA). Best management practices for pollution prevention in the textile industry. http:// infohouse.p2ric.org/ref/02/01099/0109901.pdf. (accessed: 5 April 2013).
34 European Commission Directorate. Reference document on best available techniques for the textile industry, General Jrc Joint Research Centre Institute for Prospective Technological Studies (Seville), Technologies for Sustainable Development European IPPC Bureau, 2002.

35 Onar, N. Katyonik pamuk eldesi ve uygulamaları. Tekstil ve Mühendis, 2006; 13(61), 33–40.

36 Özdoğan, E. Selüloz esaslı liflerin katyonize edilerek boyanma ve baskı özelliklerinin geliştirilmesi (PhD Thesis). Ege Üniversitesi, Fen Bilimleri Enstitüsü, Tekstil Mühendisliği Anabilim Dalı, İzmir, Türkiye, 2003.

37 Uğur, S.Ş., Sarıışık, M., Aktaş, A.H. Katyonizasyon işleminin pamuklu kumaşların bazı özelliklerine etkisi. Tekstil ve Mühendis, 2011; 18, 7–11.

38 Kazan, C. Liflerin kimyasal modifikasyonu yoluyla poliester/pamuk karışımlarının tek banyoda boyanabilirliğini sağlayacak yeni bir yöntem geliştirilmesi (MSc Thesis). Namık Kemal Üniversitesi Fen Bilimleri Enstitüsü, Tekirdağ, Türkiye, 2015.

39 Yang, Y., Carman, E.F. Non-formaldehyde nitrogen-containing fixing agent for direct dyeing. American Dyestuff Reporter, 1996; 10, 39–44.

40 Haroun A.A., Mansour, H.F. Effect of cationisation on reactive printing of leather and wool. Dyes and Pigments, 2007; 72(1), 80–87.

41 Toit, L.C.D., Pillay, V., Choonara, Y.E. Nano-microbicides: challenges in drug delivery, patient ethics and intellectual property in the war against HIV/AIDS. Advanced Drug Delivery Reviews, 2010; 62(4–5), 532–546.

42 Klajnert, B., Bryszewska, M., Dendrimers: properties and applications. Acta Biochimica Polonica, 2001; 48(1), 199–208.

43 Froehling, P.E. Dendrimers and dyes. Dyes and Pigments, 2001; 48(3), 187–195.

44 Zhang, F., Chen, Y., Lin, H., Wang, H., Zhao, B. HBP-NH2 grafted cotton fiber: preparation and salt-free dyeing properties. Carbohydrate Polymers, 2008; 74, 250–256.

45 Zhang, F., Che, Y., Lin, H., Lu, Y. Synthesis of an amino-terminated hyperbranched polymer and ıts application in reactive dyeing on cotton as a salt-free dyeing auxiliary. Coloration Technology, 2007; 123(6), 351–357.

46 http://www.masterorganicchemistry.com/2011/10/21/reagent-friday-sodium-periodate/, (accessed: 05 April 2013).

47 Demir, A., Seventekin, N. Kitin, kitosan ve genel kullanım alanları. Tekstil Teknolojileri Elektronik Dergisi, 2009; 3(2), 92–103.

48 Demir, A., Öktem, T., Seventekin, N. Kitosanın tekstil sanayiinde antimikrobiyal madde olarak kullanımının araştırılması. Tekstil ve Konfeksiyon, 2008; 2, 94–102.

49 Dilsiz, Y.F. Pamuğa kimyasal modifikasyon uygulayarak multicolor efektine ve antibakteriyellik özelliğine sahip fonksiyonel gömleklik kumaş eldesi (MSc Thesis). Namık Kemal Üniversitesi Fen Bilimleri Enstitüsü, Tekirdağ, Türkiye, 2016.

50 Liu, X.D., Nishi, N., Tokura, S., Sakairi, N. Chitosan coated cotton fiber: Preparation and physical properties. Carbohydrate Polymers, 2001; 44(3), 233–238.

51 Paul, D., Chandra Das, S., Islam, T., Siddiquee, Md. A.B., Al Mamun, Md. A. Effect of alkali concentration on dyeing cotton knitted fabrics with reactive dyes. Journal of Chemistry and Chemical Engineering, 2017; 11, 162–167.

52 Sugimoto, T. Neutral-fixing reactive dyes for cotton. Part 2: commercial reactive dyestuffs and their classification. Journal of the Society of Dyers and Colourists, 1992; 108, 497–500.

53 Dystar. Levafix®, Remazol® Reactive dyes catalogue.

54 Petry. Sequestering agent documents.

55 Çınarlı, A., Uca, R. Su kirliliğinin ve iyon tutucuların reaktif boyamaya etkilerinin spektrofotometrik incelenmesi, 2003.

56 Akıncı, R. İyon tutucuların kompleks oluşturma, dispers etme özelliği; boyama üzerindeki etkileri ve spektrofotometrik sonuçları. IX. Tekstil Teknolojisi ve Kimyasındaki Son Gelişmeler Sempozyumu, Bildiri Kitapçığı (48–59), Bursa, 2003.

57 Seventekin, N. Tekstil fabrikalarında su ve analizleri. İzmir, Türkiye: E.Ü. Mühendislik Fakültesi Çoğaltma Yayınları, 1987.

58 Akçakoca, E.P., ve Atav, R. Tekstil terbiyesinde iyon tutucuların önemi ve kullanımı. Tekstil ve Konfeksiyon, 2004; 14(2), 104–108.

59 Bozetto, G. Sequestering agents, Technical Information Service, No:58.

60 Yurdakul, A., Orzel, S., Atav, R. Numune ve üretim ölçeklerinde yapılan boyamalar arasındaki farklılıkların ve nedenlerinin incelenmesi. Proje No: TÜBİTAK-TAM 2003-01, Aralık 2004.

61 Hanxleden, Collishaw, P.S. Reaktif ve dispers boyalar pazar yönlendiriciler ve Dystar'ın cevabı, X. Uluslararası İzmir Tekstil ve Hazır Giyim Sempozyumu, Bildiri Kitabı (315), 2004.

62 Atav, R., *Boyama Teknolojisi I Ders Notları*, Tekirdağ Namık Kemal Üniversitesi, Tekirdağ-Türkiye, 2021.

63 Yurdakul, A. Klasik ve dozajlı boyama sistemlerinin karşılaştırılması. Tekstil ve Konfeksiyon, 1991, 44–51.

64 Dystar. Technical documents (Reactive continuous dyeing).

65 Babar, A.A., Peerzada, M.H., Jhatial, A.K., Bughio, N. Pad ultrasonic batch dyeing of causticized lyocell fabric with reactive dyes. Ultrasonics Sonochemistry, 2017; 34, 993–999.

66 Khatri, Z., Memon, M.H., Khatri, A., Tanwari, A. Cold pad-batch dyeing method for cotton fabric dyeing with reactive dyes using ultrasonic energy. Ultrasonics Sonochemistry, 2011; 18(6), 1301–1307.

67 Çektirme ve pad-batch maliyet. http://www.rftdye.com/index_files/Page417.htm, (accessed: 20 May 2021).

68 Ciba technical documents. Dyeing processes with CIBACRON® C Dyes.

69 Monforts-Dystar. Econtrol for new possibilities in continous dyeing, Technical Brochure.

70 Kumbasar, E.P., Yurdakul, A., Atav, R., Orzel, S. Reaktif boyama sonrası yapılan yıkama işlemlerinin optimizasyonu. Proje No: TÜBİTAK-TAM 2004-09, Haziran 2006.

71 Miljø-og Ligestillingsministeriet (2002). http://www.mst.dk./udgiv/publications/2002/87-7972-009-0/html/kap02_eng.htm#7, (accessed: 20 May 2021).

72 Yurdakul A. Reaktif boyamada yıkama efektinin klasik ve thermo-flush yöntemiyle karşılaştırılması. Tekstil ve Konfeksiyon, 1999; 6, 444.

73 Hoffman, F., Sicdow, K., ve Woydt, M. Washing off reactive dyeings in jets with bath change rinsing and continous rinsing. Melliand International, 1996; 4, 222.

74 Akçakoca Kumbasar, E.P., Atav, R., Yurdakul, A., İnce, E. Reaktif boyamalar sonrası yapılan yıkamaların optimizasyonu. XI. Uluslar Arası İzmir Tekstil ve Hazır Giyim Sempozyumu, Bildiri Kitabı (239–252). Çeşme, İzmir, Türkiye, 26–29 Ekim 2007.

75 Bradbury, M.J., Collishaw, P.S., Moorhouse, S. Optimizing processes in exhaust dyeing on cellulose. American Dyestuff Reporter, 1995; 73.

76 Aniş, P., Eren, H. Examining the effectiveness and the environmental impact of rinsing in reactive dyeing. AATCC Review, 2002; 24, 6.
77 Bradbury, M.J., Collishaw, P.S., Moorhouse, S. Controlled rinsing: a step change in reactive dye application technology. Colourage, 2000; 73.
78 Akcakoca, E.P., Ozguney, A.T., Atav, R. The efficiency of washing agents in the post-dyeing removal of hydrolyzed reactive dye. Dyes and Pigments, 2007; 72(1), 23–27.
79 Burkinshaw, S.M., Katsarelias, D. The wash-off of reactive dyes on cellulosic fibres, Part 4: the use of different alkalis with monochlorotriazinyl dyes on cotton. Dyes and Pigments, 1997; 35, 249.
80 Weible, K-H. The influence of calcium when washing off reactive dyes. Melliand English, 1990; 10, 361.
81 Kahle, U. Test of washing efficiency during washing-off reactive prints. Melliand International, 1997; 3, 162.
82 Tarakçıoğlu, I. Tekstil terbiyesinde yıkama yöntemleri ve makineleri. İzmir, Türkiye: Ege Üniversitesi Tekstil ve Konfeksiyon Araştırma-Uygulama Merkezi Yayını, 1999.
83 Akçakoca, E.P., Atav, R. Siklodekstrinlerin molekül yapısı ve özellikleri. Tekstil ve Konfeksiyon, 2005; 15(3), 131–134.
84 Akçakoca, E.P., Atav, R. Siklodekstrinlerin inklüzyon kompleksleri. Tekstil ve Konfeksiyon, 2006; 16(2), 94–99.
85 Akçakoca Kumbasar, E.P., Atav, R., Özkaya, K., Yurdakul, A. The efficiency of β-cyclodextrin in the post-dyeing removal of hydrolyzed reactive dyes. Tekstil: Journal of Textile & Clothing Technology, 2013; 62(5–6), 235–241.
86 Alat, D.V. Recent Developments in the processing of textiles using enzymes. Colourage, 2001; 48(2), 32–36.
87 Akalın, M., Merdan, N., Koçak, D., Usta, İ. Effects of ultrasonic energy on the wash fastness of reactive dyes. Ultrasonics, 2004; 42(1–9), 161–164.
88 Bradbury, M., Collishaw, P.S., ve Moorhouse, S. Smart rinsing: a step change in reactive dye application technology. Journal of the Society of Dyers and Colourists, 2000; 116, 144.
89 Brennich, W., Beitlich, R. Moderne nachbehandlungsmittel zur verbesserung der nassechtheiten von cellulosefarbungen. Melliand Textilberichte, 2001; 9, 711–716.
90 Çoban, S. Bitim işlemlerinde yumuşak tutum ve yumuşatıcı maddeler. Tekstil ve Konfeksiyon, 1997; 3, 167–173.
91 Nostadt, K., ve Zyschka, R. Softeners in the textile finishing industry. Colourage, 1997, 53–58.

10

Dyeing with Other Dyes

10.1 Indosol Dyes

The dyeing systems available to the dyer for cellulosic fibres are well-known and are classified as direct, reactive, sulphur, vat and azoic dyes. Undoubtedly, the simplest technique is, of course, direct dyeing, but with important fastness limitations. It therefore seems obvious that if direct dyes can be used to achieve acceptable fastness properties, by a suitable aftertreatment fixation, production cost savings should be realised when compared with the other high-fastness dyeing systems. In this context, the term "reactant fixable dyes" refers to direct dyes that are suitable for aftertreatment with certain developed products to give exceptionally high wash fastness.[1] A dye group with the trade name Indosol has been introduced to the market by the Sandoz company for the purpose of obtaining dyeings with very good wet fastness values with direct dyes with the help reactive fixing agents.[2] The range of Indosol SF dyes consisted of twelve dyes based on copper complexes which give good light fastness when post-treated with suitable auxiliaries.[3] Indosol dye gamut is limited, a red Indosol dye is not available, mostly blue-coloured dyes are present.[2] The dye gamut is insufficient in the range from yellow to violet. Although the nature of very few of these dyes has been disclosed, CI Direct Violet 66, marketed as Indosol Violet SF-B gives us an idea of the structure of these dyes (Figure 10.1).[3]

10.1.1 Fastness Properties of Indosol Dyes

Indosol dyes have very good standard wet fastness values as well as good chlorine and perborate fastness values. Thus, in the dyeing of cellulose fibres, fastness values that could only be achieved with vat and reactive dyes in the past can be achieved with this dye class.[2]

10.1.2 End-uses of Indosol Dyes

If the material dyed with vat or reactive dyes is to undergo a crease-resistant treatment afterwards, then it is much more economical and easier to work with Indosol dyes. After dyeing with these dyes, the tensile and friction strength of the fabric decreases, so it is impractical to use Indosol dyes for products such as towels or products with low strength. Indosol dyes have not found a significant use in knitwear due to the difficulty of properly

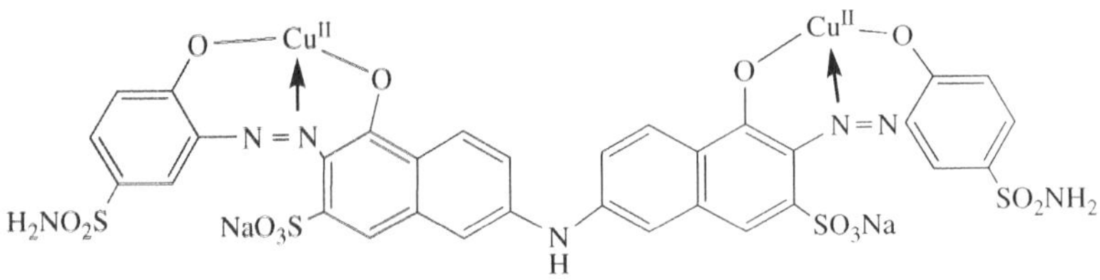

Figure 10.1 Structure of Indosol Violet SF-B (CI Direct Violet 66). Ref. [3]/Society of Dyers and Colourists.

impregnating the fabric without stretching it too much, the decrease in the ability of underwear to absorb perspiration and the slight hardening of the handle.[2]

10.1.3 Dyeing with Indosol Dyes

In the Indosol system, first dyeing is done as with a normal direct dye, then the rinsed and dried fabric is impregnated with a reactive fixing agent solution called Indosol CR, dried and condensed. The reactive fixing agent used in this system is a polyfunctional (polyvalent) compound and has a cationic structure.[2] Its structure consists of three components:

1. A polybasic condensate consisting of diethylenetriamine and an amide such as cyanamide, dicyandiamide, guanidine or biguanide.
2. A N-methylol precondensate of the dicyandiamide-formaldehyde type.
3. An acid catalyst such as lactic acid.[3]

As a result of such a molecular structure, it acts in three different ways that complement each other:

1. By forming bridging bonds between cellulose macromolecules, it not only binds itself to the fibre with covalent bonds but also reduces the swelling of the fibres (therefore, the pores that allow the dyes to enter and exit the fibres).
2. Since it has a cationic structure, it blocks the water-soluble anionic ($-SO_3^-$) groups in the dye molecules.
3. By reacting with dye molecules, it can bind to dye molecules via covalent bonds over the reactive group.[2]

First, as with a normal direct dye, the dyeing is started at around 40 °C and the boiling temperature is reached within 30 minutes, and if necessary, 0–15 g/L calcined sodium sulphate (or common salt) is added in 2–3 portions. After dyeing is continued for 45–60 minutes at the boiling temperature, the levelness of the dyeing is checked, and if sufficient levelness is achieved, the dyeing liquor is cooled to 80–90 °C within 15 minutes and the dyeing is finished. If the levelness is not good, it is useful to continue the dyeing at boiling temperature for a while and wait for the dyeing to become level by migration. This is because it is not possible to level the dyeing after the fixation process. At the end of the dyeing, the material must be thoroughly rinsed and dried. Adding 1–5 g/L of Indosol CR to the final rinse water is beneficial in terms of preventing migration of the dye that has not yet been fixed.[2]

After Indosol CR is dissolved in cold water and diluted to the desired concentration, other substances are added to the liquor in diluted form.

80–130 g/L Indosol CR
10–30 g/L Cationic or non-ionic suitable softener
20–40 g/L Additive (polyethylene dispersion)
10–30 g/L Additive (It can be omitted if desired. It can be based on polyvinylalcohol or silicone elastomer.)
1 g/L Cationic or non-ionic wetting agent
pH 4–4.5 (Addition of 0.5–1 g/L acetic acid is beneficial.)
Pick-up: 70–100%
Drying: at 100–120 °C
Condensation: 30 seconds at 180 °C or 1 minute at 170 °C or 3 minutes at 160 °C.

If suitable equipment is available, drying and condensation can be done in a single step if desired. In this case, passing the fabric at 180 °C for 60 seconds is usually sufficient.[2]

Indosol CR acts as a reactant-type anti-crease agent as well as dye fixative. As a result, the fabric also gains good crease-proof, non-swelling and non-shrinking properties. However, as in all crease-proof finishing processes, the tensile and abrasion strength of the fabric decreases and formaldehyde is released. However, if a suitable softener is added to the liquor, there is no adverse effect on the handle of the product.[2]

The polybasic condensate component in the structure of Indosol CR was subsequently marketed alone as Indosol E-50 (S) for post-treatment of Indosol dyeings by exhaust method. In order to achieve even better fastness to washing, a third type of Indosol after-treatment agent was introduced to the market. Indosol EF forms the dye-aftertreatment complex in the same way as Indosol E-50, but this complex is more securely attached to the fibre thanks to the covalent bond between the auxiliary and the fibre, similar to the dye-fibre bonds in conventional reactive dyes. Indosol E-F is designed for application by exhaust method, and final fixation in alkaline solution at 40 °C is required.[3]

When the boiling method with dimethylformamide (DMF):water 1:1 mixture is utilized to extract the dye that is not covalently bonded to the fibres from the fibres is applied to Indosol dyes, dye does not bleed into the DMF:Water mixture even after a long time (2–8 hours) of boiling. This indicates that the dye molecules are also attached to the fixing agent by covalent bonds. Removal of the fixed dye from the fibres can only be achieved by treating the material in an acidic (hydrochloric acid or phosphoric acid) bath containing auxiliaries for 30 minutes at 80 °C. This shows that the covalent bonds between dye/fixing agent and fibre/fixing agent have good hydrolysis resistance.[2] Compared to reactive dyes, Indosol dyes provide economic advantages due to reduced dye, chemical and water consumption and processing time.[3]

Indosol SF dyes have also proven to be highly suitable for dyeing by pad-batch method if urea and appropriate auxiliaries are used to aid solubility and diffusion during the 12-hour holding period.[3]

10.2 Diazotisation Dyes

Most of the direct dyes are azo dyes, and the biggest drawback of these dyes is their poor wet fastness. During the studies carried out to improve their wet fastness, diazotisation dyes were developed. Basically, diazotisation dyes are also substantive dyes, but wet fastness values can be increased by the processes performed after dyeing (diazotisation and coupling).[2] Direct dyes containing aromatic amino groups are diazotised on the fibre after dyeing and then coupled with a developer (a phenol, naphthol or aromatic amine). Wet fastness in particular is improved by this increase in molecular size, and the colour shade is also changed.[4]

10.2.1 Fastness Properties of Diazotisation Dyes

As a result of the processes applied in dyeing with these dyes, although the number of water-soluble groups does not increase, wet fastness values also improve as the molecular size increases. However, the wet fastness values of diazotisation dyes are inferior to the wet fastness values of vat, naphthol, reactive or sulphur dyes.[2]

10.2.2 End-uses of Diazotisation Dyes

In the past, diazotisation dyes were widely used in terms of obtaining dyeings with very good wet fastness values in a cost-effective manner.[2]

10.2.3 Dyeing with Diazotisation Dyes

The material to be dyed is dyed with a special dye containing one or more amino groups that can be diazotised (diazotisation dye) according to the principles described for the direct dyes. In a second bath, the amino groups of the dye attached to the fibre are diazotised. In a third bath, the diazonium groups formed as a result of diazotisation are coupled with a suitable phenol, naphthol or aromatic amine (Figure 10.2). As a result of diazotisation and coupling, there is usually a change in the colour shade of the dyeing, typically a darkening. On the other hand, although the number of groups providing solubility in water does not increase, wet fastness values also improve as the molecular size increases.[2]

The material to be dyed is first dyed with diazotising dyes according to the principles and methods applied to a normal direct dye, and after being washed thoroughly, it is processed in a cold diazotising bath for 15–20 minutes. Depending on the depth of the dyeing, the diazotisation bath contains:

1.5–3% Sodium nitrite
5–1.5% Hydrochloric acid (20 °Bé) or
3–5% Sulphuric acid (96% = 66 °Bé).[2]

After diazotisation, the material is thoroughly rinsed. Since diazotised dyeing is sensitive to alkalis, care must be taken that the water used for rinsing is not alkaline. It is recommended to perform coupling immediately at the end of the rinsing, because diazotised dyes are very sensitive to air and especially to light. As a result of sunlight or drying, diazonium groups can be partially broken down, causing a stained, uneven dyeing.[2]

Coupling is done in the cold within 20–30 minutes with a suitable phenol, naphthol or amine. The substance that reacts with diazonium groups during this coupling is called a developer. Depending on the type of developer used and the depth of the dyeing, 0.5–1.5% developer is added to the coupling bath. Since the dissolution abilities of the various

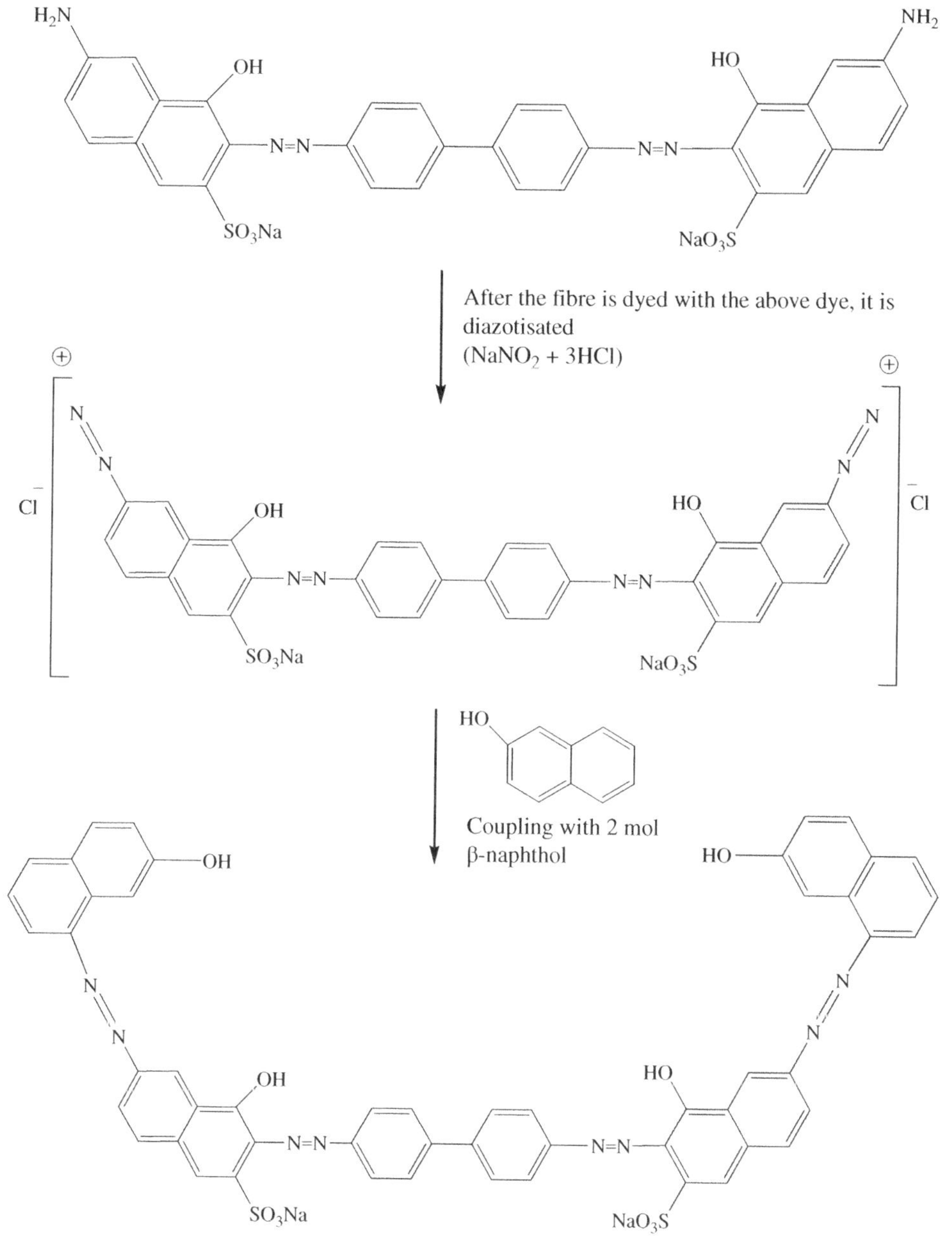

Figure 10.2 Mechanism of dyeing with diazotising dyes. Ref. [2].

developers vary greatly, the dissolution conditions specified in the catalogues must be observed.[2]

In case of dyeing large lots, diazotisation and coupling (development) can also be done continuously. In this case, the fabric is treated successively with sodium nitrite and hydrochloric acid to prevent the formation of irritating nitrogen oxides.[2]

10.3 Developing (Naphthol) Dyes

Although diazotisation dyes have a very good washing fastness due to their large molecules, their washing fastness is still limited because they contain sulpho groups. Therefore, an azo dye with good washing fastness should not contain water-soluble groups. On the other hand, since the dye that does not contain such a group is insoluble in water and has no affinity for the fibres, dyeing cannot be achieved with it. In order to solve this problem, if the water-soluble components are brought together and coupled on the fibre, an azo dye that does not contain water-soluble groups is formed in the fibre, which has good wet fastness values. An example of this is given in Figure 10.3.[2]

As can be seen in Figure 10.3, the molecules of the naphthol dyes are formed as a result of the coupling of a naphtholate component and a diazonium salt in the fibres. Accordingly, dyeing with naphthol dyes consists of the following stages:

1. Grounding with naphtholate
2. Intermediate processing
3. Treatment with diazonium salt and coupling of naphtholate and diazonium salt to develop dye
4. Aftertreatment.[2]

Azoic components are defined in the Colour Index (CI) as those used to produce "insoluble azo dyes in situ, usually on a textile substrate". Nowadays the substrate is almost always cotton; corresponding azoic components were formerly available for dyeing the ester fibres, but these are now obsolete because of environmental hazards and time-consuming application methods. Azoic coupling components are mostly naphthols and azoic diazo components are primary amines or stabilised diazonium compounds, often called fast bases or salts, respectively. So-called azoic compositions are described in the CI as "mechanical mixtures of an azoic coupling component and an azoic diazo component". The first patent for the production of insoluble azo dyes on cotton was granted in 1880 to Thomas and Robert Holliday of Read, Holliday & Co., for the preparation and coupling of 2-naphthol with diazotised naphthylamines to obtain red or bordeaux colours.[5]

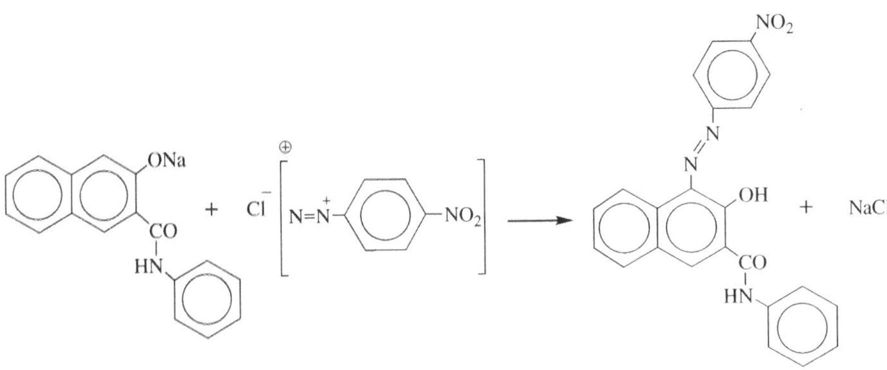

Figure 10.3 Formation of the naphthol dye. Ref. [2].

Some of the advantages of azoic dyeing are the following:

- They are particularly strong in the orange, red and bordeaux ranges, but the range also includes dark blue and black.
- Dyeings can be produced in certain bright full depths.
- In general, the standard of fastness is high.
- Most combinations are dischargeable.
- They are economically viable.

The main disadvantages of this dyeing method include the following:

- Difficulties in handling the component chemicals under typical dyehouse conditions, particularly the diazotising of arylamine fast bases.
- Complicated and time-consuming application procedures.
- The limitations of hue selection imposed by the need to avoid the crosscoupling that occurs if more than one azoic combination is applied.[5]

10.3.1 Fastness Properties of Developing Dyes

Since the naphthol dyes formed in the fibre do not contain groups that provide water solubility in their molecules, their fastness to aqueous liquors is good. Although not equal to vat dyes, their fastness to washing, mercerisation and partial bleaching is very good. Some suitable combinations can qualify for the label "Indanthren". Colour shades are vivid.[2]

The dark shades exhibit excellent light fastness. However, in medium and pale shades, light fastness is good to fair. Like many other dyes, the light fastness of azoic-dyed ground fabric or yarn is also affected by environmental humidity. Unlike sulphur dyes, most of the azoic pigments withstand chlorine bleaching. Most of the azoic combinations have excellent perspiration fastness as well.[6]

Although rubbing fastness values can be improved to some extent by paying attention to dyeing conditions, using suitable dispersing agents and performing a good aftertreatment, they are in many cases low. Some combinations are unstable to trichloroethylene, paraffin, wax and some synthetic softeners.[2]

10.3.2 End-uses of Developing Dyes

Due to the above-mentioned features, it was a dye class that was widely used in the dyeing of clothes, aprons, shirts, bed covers and flags that required frequent washing in the past. Today, naphthol dyes have lost their importance.[2]

10.3.3 Dyeing with Developing Dyes

10.3.3.1 Dyeing by Exhaust Method

10.3.3.1.1 Grounding with Naphtholate

In naphthol dyes, the initial difficulty was finding the naphthol component with affinity for cellulose fibres. With the discovery of Naphthol-AS (2-hydroxy 3-naphthoic acid anilide), this problem was solved. Now, besides naphthol-AS, many compounds that are derivatives or analogues can be used as naphthol components. For this reason, naphthol dyes are also called "Naphthol-AS" dyes.[2]

The affinities of the various naphthols available in the market for cellulose fibres vary widely. Some show almost no affinity, while others show high affinity.[2] Naphthols with high substantivity are generally preferred for application by the exhaust method, while naphthols with low substantivity are used for continuous applications. The affinity of low- or medium-substantivity naphthols can be improved by adding electrolytes during application by the exhaust method. The affinity of most naphthols decreases with increasing application temperature. Since the affinity of naphthol decreases with increasing temperature, applications by the exhaust method are normally carried out at 30–40 °C. If better wetting or penetration is required, the temperature can be increased to 50 °C. Since the stability of the naphthol solution and the naphtholated material is limited, it is essential that the naphthol solution is prepared immediately before use. The processing time usually required is 20–30 minutes.[6]

In the naphthol dyes, the dye in the desired colour and nuance is obtained in the fibres during dyeing as a result of the combination of two separate components. Thus, the colour and nuance of the dyeing depend on the type of naphtholate and diazonium salt chosen. Furthermore, since certain naphtholates give particularly good results in obtaining certain colours, they are grouped by the dye manufacturers according to the colours they give. However, a red naphthol does not necessarily have to be used in red and similar colours. A green colour can also be obtained by combining it with a suitable diazonium salt. However, if a green naphthol is selected, the number of green-colouring diazonium salts combined with it increases considerably, making it easier to find a combination with the desired nuance and fastness. Therefore, a suitable naphtholate/diazonium salt combination must be selected before dyeing with naphthol dyes. While making this choice, the following matters should be taken into account:

- Desired colour and nuance
- Dyeing method (exhaustion, impregnation)
- Fastness
- Price
- Compatibility of the naphtholate and diazonium salt.

Not every combination of naphtholate and diazonium salt works well. As the nuances and fastness values obtained from combination to combination vary, for example, the combination of a naphtholate and diazonium salt with very different coupling pH regions causes various technical difficulties. For this reason, dye factories also specify in their catalogues which of the various combinations of their naphthol and diazonium salts give very good results, which give moderate results, and which are not recommended.[2]

Sodium naphtholate is the sodium salt of naphthol compounds that form the naphthol dye by coupling with the diazonium salt. Since the naphthol compound dissolves in the caustic soda, naphtholate is formed during dissolution. It should be noted that at any stage before coupling (development of the dye), naphtholate must not revert to naphthol, because the part that reverts to naphthol does not undergo a coupling reaction.[2]

Dissolution of naphthols can be done according to "cold dissolving" or "hot dissolving" methods. Since grounding with naphtholate, which is performed by exhaust method, is done in the cold, the "cold dissolving" method is preferred in dyeing according to this method. Ethyl alcohol is required in cold dissolving, and some companies have also brought cold-soluble naphthols to the market that do not require ethyl alcohol (or use a small amount). In cold dissolving, the naphthol compound is pasted with technical ethyl alcohol in the amount specified in the catalogues. Then the required amount of caustic soda (38 °Bé) is added and diluted with cold water and dissolved by stirring. After the addition of formaldehyde (33%), the solution is left to stand for 5–15 minutes.[2]

The naphtholate solution prepared as described above is added slowly to the liquor containing 6 mL/L caustic soda (38 °Bé) (4 or 8 mL/L in some naphthols), 2–6 g/L protective colloid and 0–30 g/L common salt while stirring well. Grounding with naphtholate liquor is generally done at 30 °C for 30 minutes. While there is no harm in a slightly longer time, it may be recommended to ground for 45 minutes during bobbin dyeing, especially on apparatus with poor liquor movement.[2]

Depending on the type of naphthol, 0–30 g/L common salt or Glauber's salt is added to the grounding liquor. The purpose of salt addition is to increase the substantivity of naphtholates, thus increasing the amount of naphtholate uptake by the fibre. Thus, the amount of naphtholate remaining in the liquor, which can be mechanically bonded to the fibres, is reduced. In this way, the rubbing fastness is also better. There is no need to add salt when grounding with naphtholates, which show a high degree of substantivity.[2]

Most naphtholates are sensitive to humid air. On the one hand, carbon dioxide in the air, as a weak acid, can cause some of the naphtholate to be converted to naphthol, which itself does not undergo a coupling reaction. On the other hand, as shown in Figure 10.4, dinaphthols that do not give a coupling reaction can be formed from naphthols under the influence of atmospheric oxygen.[2]

Therefore, with a few exceptions, formaldehyde is added to the naphtholate solution when working by exhaust method. Formaldehyde forms 1-methylol compounds with naphtholates, as shown in Figure 10.5, which prevents the conversion of naphtholates to naphthol or the formation of dinaphthols. In the next step, formaldehyde is released again during the coupling (development) reaction.[2]

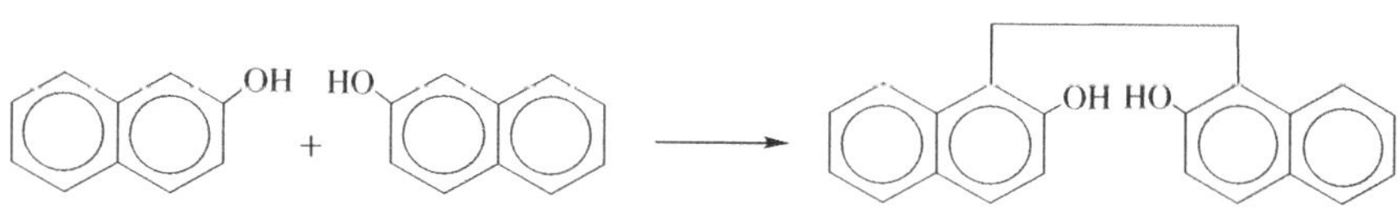

Figure 10.4 Formation of dinaphthols from naphthols under the influence of atmospheric oxygen. Ref. [2].

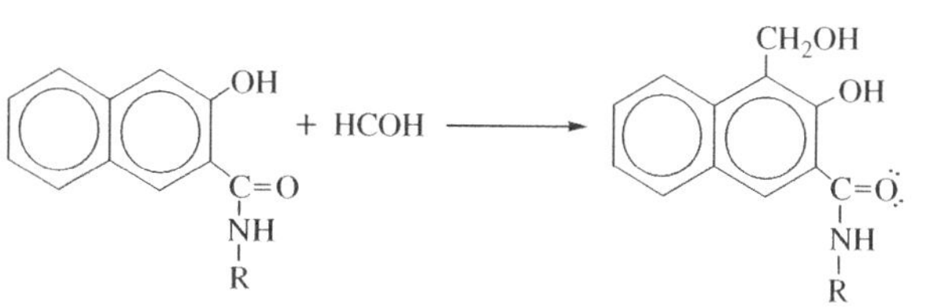

Figure 10.5 1-methylol compound formation of formaldehyde with naphtholate. Ref. [2].

10.3.3.1.2 Intermediate Processing

After the material to be dyed is grounded with naphtholate, excess naphtholate solution must be removed from the material. Otherwise, when the grounded material is introduced into the development bath, a water-insoluble naphthol dye is formed in the liquor, and some of it is mechanically bonded to the surface of the material to be dyed. Since this part, which is mechanically bonded, will be removed by friction, a dyeing with poor rubbing fastness is obtained.[2]

Depending on the type of material being dyed, the excess naphtholate solution can be removed by squeezing, suction or centrifugation. It should be noted that naphtholates are not very resistant to air (despite the addition of formaldehyde), especially in suction processes applied to packages, and therefore the suction process should not be prolonged. If a centrifuge is used, it is useful to rinse the centrifuge with naphtholate solution before loading the material. During squeezing and centrifugation, the extracted solution can be collected and reused in the naphtholate liquor.[2]

Naphtholate-grounded material is very sensitive to factors that can cause hydrolysis, especially acid or water, because naphthols formed as a result of hydrolysis of naphtholates do not give a coupling reaction. For this reason, it is necessary to ensure that acid vapours or water drops do not contact the grounded material in any way; otherwise, the shade of the dyeing will be lighter, as the coupling will be less at the points they contact. For the same reason, one part of the material should be prevented from drying out more than the other part, since the equilibrium shown in Figure 10.6 will shift to the right and left to varying degrees depending on the amount of water, resulting in an uneven dyeing.[2]

In addition to the possibilities mentioned above, excess naphtholate can be removed by rinsing with a solution containing caustic soda and common salt. Some dye factories specifically recommend this method.[2]

10.3.3.1.3 Treatment with Diazonium Salt

Most diazonium salts are unstable at room temperature. Although specially stabilised diazonium salts have been introduced to the market by dye manufacturers, they are not always recommended because they are expensive and do not last long in a very hot and humid climate. Generally, an amine compound suitable for diazotisation is selected, and the diazonium salt is obtained by diazotising in the dyehouse. Since diazonium salts are not very durable, they should be used immediately after they are prepared.[2]

Amine compounds suitable for diazotisation are sold in the market under the name "Base". The most suitable diazotisation recipe for each Base is specified in the catalogues. Generally, Bases are dissolved in hydrochloric acid. In this case, the Base is pasted with water and the specified amount of hydrochloric acid (20 °Bé) and dissolved by adding

Figure 10.6 Conversion of naphtholate to naphthol by hydrolysis of water. Ref. [2].

cold water. Sodium nitrite solution dissolved in cold water is added with stirring. Since diazonium salts are sensitive to heat, the diazotisation temperature should not exceed 5–15 °C. The decomposition of diazonium salts is evident from the formation of gas bubbles and foam. Besides heat, light and metals such as copper, tin and iron facilitate the breakdown of diazonium salts. Therefore, copper and iron vessels are not used during the production and storage of diazonium salts. The diazotisation time is 15–30 minutes.[2]

During diazotisation reactions, excess acid must be present in the medium. Otherwise, there will be unpleasant side reactions. After the diazotisation is over, the excess acid is neutralised or buffered, and liquor is adjusted to the optimum pH values for the coupling reaction. The coupling reaction of naphtholate and diazonium salt depends on the pH of the medium. For example, in a strongly alkaline environment, diazonium salts can transform into the *trans*-diazotate form, which does not give a coupling reaction. Generally, coupling reactions occur best around pH 4–7. The most suitable coupling pH for each naphthol and Base is specified in the catalogues. Especially when working with Bases with high coupling ability, good results can be obtained with acetic acid for pH around 4–5 and sodium acetate/acetic acid combination for pH around 5–5.7. Disodium phosphate/monosodium phosphate, on the other hand, is a buffer mixture used to adjust the pH to 6–6.8, especially when working with Bases that do not have a strong coupling ability. Sodium bicarbonate/chromacetate S grain is a recommended combination, mostly for buffering the coupling pH to 7. In addition to the above-mentioned additions, 20–50 g/L common salt can be added to the development liquor. When the naphtholate-ground material is introduced into the development liquor, some naphtholate will tend to bleed into the liquor. The purpose of adding salt is to reduce the amount of naphtholate bleeding into the liquor.[2]

10.3.3.1.4 Aftertreatment

The formation of some naphthol dye in the liquor and its mechanical bonding to the material cannot be completely avoided. For this reason, the material coming out of the development bath should be washed well. When ready-made diazonium salts are used instead of Base, the first aftertreatment is acidification. Ready-made diazonium salts contain $Al_2(SO_4)_3$ salts as stabilisers. The hydroxides of these salts may precipitate during development, which requires acid treatment to remove them from the material.[2]

- **Acidification:** The material is processed at 30 °C for 10 minutes with a liquor containing 3–4 mL/L 20 °Bé hydrochloric acid. Then it is rinsed thoroughly with cold water. No acidifying is required when working with Bases.
- **1st soaping:** To remove the mechanically bound dye, processing is carried out at 60 °C for 10 minutes with a liquor containing:

  1–3 g/L washing agent
  1–2 g/L calcined soda ash
  1–2 g/L complexing agent.

  Then the material is rinsed thoroughly with hot and cold water.
- **2nd soaping:** The aim here is to ensure the growth (aggregation) of the dye particles in the material. As the total surface area all dye particles will be reduced as a result of the small particles combining to form larger particles, light and wet fastness values

improve, and a change in the colour shade of the dyeing is observed. The temperature and duration of the second soaping are different for each combination and are indicated in the dye catalogues. Generally, the temperature varies between 80 and 100 °C and the time varies between 5 and 20 minutes. Liquor contains:

1 g/L dispersing agent
1 g/L soda ash
1 g/L complexing agent.

After the second soaping, the material is rinsed with hot and cold water, and the dyeing is finished.[2]

10.3.3.2 Dyeing Impregnation Method

10.3.3.2.1 Grounding with Naphtholate

Naphtholates to be used in continuous dyeing should have no or low substantivity. In this way, the concentration of the liquor does not undergo a significant change during the impregnation in the padder, and it is possible to obtain dyeings with the same depth at the beginning and the end. Since the substantivity of naphtholates decreases with increasing temperature, impregnation is usually done at 95 °C. Since grounding is done with hot liquor, the dissolution of naphthol in continuous dyeing is performed by the "hot dissolution" method. For this purpose, the naphthol compound is poured into water at at least 90 °C containing caustic soda (38 °Bé) and protective colloid and dissolved by boiling. Some naphthols are not recommended to be dissolved while hot, so they are dissolved in the cold and poured into a liquor containing 5–10 mL/L 38 °Bé caustic soda.[2]

As mentioned before, naphtholates are sensitive to air when they have moisture, and therefore it is useful to add formaldehyde to the grounding liquor. In the continuous dyeing method, there is no need to use formaldehyde since the material grounded with naphtholate is immediately subjected to an intermediate drying, that is, it is not kept damp. In fact, it is harmful to use formaldehyde in continuous dyeing since the impregnation temperature is high and formaldehyde will cause undesirable side reactions at high temperatures. Grounding with naphtholate can be done on two or three roller pads. It is useful to have the vessel of the padder quite small so that there are no large concentration changes, as the liquor will be renewed in a short time. After the material passes through the padder, some time is required for the naphtholate taken to diffuse into the material. Therefore, after exiting the padder, it is recommended to pass the material through an air passage for 20 seconds before entering the drying unit.[2]

10.3.3.2.2 Intermediate Drying

Intermediate drying of naphthalate grounded material is usually done in a hot-flue. Roller dryers can also be used when hot-flue is not available. In this case, the critical point is that the temperature of the first two rollers does not exceed 60 °C and the material is passed between the rollers in such a way that it contacts the rollers on both surfaces. If a frame dryer (stenter) is to be used, pin frames should be preferred. Clip frames leave clip marks. The drying temperature is adjusted according to the fabric speed. The important thing is to prevent over-drying. Up to 8% moisture should remain in cotton material and up to 14% in viscose.[2]

10.3.3.2.3 Development

Either the Base is taken and converted to diazonium salt by diazotisation as described before, or the ready-made diazonium salt is used. The usage rate of ready-made diazonium salt in continuous processes is higher than that in dyeing performed by the exhaust method. Here, too, the pH of the development liquor needs to be adjusted. Generally, sodium acetate/acetic acid combination gives good results. The material coming out of the development liquor is passed through an air passage of at least 30 seconds and then through a hot water passage (overflow washing) at 70–80 °C.[2]

10.3.3.2.4 Aftertreatment

Soaping and rinsing can be done continuously in open-width washing machines. Generally, a washing machine with 7–8 compartments is sufficient. An example of post-processing is given below.

- **1st compartment:** Hot water passage at 70–80 °C
- **2nd compartment:** Rinsing with soft water of 60 °C
- **3–6th compartments:** Soaping with a boiling liquor containing
  1 g/L Soda ash
  1–2 g/L Dispersant washing agent
  1–2 g/L Complexing agent
- **7th compartment:** Rinsing with 60 °C water
- **8th compartment:** Rinsing with cold water.[2]

10.4 Oxidation Dyes

The basis of oxidation dyes is to treat the material to be dyed with a suitable aromatic amine and then oxidise this amine on the material to form the dye. Among the dyes belonging to this group, the most important is aniline black. This dye is formed as a result of the oxidation of anilinium chloride in an acidic environment. Diphenyl black is also included in this group and is formed as a result of the oxidation of p-amino-diphenylamine.[2] The importance of oxidation dyes has declined rapidly because aniline and other aromatic amines, as well as bichromate used for oxidation, are toxic.[4]

10.4.1 Fastness Properties of Oxidation Dyes

The colours produced by aniline black are characterised by a full bluish-black shade and excellent fastness.[4]

10.4.2 End-uses of Oxidation Dyes

Special aromatic amines (e.g. water-soluble diphenylamine derivatives) have been released to the market by some dye factories due to the complexity of dyeing with aniline or diphenyl black base and the high risk of faulty dyeing. They can be used especially in printing and also in continuous dyeing. Black dyeings obtained with oxidation dyes can be labelled as "Indanthren" due to their high fastness. In the past, they were widely used in ground dyeing of prints, dyeing of black work clothes and linings.[2]

10.4.3 Dyeing with Oxidation Dyes

10.4.3.1 Dyeing with Aniline Black

Aniline black dyeing can be done according to three different methods.

a. ***Single bath black:*** The material to be dyed is treated in a bath containing aniline salt (anilinium chloride), inorganic acid and potassium bichromate.
b. ***Oxidation black:*** The basis of this method is to impregnate the product to be dyed several times with a liquor containing aniline salt, oxidation salt, oxidation agent (e.g. sodium chlorate) and oxidation catalyst (e.g. ammonium vanadate or copper (II) sulphate). Then the material is dried in hot-flue and oxidised. The dried and oxidised material is treated with a sulphuric acid and potassium bichromate solution to complete the oxidation.
c. ***Vapour black:*** The principle of this method is that the product to be dyed is impregnated with a liquor containing aniline salt, sodium chlorate and potassium ferrocyanide, then dried in hot-flue and oxidised in a continuous steamer. Finally, it is necessary to treat it with potassium bichromate to complete the oxidation.[2]

10.4.3.2 Dyeing with Diphenyl Black

Although it is a bit more expensive than aniline black, it is advantageous because there is no risk of the colour shade turning green over time, and it causes less damage to the fibres.[2] To produce diphenyl black (CI Oxidation Base 2, CI 76085), cotton is impregnated with a solution containing diphenylamine, acetic acid, lactic acid, sodium chlorate, aluminium chloride, copper sulphide, ammonium vanadate and some tragacanth.[7] The product is then dried in a hot-flue and then steamed.[2]

10.5 Vat-leucoester (Indigosol) Dyes

Vat dyes are not soluble in water but can be rendered soluble only after special vatting process. In order to overcome this difficulty, some vat dyes have been rendered water-soluble and introduced to the market in a form resistant to air oxidation. These are generally called vat-leucoester dyes.[2] Solubilised vat dyes are commonly known as indigosol dyes as BASF (Germany) first introduced these to the market with the brand name – indigosol.[7] Vat-leucoester dyes are produced in factories by treating the reduced form of normal vat dyes with chlorosulphonic acid and then converting it into sodium salt using caustic soda. In other words, these are the sodium salts of the esters formed by reaction between sulphuric acid and the leuco form, which is the reduced form of vat dyes.[2] In another method, parent vat dye is treated with copper-pyridine complex and finally converted into the sodium salt of the leuco sulphate ester.[7] Preparation of vat-leucoester dye is shown in Figure 10.7.

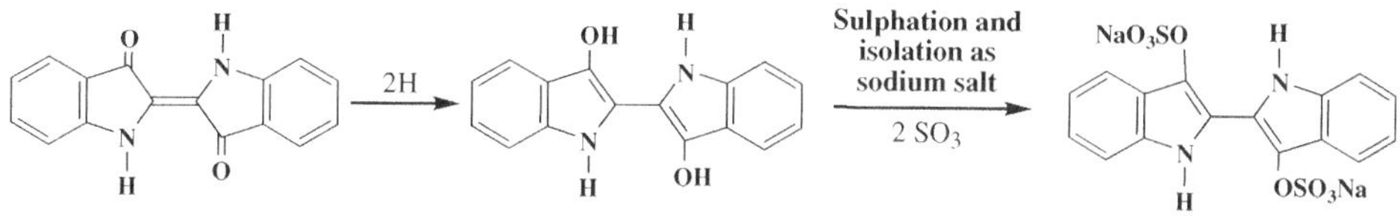

Indigo (CI 73000) **Leuco indigo (CI 73001)** **Indigosol O (Durand and Huguenin) Anthrasol (Hoechst)**

Figure 10.7 Preparation of vat-leucoester dye. Ref. [8]/John Wiley & Sons.

Vat-leucoester dyes are soluble in water. The material to be dyed is treated with this solution. By saponification and oxidation, the dyes that have penetrated into the fibres are converted back into water-insoluble vat dye pigments. Vat-leucoester dyes are easy to work with because they are water-soluble. However, their main advantage is their lack of affinity for cotton fibres. Therefore, unlike vat dyes, it is easier to achieve uniform dyeing with them, especially in light shade dyeing.[2]

10.5.1 Fastness Properties of Vat-leucoester Dyes

The advantages of dyeing with vat-leucoester dyes are not only good levelness and penetration but also excellent fastness properties associated with vat dyes.[4]

10.5.2 End-uses of Vat-leucoester Dyes

One of the major drawbacks and difficulties of working with vat dyes is that it is difficult to obtain uniform dyeing with them, and it is only possible by special methods. For this reason, indigosols, which are more expensive than vat dyes, can often be used for dyeing fabrics in pale colours which are difficult to dye properly. As they are expensive, they are not preferred for dyeing in dark colours.[2] They are mainly used for dyeing high-quality products made of cellulose fibres in light shades and polyester/cellulose blends in medium shades.[4] Vat-leucoester dyes can also be used in printing.[2] As a result of their high price, low substantivity and toxicological problems during production, the importance of the leuco esters of vat dyes has decreased.[4]

10.5.3 Dyeing with Vat-leucoester Dyes

Cellulose fibres can be dyed with vat-leucoester dyes by both exhaust and impregnation methods. Since most of the vat-leucoester dyes can be easily developed (converted into vat dyes), their development can be carried out in cold. Some of the difficult-to-develop ones should be treated at higher temperatures (70 °C). Although the affinity of these dyes for cellulose fibres is generally low, it can vary greatly among themselves. The temperatures at which they exhibit maximum affinity (maximum uptake by the fibres) also differ. Most of them exhibit maximum uptake at low temperatures (25–30 °C), while some are best absorbed at hot temperatures (55–70 °C). Addition of salt increases the amount of dye taken up by fibres.[2]

Since vat-leucoester dyes are soluble in water, it is not necessary to boil them during the preparation of the liquor. They can be dissolved by pouring water heated to a maximum of 70 °C. It is necessary to protect their solutions from strong rays and the effects of acids and oxidising agents. Similarly, material impregnated with dye solution, but not yet developed, should be protected from sunlight and acid vapours.[2]

In the development of indigosols, sulphuric acid is used for saponification and sodium nitrite is usually used as oxidant. It is not common to use chemicals other than sulphuric acid and sodium nitrite for the development of these dyes. In some cases, sodium bichromate or iron (III) salts can also be used as oxidising agents. For example, these can be used instead of sodium nitrite in cases where partial insolubility occurs due to sodium nitrite in the liquor when working at high dye concentration. Although the development with sulphuric acid and potassium bichromate had gained a certain importance in the dyeing of woollen products with vat-leucoester dyes, given that there is no significant use of vat-leucoester dyes in wool dyeing today, the development with bichromate has also lost its importance. The use of potassium bichromate instead of sodium nitrite in the dyeing of cellulose fibres does not provide much benefit other than preventing the formation of irritating nitrous gases, but it causes drawbacks such as decreasing the dyeing efficiency, not being suitable for use in all vat-leucoester dyes, causing embrittlement of the fibres and dulling of the dyeing, and releasing of chromium ions into the wastewater.[2]

The possibility of using other acids instead of sulphuric acid has also been investigated in order to reduce the corrosion hazard, especially in the continuous dyeing method. The best results are obtained by using oxalic acid. In this case, a liquor containing 25–35 g/L of oxalic acid at 80 °C is used instead of a developing liquor containing 20 mL/L of sulphuric acid at 70 °C. Formic acid also gives very good results, and if it is used, the concentration should be 35 mL/L (for 85% formic acid) and the temperature of the developing liquor should be 80 °C.[2]

10.5.3.1 Dyeing by Exhaust Method

10.5.3.1.1 Single Bath Nitrite Method

In this method, which is mostly preferred for dyeing in light shades, dyes that can be easily developed and are well absorbed by the fibres are used. As the name suggests, in this method, the grounding with the dye and the development of the dye are carried out in the same bath. After adding 0.5–1 g/L wetting agent, 1–2 g/L dispersing agent and 1 g/L calcined soda ash to the liquor heated up to 40–50 °C and after thoroughly wetting the fabric, the dye solution is added in portions within 10–20 minutes. Previously dissolved 5–30 g/L sodium sulphate is added in portions within 20 minutes and 0.5 g/L sodium nitrite is added (if oxidation-sensitive dyes are used, a reducing agent such as urea or thiourea should be added before the addition of sodium sulphate). In the meantime, the temperature of the liquor should have dropped to 35 °C. Then 5 mL/L sulphuric acid (96%, i.e. 66 °Bé), diluted by pouring it to four times its volume of cold water, is slowly poured into the liquor and allowed to develop at 30–35 °C for 15 minutes. After developing, the fabric is rinsed and completely neutralised with warm soda ash solution. Soaping is performed for 10 minutes with boiling liquor containing a suitable detergent, and the dyeing is finished by rinsing.[2]

10.5.3.1.2 Two-bath Nitrite Method

There are different variations of this method.

a. ***When working with hot-developed dyes:*** the material is treated (grounded) for 1–2 hours at 60–20 °C (usually starting hot and cooling down) with a liquor containing dye, sodium sulphate (5–50 g/L) and sodium nitrite (3–15 g/L). After grounding, the product is squeezed or sucked out if possible (especially in the case of bobbins) before treatment with the developing liquor. The development liquor contains 25–30 mL/L sulphuric acid (66 °Bé). The development time is 10 minutes and the temperature is 40–70 °C. Aftertreatment of the dyeing is as described in the first method.[2]
b. ***When working with cold-developed dyes:*** after the material is treated with 30 °C liquor containing dye, calcined soda ash and wetting agent (in bobbin dyeing) for 15–20 minutes (two passages in the jigger), 5–30 g/L calcined sodium sulphate is added in two portions (at the beginning of the third and fourth passages in the jigger) and grounding is continued for 25–30 minutes. After the grounding liquor is drained and the product is squeezed or sucked if possible (in bobbin dyeing in apparatus), the treatment with the development liquor is started. In the development liquor, 5–20 mL/L sulphuric acid (66 °Bé) and 0.5–1 g/L sodium nitrite are present. Liquor temperature is 20–25 °C and development time is 15 minutes.[2]

10.5.3.2 Dyeing by Impregnation Method

Since the affinity of vat-leucoester dyes for cellulose fibres is generally not very high, it is possible to work with them in a continuous process and to obtain dyeings with the uniform depth at the beginning and end of the fabric. For preparing the impregnation liquor:

- X g/L dye
- 1–2 g/L wetting or impregnation auxiliary and
- 1 g/L calcined soda ash

are dissolved in boiling water and added in a liquor vessel. Then water is added. After 6–10 g/L sodium nitrite is dissolved elsewhere, it is added to the liquor tank while stirring and the volume of liquor is made up with water.[2]

a. ***Pad-batch method:*** The fabric is impregnated with a cold liquor containing:
   - X g/L vat-leucoester dye
   - 1 g/L soda ash
   - 2 g/L wetting agent
   - 1–2 g/L auxiliary (reducing the substantivity of the dye).

   Then, after the fabric is squeezed, it is wrapped properly and covered with polyethylene foil. After rotating slowly for 6 hours, the fabric is impregnated with the development liquor containing dispersing agent, sodium nitrite, thiourea and sulphuric acid (96%) and then squeezed. The liquor pick-up in second impregnation should be 15–30% more than the liquor pick-up in the first impregnation. After impregnation with the development liquor, the batch must be kept for another 15 minutes before starting the aftertreatment process consisting of rinsing, neutralisation and soaping.[2]

b. ***Pad-jig method:*** After the fabric is impregnated in the padder with the liquor prepared as described above, it is advisable to wait at least 2 hours before it develops in the jigger. However, if this waiting time will be much longer, it would be beneficial to perform intermediate drying and cover the product thoroughly while waiting. Then, it is treated with the development liquor containing 20 mL/L sulphuric acid for two passages (at least 15 minutes) in the jigger. Liquor temperature is 20–25 °C. Aftertreatment of the dyeing (rinsing, neutralising, soaping and rinsing) is done in the same jigger.[2]

c. ***Classical vat-leucoester continuous dyeing method:*** It is the most common continuous dyeing method. The fabric is impregnated with a dye liquor in the first vessel, passed through an air passage (10 seconds) and impregnated with a development liquor in the second vessel. The development vessel usually contains 20 mL/L sulphuric acid (66 °Bé). When working with dyes that are difficult to develop, this amount can be increased to 35 mL/L.[2]

   The temperature of the developing liquor is 70 °C. Immersion time through the liquor is 8–10 seconds or 2–3 seconds + air passage. The second method is generally preferred. Aftertreatment of dyeing is usually carried out continuously in open-width washing machines. Care should be taken to ensure that there are enough compartments for rinsing, neutralising, soaping and rinsing steps.[2]

d. ***Pad-steam method:*** In this method, the fabric impregnated with dye is steamed at 103–105 °C for 10 minutes and then passed through the developing liquor. Generally, the dyeing efficiency is higher in dyeing by this method compared to the classical continuous dyeing method.[2]

10.6 Basic (Cationic) Dyes

The common characteristic of basic dyes is that they all show basic properties, as can be understood from their names. Apart from this, the dyes under this name are chemically very different from each other and belong to different groups (such as diphenylmethane, triphenylmethane, thiazine, oxazine, thiazole, quinoline and polymethine dyes). Basic dyes are available in the market as salts (chloride, acetate, zinc (II) chloride, etc.).[2]

10.6.1 Fastness Properties of Basic Dyes

The light fastness of basic dyes is very low. They have low washing fastness even on cotton fibres mordanted with tannin.[2]

10.6.2 End-uses of Basic Dyes

Since the light fastness of basic dyes is very low, they can only be used for dyeing materials that do not require light fastness. They are inexpensive, and intense colours can be obtained by using very small amounts of them. Their colour shades are very vivid. For example, they can be used for dyeing theatre decor and costume fabrics.[2]

One of the main uses of basic dyes is as a supplement to other dyes. By adding a very small amount of basic dye to normal substantive, vat or sulphur dyes, a vivid colour shade can be achieved. Here, in addition to the main dye, the small amount of basic dye molecules bound to the fibre does not impair the fastness of the dyeing.[2]

10.6.3 Dyeing with Basic Dyes

Basic dyes do not show any substantivity for cellulose.[4] They cannot bind directly to cellulose fibres because they do not contain acidic groups. First, the material to be dyed is treated with tannin. During this treatment, tannin is bound to cellulose by hydrogen bonds and van der Waals attraction forces.[2] The —COOH group in the structure of tannic acid reacts with the basic dye, and thus tannic acid acts as a mordant in the dyeing of cellulose with basic dye (Figure 10.8).[7] However, tannin itself does not bind to cellulose very well and partially bleeds, especially during washing. In order to prevent this, after the product is treated with tannin, it is treated with potassium antimonyl tartrate, and the antimonyltannate thus formed is hardly soluble and no longer bleeds during washing. Various commercially available products can also be used instead of tannin for the binding of basic dyes to cellulose fibres. Their advantage is that they do not require a second treatment with potassium anthimonyltartrate as in tannin. If these are used, the washing fastness values obtained are also slightly better.[2]

Basic dyes are dissolved by pasting with acetic acid and then pouring hot water over them. A critical point is that there should be no iron ions in the water used. Otherwise, they form dark-coloured compounds with tannin. The material to be dyed is first treated with a solution containing 1–6% tannin at 60–70 °C for 1/2–1 hour, and then the material is left there for 2–4 hours while the liquor cools. The material is then taken, squeezed and treated in the cold with a second bath containing half the amount of potassium antimonyl tartrate used in the first bath.[2] The amount of tannic acid must be optimised for a specific depth of shade. Insufficient tannic acid application will result in the dye bath not being fully exhausted, while excessive addition will result in material waste and the dye in the mordanted fibre will rust. Generally, twice the amount of tannic acid relative to the weight of the dye is used in mordanting.[7]

After 2–3% acetic acid is added to the main dyeing bath, the dye dissolved as described above is added. The product, which has been treated with tannin or similar products, is started to be dyed with this liquor in the cold and the temperature of the liquor is raised up to 50–70 °C. The dyeing time is about 1 hour. The biggest drawback of basic dyes is their

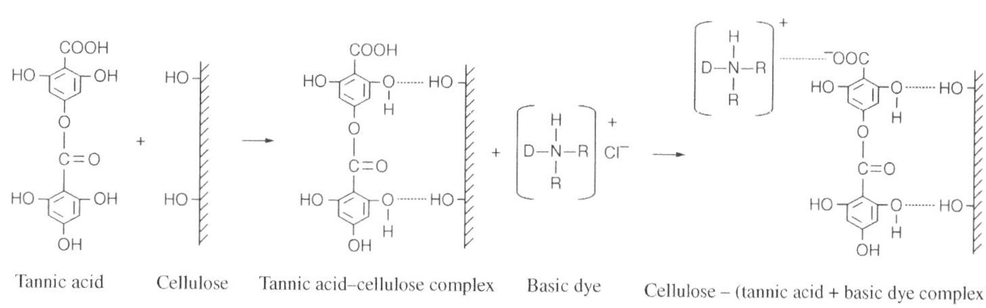

Figure 10.8 Fixation of basic dye on mordanted cellulose. Ref. [7]/Elsevier.

low light fastness. If the dyed material is treated with sodium phosphate molybdenum wolframate, light fastness is improved. However, this effect disappears after washing.[2]

10.7 Phthalocyanine Dyes

Dyes based on phthalocyanine are mostly blue, turquoise and green-blue, and are known for their vivid shades and high light fastness.[2] Phthalogenetic dyes are the precursors of copper phthalocyanine, which is produced by a complex chemical reaction in the fibre (Figure 10.9). For example, impregnation of cotton with di-imino-isoindoline in the presence of a special reducing solvent and a copper salt or copper complex and subsequent heating produces copper phthalocyanine pigment.[9] Phthalic anhydride, in combination with other chemicals, especially copper complex, forms a blue dye in which four phthalocyanine molecules are attached to a copper atom. The dyeing process consists of liquor preparation, impregnation, drying, fixation and development. Therefore, the dyeing technology is complex and requires great care.[7]

10.7.1 Fastness Properties of Phthalocyanine Dyes

Dyeings with phthalocyanines are characterised by high brightness and excellent fastness.[4] Copper phthalocyanine is highly resistant to the effects of chemicals and is characterised by a bright blue colour.[7]

10.7.2 End-uses of Phthalocyanine Dyes

They are suitable for garments that are resistant to washing and weather conditions.[7] Since it is easier to use reactive dyes containing phthalocyanine groups for blue, turquoise and green-blue colours in dyeing, these dyes have not found widespread use.[2]

10.7.3 Dyeing with Phthalocyanine Dyes

As in oxidation dyes, in the phthalocyanine process, the actual dye (insoluble phthalocyanine pigment) is formed in the fibre itself. The dyeing process is carried out with precursors. There are two main processes: Aminoiminoindolene and its derivatives are low molecular phthalogenetic developers, applied by impregnation with heavy metal donors

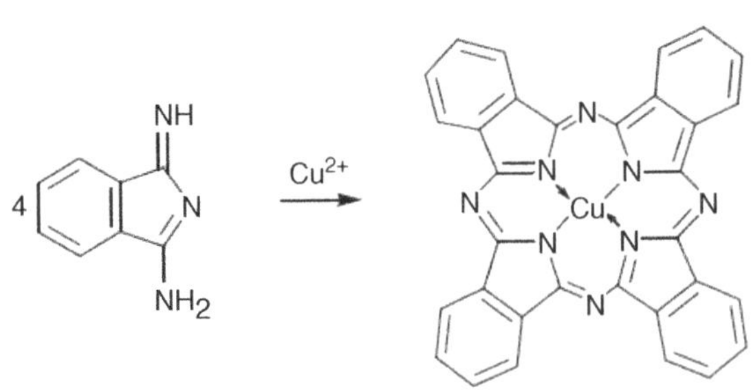

Figure 10.9 Formation of copper phthalocyanine from suitable precursors. Ref. [9]/Society of Dyers and Colourists.

(preferably Cu or Ni). The most important aminoiminoindolene is CI Ingrain Blue 2:1, 74160. Polyisoindolenes, which are high molecular phthalogenetic developers and complexes of indolene with heavy metals, are developed with reducing agents in a wet process. Polyisoindolene complexes can be applied using the exhaust process. Reductive development is carried out with hydrosulphite in an alkaline medium.[4] Phthalogenetic blue is obtained by developing on cotton by reaction of several chemicals and subsequent development under suitable conditions. The component chemicals applied for this purpose have no affinity for cotton and are therefore applied primarily by impregnation. Then, controlled drying, fixation or polymerisation, development and then soaping and washing are performed. The main component formed by the reaction between phthalic anhydride and urea, the water-insoluble amino-imino-isoindole, is emulsified with a solvent as a fine dispersion and then a copper complex is added. Cotton fabric is impregnated with this emulsion, dried and fixed at 150 °C for 4–5 minutes to form a chelated cyclic product, copper phthalocyanine. The dyed fabric is treated with $NaNO_2$ and HCl to remove the various by-products formed and to reveal the true colour tone.[7]

The old Alcian dyes (ICI) were water-soluble isothiourea derivatives of copper phthalocyanine carrying chloromethyl substituents. After application to cotton, post-treatment with an alkaline solution formed the insoluble pigment.[9]

10.8 Mordant Dyes

The fibre is first treated with metal salts (mordanted). Extremely sticky, basic metal compounds are formed on the fibre. These compounds can form insoluble coloured complexes with certain azo and anthraquinone derivatives. Alizarin is the best known anthraquinone derivative for this process. It was formerly isolated from the root of the madder plant but has now been replaced by synthetic products. Suitable azo dyes contain hydroxyl or carboxyl groups in one of the aromatic nuclei or in the ortho position to each azo group. The shade of the dye depends on the type of metallic mordant used.[4]

10.8.1 Fastness Properties of Mordant Dyes

They have good washing and light fastness.[2]

10.8.2 End-uses of Mordant Dyes

Despite the very complex and difficult dyeing processes, they have played an important role in textile dyeing for a long time because no other dye groups with good light and washing fastness were known in the past. In the past, they were used in the production of flags, mattress covers, etc. Mordant dyes, which play an important role in dyeing, have lost their importance with the discovery of vat, development and reactive dyes.[2]

10.8.3 Dyeing with Mordant Dyes

The most widely known and used of mordant dyes is Alizarin red (flag red, Turkey red).[2] The Alizarin dyeing process is very detailed. The old Turkey red dyeing process required at least ten working steps: boiling, treatment with rancid olive oil, washing, treatment with tanning agent, mordanting with aluminium, dyeing in hard water, softening, acidification and soaping. In the new process, treatment with Turkey red oil (sulphorycinoleic acid) has replaced the oiling step. The light and wet fastnesses obtained are excellent.[4]

10.9 Pigments

Pigments are organic or inorganic solids with coloured, black, white or fluorescent particles.[10] Organic and inorganic pigments are not considered true textile dyes because they have no affinity for any textile fibre. In addition to organic pigments, inorganic coloured oxides can be used to colour textiles, especially those made of cellulosic fibres.[4] Pigments are generally insoluble in the materials they interact with and are not physically or chemically affected by them. As a result of the physical and chemical structure of pigments, their application is different from that of dyes. When dyes are applied to the fibre, they penetrate into the fibre in a soluble state and then become insoluble or remain in the same state. Pigments, on the other hand, do not have any affinity for textile materials. They are very well dispersed in water and maintain this state during dyeing.[10] Since they have no affinity for textile fibres, they are bound to the fibres by means of adhesives called "binders" added to the dyeing liquor.[2] Usage areas of pigments are as follows:

- Melt dyeing
- Dyeing by impregnation method
- Dyeing by exhaust method
- Printing.

The most important ones among these are melt dyeing and printing processes. In other words, the use of pigments in normal dyeing processes is not common. However, in recent years, bleeding dyeings with a vintage-look achieved by pigment application by exhaust method in obtaining current fashion effects has attracted increasing interest.[10] Despite the various advantages of pigments, the following limitations hinder their widespread use in textile dyeing:

- Rubbing and mechanical washing fastnesses are low.
- The handle of the material becomes harder.
- Dark shades do not give good results.
- There is a risk of uneven dyeing due to migration.[2]
- Correction of faulty dyeing or lightening of the colour is very difficult.[11]

It is not recommended to obtain dark shades with pigments. This is because high binder concentrations are required to bind the large amount of pigment which results in hardening of the handle. In addition, up to a concentration of about 10 g/L impregnation liquor, the depth of the dyeing obtained increases with increasing pigment amount as with other dyes, while the increase in depth slows down at higher concentrations. In other words, in order to obtain a darker dyeing, the amount of pigment to be used must be increased considerably. This is also not economical. As for the migration risk, this risk, which can occur in all dye classes during drying before the pigment is fixed in the fibres, can be reduced by using suitable antimigration agents and some other measures.[2] Some advantages of pigments can be listed as follows:

- The dyeing methods are simple. Since the pigment is bound to the fibre through the binder, drying and condensation processes after impregnation are sufficient to complete the dyeing. There is no need for washing after dyeing.
- Since the pigment has no affinity for fibre, the concentration of the liquor does not change during dyeing, so they are very suitable for continuous dyeing.
- Since there is no affinity between the pigment and the fibres to be dyed, fibre blends can also be dyed with the same dye in one bath.
- Pigments have a fairly good gamut, and some selected pigments have high light, dry cleaning, chlorine and perspiration fastness.
- By selecting the appropriate binder from the various products on the market and adjusting its usage amount, the desired handle finishing effect can also be achieved during dyeing, and this finishing will be wash-resistant.
- With pigments, dyeing and crease-resistant finishing can be done in the same step. Thus, one process step is saved and the cost can be reduced considerably and the wet fastness of the dyeing is improved.[2,10]

10.9.1 Fastness Properties of Pigments

Pigments have good light fastness and general fastness properties. However, rubbing fastness is low in dark shades and washing fastness is limited.[11] On the other hand, some selected pigments have high light, dry cleaning, chlorine and perspiration fastness.[10]

10.9.2 End-uses of Pigments

Pigments can be used for economical dyeing of products such as tablecloths, t-shirts, dresses, linings, curtains, decoration materials and beddings made of cellulose fibres or their blends in light and medium shades when high washing and rubbing fastness values are not required.[2]

10.9.3 Dyeing with Pigments

10.9.3.1 Dyeing by Impregnation Method

Since they are suitable for dyeing different fibres and fibre blends, pigments, which are called universal dyes, are very suitable for continuous dyeing since they have no affinity for fibres. The dyeing process is quite simple, and the fabric impregnated with a liquor containing suitable pigment, binder and catalyst (acidic salt) is dried and then condensed at 140–150 °C for 5–4 minutes.[11] Pigments may belong to azo, diazine, anthraquinone, indigoid, phthalocyanine, metal oxides, etc. groups in terms of their chemical structures. Since the pigment has no affinity for fibres, the structure of the pigment molecule (to which group it belongs) does not play an important role. Pigments, together with the dispersing agent, are made ready for use by grinding or milling well when wet. This pigment finish has a role in the distribution of pigments within the fibres. It is difficult to obtain high fastness values with pigments without a good finish.[2]

Since the binding of the pigment molecules to the fibres is through the binder, the importance of the binder for obtaining a good dyeing is obvious. Binder is the substance that adheres to the fibres and forms a film layer surrounding the pigment particles. Thus, it ensures that the pigment particles it envelops also bind to the fibres. The binders themselves bind to the fibres mainly by hydrogen bonds and physical adsorption forces. Some binders also contain reactive groups and can bind covalently to fibres via these groups.[2]

Binders are high-molecular-weight substances obtained by polymerisation of monomers. Products such as various esters of acrylic acid, acrylonitrile and styrene are widely used as monomers. Binders are required to fulfil many properties simultaneously. Binders obtained as a result of polymerisation of a single type of monomer cannot provide this. For this reason, better quality binders are generally obtained by copolymerisation of mixtures of various monomers in certain ratios. The binder should form a film layer on the fibres. For this, the binder must be homogeneously dispersed in the dyeing liquor. For this reason, the binder should either be soluble in water or be available as aqueous dispersion. Aqueous dispersed binders are generally used in the market today. The binder, which is dispersed homogeneously in the liquor, is absorbed into the fabric together with the pigment, and when the water evaporates during drying and condensation, it deposits on the fibres and forms a film layer. The binder macromolecules in chain form bind to each other through the reactive groups they contain and form macromolecules that can no longer be dispersed. Properties of a binder depend on type and proportion of monomers used, molecular weight (DP), type of reactive group in molecules, and type of emulsifier (dispersing agent) used to produce the dispersion. The properties expected from a good binder used in textile dyeing can be listed as follows:

- It should have a high ability to adhere pigments to fibres.
- The film layer they form should be colourless, clear and of the same thickness everywhere.
- The film layer should be soft and elastic, not sticky after drying.
- The film layer must be as resistant as possible to light, mechanical effects (rubbing, scrubbing), chemicals and organic solvents (dry cleaning).
- The binder must not react adversely with other auxiliaries in the liquor (e.g. form precipitates).
- Vessels and devices should be easily cleaned from liquor and thus binder residues after use.[2]

When dyeing with pigments, it is desirable to add as little binder as possible to the liquor in order to reduce the cost and to avoid a hard handle. On the other hand, since the binding of the pigments to the fibres is only possible through this binder, using less binder than necessary causes a decrease in fastness. For this reason, using insufficient binder should be avoided, and the usage amounts recommended by the binder manufacturers for their binders in their catalogues should be followed.[2] Although the binder concentration increases as the amount of pigment increases, it generally varies between 20 and 60 g/L.[10]

Although some of the binders on the market can be used directly, some require the addition of a catalyst (e.g. diammonium phosphate) or even polyfunctional compounds (e.g. N-methylol compounds) to the liquor. After drying, depending on the type of binder, 5–2 minutes at 120–150 °C is usually sufficient for the formation of the condensation film layer. It is also possible to add finishing agents and softeners to the liquor. It is important that pigments, binders and additives do not interact adversely and do not form precipitates. Although some binders are produced to reduce migration, it is beneficial to add a suitable antimigration agent to the liquor.[2] As an antimigration agent, natural- or synthetic-based substances that increase the viscosity of the liquor are used. Wetting agent and sometimes an antifoaming agent can also be added to the impregnation liquor. A wetting agent to be used should be of the non-ionic and non-foaming type.[10]

Pigment and binder are diluted in separate containers by adding water and mixing. Especially in combinations prepared by mixing more than one pigment, care should be taken to ensure thorough. Finishing agents and auxiliaries are also dissolved in separate containers, and after cooling, they are added to the solution containing pigment and binder, mixed.[2]

The product to be dyed with pigment must be well prepared for dyeing and its hydrophilicity must be sufficient. Inadequate or improperly cleaned product causes stained and faulty dyeing. The product to be dyed should be free of sizing and preparation residues, and furthermore, the product should be subjected to singeing. Since the protruding fibres are dyed in a lighter shade compared to the ground, dyeing the fabrics which were not subjected to singeing with pigment gives an appearance of unevenness. In addition, care should be taken to ensure that all parts of the product are at the same dryness. Otherwise, since the wet parts will absorb less pigment liquor, they appear as stains at the end of dyeing.[2]

One of the most important issues to be considered in pigment dyeing is migration. In order for surface migration to occur, there must be a liquid layer (film) on the surface of the fabric in which the pigments can move. In other words, the risk of migration exists until the surface water evaporates and capillary water remains (30% moisture in cellulose fibres). The less surface water to be evaporated, the shorter the first drying phase in which the risk of migration exists. Therefore, the pick-up value will be low if the impregnation time (immersion length) is short and the fibres are squeezed under high squeezing pressure before they have a chance to swell fully. If the product is also passed through an air passage for 30–60 seconds before drying, the fibres have time to swell and use 30% of the water in the picked-up liquor as swelling water, in which case the amount of surface water is lower (30–40%). Drying can be carried out in a hot-flue, stenter or dyeing abrage dryers, and it is notable that an IR channel can be placed in front of the dryer (especially in stenter dryers) to prevent migration. Condensation can normally be carried out in all devices where other curing processes are made.[2]

It is also possible to perform pigment dyeing and crease-resistant finishing in one step. A crease-recovery agent, catalyst, binder, auxiliaries (antimigration agent) and pigment are added to the liquor. The washing fastness of the dyeings obtained in this way is much higher compared to standard pigment dyeing. This is because the crease-recovery agents act as a bridge between the hydroxyl groups of cellulose and the reactive groups of the binder, allowing the binder to bind covalently to the fibres via the anti-creasing agent bridge as well as a physical adsorption forces and hydrogen bonds. As a result, the binding of the pigments to the fibres via the binder is more robust. For this reason, it is a simple and inexpensive process that can be used in the dyeing of shirting, t-shirts, women's dresses, linings and terry fabrics in light and medium shades. They are especially recommended for ground dyeing in pale shades of fabrics to be printed with pigments.[2]

It is not recommended to use resin-forming urea-formaldehyde and melamine-formaldehyde products as crease-resistant finishes. This is because neither the washing fastness of the dyeings nor the washing resistance of the crease-resistant effect is very high. Less resin-forming products (cyclic N-methylol compounds and carbomates) give better results. As catalysts, magnesium chloride, ammonium chloride, monoammonium phosphate and, particularly, mixtures of magnesium chloride and ammonium chloride give good results.[2] The problem that may occur here is that the anionic dispersing agent required for pigment dyeing interacts with the (+) charged cations (Mg^{++}, Al^{+++}, etc.) of the catalyst necessary for crease-resistance finishing. The same problem can arise if a cationic softener is used, however unlike the catalyst, it is possible to use softeners with a different ionic character.[10] In addition to a suitable antimigration agent (antimigrating agents such as alginate, which harden the handle, are not recommended for pigment dyeing), auxiliaries should also be added to the liquor to prevent the formation of a layer on the pad rollers. First, the crease-recovery agent diluted with a little cold water is added, followed by the binder, antimigration agent, softening agent, pigment and finally the catalyst solution diluted with cold water. It is recommended to dry at 120–130 °C (until a residual moisture content of 6–7% is reached). The conditions for curing are the same as when the crease-recovery agent is used alone.[2]

10.9.3.2 Dyeing by Exhaust Method

Pigments are used in garment dyeing by the exhaust method. Although garment dyeing has been used for many years, interest in this technique has increased in recent years. One of the important reasons for this is the fashion for creased and vintage-look clothes in daily wear. In fact, this fashion aims to provide clothing without ironing, ease of use and cost advantage. This appearance can be easily achieved with garment dyeing processes on 100% cotton fabric. In addition, this method allows small batches to be dyed in fashionable colours and effects. Direct, reactive, sulphur and vat dyes, and pigments can be used for garment dyeing of cotton fabric. Special dyeing methods can be determined depending on the type of piece goods, desired colour, machine and dyeing process.[12]

Pigments have no affinity for fibres. Therefore, in order to be able to dye by the exhaust method, it is necessary to ensure that the pigment particles go onto the fibre and fix there. Pigments are present as (−) charged particles in the liquor because they are surrounded by the anionic dispersing agent used in dyeing.[10] For this reason, the amount of pigment to be absorbed by the fibre can be increased by pretreatment with a cationic substance before dyeing. Colour depth directly depends on the amount of cationic substance on cotton. Dark colours cannot be obtained in a fabric with a low level of cationisation. However, there is a change not only in the depth of shade but also in the nuance of cationic cotton compared to untreated cotton. Therefore, it is not an easy to adapt a conventional dyeing recipe to the dyeing of cationic cotton. Fixation is provided by the binder. The cationisation and dyeing of cotton with pigments can be done in two different ways:

a. ***Two-step method:*** The cationic pretreatment and acrylic-based binder are applied first and then the garment is dyed with pigment in the second step.
b. ***Three-step method:*** The garment is first cationised, then dyed with a pigment and then treated with an acrylic-based binder.[13]

Colour yield in pigment dyeing process is slightly affected by temperature and time. For a good colour uniformity, 30 minutes at 65 °C is sufficient, and the liquor ratio can be reduced to 1:10.[14] After dyeing, post-treatment with cationic softener can be performed, but yellowing problem is generally experienced after drying in cationic softening processes. Particular attention should be paid to unacceptable colour changes in white and pastel shades. If desired, the material can be subjected to enzymatic treatment with cellulases. With this process, undyed fibre ends on the material are removed and a smoother surface appearance is obtained.[13] After post-treatment, the material is removed and dried by conventional methods. Drum dryers give the fabric a soft texture with the mechanical effect they provide. Another option is to hang the garments on hangers and dry them in a transport dryer. During drying, it is necessary to ensure that the fabric temperature does not exceed 115 °C. It is preferred to leave at least 5–10% moisture in the material. If the dyed garment needs to be ironed, it can be done on shaped ironing boards or hot presses to ensure a very smooth surface.[12]

Before concluding the subject of dyeing with pigments by exhaust method, it is useful to briefly mention the "bleeding dyeing" process, which has been in fashion for some time.

The essence of the bleeding dyeing process, which is applied to obtain a used, vintage appearance on the material, is based on reducing the amount of binder used to bind the pigment to the fibre. At the end of bleeding dyeing, the material must be washed before being delivered to the consumer. The recipes and washing methods applied for this purpose vary from company to company. If it is desired to stop the bleeding after obtaining the vintage effect, post-treatment can be done with a fastness-enhancing product.[14]

10.10 Minerals

Mineral colours are formed on the fibre by applying water-soluble metal salts, which are then converted into insoluble metal oxides. The colour of a particular metal oxide, alone or in combination, forms the final colour. Metal oxides occupy free space within the fibre, leading to hardening of the material. They are over-dyed with vat or sulphur dyes. An important member for military uniforms is mineral khaki. Other examples are chrome yellow, chrome red, chrome orange, chrome green, manganese brown, Prussian blue, iron oxide, etc. Mineral colours are not bright. Viscose is not dyed with mineral colours due to the problem of brightness loss.[7]

10.10.1 Fastness Properties of Minerals

Mineral colours have good washing and light fastness, but they are sensitive to acid.[7]

10.10.2 End-uses of Minerals

Brown and khaki tones (mineral khaki) are obtained with chrome and iron salts. This dyeing is inexpensive and resistant to light, weather conditions and decay. Its disadvantages are wastewater pollution and hardening of the fabric. This method is still important in some countries for dyeing tarpaulins and uniform materials.[4]

10.10.3 Dyeing with Minerals

Metal salts can be applied to cotton alone or in combination, then treated with alkali to obtain metal hydroxides and finally metal oxides through oxidation, thus forming coloured pigments on the fibre. This cannot be called dyeing but rather the formation of coloured metal oxide complexes. Metal oxides such as TiO_2, ZnO, etc. are used as white pigments, while Fe_2O_3 is reddish-brown. Metal oxides that are insoluble in water can be formed, alone or in combination, in a certain proportion to achieve a specific colour hue. The metal salts used do not have an affinity for cotton, and the impregnation method is suitable for application.[7] By impregnation, inorganic metallic salts can be applied to cotton and then converted to the corresponding oxide by steam treatment in an alkaline medium.[4]

The most important colour in this category, mineral khaki, is a mixture of chromium oxide and iron oxide in the right proportions for colouring military uniforms. Cotton is impregnated with chromium sulphate and iron sulphate solution, dried and passed through an alkaline solution at the boiling point to form the respective hydroxides. When dried, chromium hydroxide is oxidised to green chromium oxide (Cr_2O_3) and iron hydroxide to red iron oxide (Fe_2O_3), finally obtaining khaki. The insolubility of these two oxides in water provides excellent washing fastness. Aftertreatment with Na_2SiO_3 also increases perspiration fastness.[7]

10.11 Natural Dyes

The production of dyes using naturally obtained materials has been known as an art since ancient times. Until the middle of the nineteenth century, when synthetic dyes were discovered, all coloured materials were supplied from natural dyes obtained from plant or animal sources.[15,16] Although natural dyes can also be used in cotton dyeing, they are more common in wool dyeing, because cotton fibres are more difficult to dye with natural dyes than wool. However, some of the plants used in wool dyeing (buckthorn, indigo, madder, etc.) give good results in cotton dyeing.[16] Although the natural dyes that provide good results on cotton are limited, colours such as yellow with pomegranate peel and turmeric, blue with indigo, red with madder, yellowish-brown with onion peel and reddish-brown with catechu can be obtained.[17] Since natural dyes have gained importance, especially in the dyeing of protein fibres such as wool and silk, detailed information about their chemical structure, classification, fastness properties, end-uses and dyeing methods is given in the book, namely "Dyeing of Proteinous Fibres".

References

1 Hook, J.A., Welham, A.C. The use of reactant-fixable dyes in the dyeing of cellulosic blends. Journal of the Society of Dyers and Colourists, 1988; 104, 329–337.

2 Tarakçıoğlu, I. Tekstil Boyacılığı I Teksiri. İzmir, Türkiye: Ege Üniversitesi Mühendislik Fakültesi Çoğaltma Yayınları, 1982.

3 Shore, J. Dyeing with direct dyes. In Cellulosics Dyeing (Ed. J. Shore, Chapter 3, 152–186). Bradford, West Yorkshire, UK: Society of Dyers and Colourists, 1995.

4 Leube, H. Textile dyeing, In Industrial Dyes: Chemistry, Properties, Applications (Ed. Dr. K. Hunger, Chapter 4, 339–426). Wiley-VCH Verlag GmbH & Co. KGaA, 2003.
5 Shore, J. Dyeing with azoic components. In Cellulosics Dyeing (Ed. J. Shore, Chapter 7, 321–350). Bradford, West Yorkshire, Birleşik Krallık: Soceity of Dyers and Colourists, 1995.
6 Chattopadhyay, D.P. Azoic dyeing. In Handbook of Textile and Industrial Dyeing, Volume 1: Principles, Processes and Types of Dyes (Ed. M. Clark, Chapter 19, 604–626). Cambridge, Birleşik Krallık: Woodhead Publishing Limited, 2011.
7 Chakraborty, J.N. Fundamentals and Practices in Colouration of Textiles. Woodhead Publishing India Pvt Ltd, 2010.
8 Pink, T. Leuco esters of vat dyes. In Ullmann's Encyclopedia of Industrial Chemistry (Ed. F. Ullmann). Wiley-VCH Verlag GmbH & Co. KGaA, 2000.
9 Broadbent, A.D. Basic Principles of Textile Coloration. Bradford, England: Society of Dyers and Colourists, 2001.
10 Aspland, J.R. Pigments as textile colorants: pigmenting or pigmentation. Textile Chemist and Colorist, 1993; 25(10), 31–37.
11 Yurdakul, A., Atav, R. Boya baskı esasları. Bornova, İzmir: Ege Üniversitesi Tekstil ve Konfeksiyon Araştırma Uygulama Merkezi Yayını, 2006.
12 İnce, E., Arabacı, A., Yoğurtçu, F.E., Eksen, D., Hovardakaya, D. Antik parça boyama yönteminde koyu renklerde yaşanan renk verimi ve haslık sorunlarının giderilmesi, Proje No: Tübitak TAM 2006-06, 2006.
13 Tyndall, R.M. Garment wet processing-an international update. Textile Chemist and Colorist, 1996; 28(3), 25–27.
14 Pigmafast Boya Kataloğu, 2006.
15 Eyüboğlu, Ü., Okaygün, I., Yaraş, F. Doğal boyalarla yün boyama, uygulamalı ve geleneksel yöntemler. İstanbul, Türkiye: Arkeoloji ve Sanat Yayınları, 1983.
16 Öztürk, İ. Doğal bitkisel boyalarla yün boyama. Ankara, Türkiye: Ürün Yayınları, 1999.
17 Karabulut, K. Pamuklu örme kumaşlara doğal boyalarla boyama yoluyla tek adımda renk, UV koruyuculuk ve antibakteriyellik kazandırılması (MSc Thesis). Tekirdağ Namık Kemal Üniversitesi, Fen Bilimleri Enstitüsü Tekstil Mühendisliği Anabilim Dalı, 2015.

11

Dyeing of Other Cellulosic Fibres

11.1 Dyeing of Natural Cellulosic Fibres Other Than Cotton

When natural cellulosic fibres other than cotton are mentioned, fibres such as flax, hemp, jute, ramie and sisal come to mind. Although the dyeing principles of all these fibres are basically similar to those of cotton fibre, there are slight differences between them.[1] Especially in recent years, it is noteworthy that hemp, jute and ramie fibres have started to be used in the field of clothing fabrics as a blend with various fibres in textiles, but other bast fibres other than flax (hemp, jute, ramie, etc.) and leaf and fruit fibres are generally used outside the field of clothing, such as carpet backing, bagging, etc., so the finishing processes and therefore the dyeing of these fibres have not gained sufficient importance. For this reason, this section will focus on the dyeing of cellulosic fibres other than cotton, especially flax, and then other bast fibres will be mentioned very briefly. Leaf and fruit fibres will be excluded.

Since the basis of flax fibres is also cellulose macromolecules, they can be dyed with dyes (such as direct, vat and reactive) generally used for dyeing cotton fabrics, and they are dyed under similar conditions.[2] The residues of the woody core of the plant stem in flax fibres make it difficult to dye these fibres properly. This is because flax fibre and these residues differ in dyeability, just like cotton and seed coat residues.[3] Therefore, as with all fibres, a correct pretreatment is of critical importance for flax fibres.

As it is known, flax is not composed of fibres alone but of fibre bundles, and these bundles are bound together by pectin. Since this pectin substance will decompose under the influence of strong alkalis, the fibre bundles also decompose in a strong alkali environment, especially at high temperatures. Therefore, working with hot alkali liquors should be avoided during the dyeing of flax. Another point to be considered is that due to the stiffness of flax fibres, it is difficult for the dye to penetrate them completely and properly. Therefore, during the dyeing of these fibres, the following points should be taken into consideration:

- For tightly woven fabrics to be dyed in one colour, dyeing the yarns first instead of the fabric and weaving the fabric with these yarns should be preferred.
- Auxiliaries that increase the ability of the dye to penetrate into the fibres and promote uniform dyeing should be used.
- Dyes that do not have a high affinity for fibres should be used.

- Special dyeing methods should be used when working with dyes with high affinity for fibres.
- In order not to cause hardening of the fibres during drying, the temperature should not exceed 80 °C.[2]

Flax can be dyed as bobbin or fabric. Dyes, dyeing method and the device to be used are selected according to the place where the finished product will be used. Flax yarns are dyed as bobbins to obtain yarn-dyed woven fabric. The bobbins must be loosely wound. Otherwise, the dye liquor cannot penetrate the inside of the bobbin sufficiently, and as a result, it cannot be dyed. Also, the winding density of the bobbins in the same batch must be homogeneous.[4] It is recommended to use special (telescopic or elastic) bobbins for crossed bobbin dyeing. If the crossed bobbins are subjected to an intermediate drying period after the carding and then dyed, the dye penetrates into every point of the yarns better.[2] In bobbin dyeing, it is difficult to ensure good penetration into the yarn, especially when conical bobbins wrapped in rigid bobbins are used. For this reason, the use of dyes with good levelling ability and a sufficient amount of levelling agent in combination with flexible bobbins provides the best results. All the dyeing processes described for cotton yarn can also be used for linen.[5]

Different methods and devices can be used for dyeing linen fabrics.[4] The only problem in dyeing linen fabrics is to obtain a satisfactory penetration. In this respect, the best results are obtained by beam dyeing at high temperatures.[5] Jet and overflow dyeing machines are not recommended due to the risk of abrasion and breakage of the fabric. The most commonly used device is the jigger. However, the impregnation method is preferred for dyeing thick fabrics.[4] Pad-jig and pad-batch processes are suitable for linen.[5]

Flax fibres can be dyed with dyes that dye cellulose (direct, reactive, etc.) to give good wet fastness. Dyes with high durability and good penetration properties should be used in dyeing. Different types of dyes can be used, but the performance of each is different. The place where the textile product will be used should be taken into consideration in the selection of dye.[4] Linen fibres are more easily affected by hot alkalis than cotton. Caustic soda is necessary for the dissolution of naphthols and the reduction of vat and sulphur dyes to leuco compounds. Therefore, it is preferred to use direct and reactive dyes for flax fibres.[6] The properties of the dye groups that can be used in the dyeing of these fibres are described below.

a. ***Direct (substantive) dyes:*** These dyes are water-soluble and have low colour fastness. Therefore, direct dyes can be preferred when colour fastness is not very important.[4]
b. ***Development (naphthol) dyes:*** Naphthol dyes are used for dyeing in dark shades. When working with naphthol dyes, grounding is usually carried out at 60–80 °C in order to increase the penetration of naphthol into the fibres. Soda ash is first added to the grounding liquor and the required caustic soda is added in portions. Generally, the development (coupling of naphtholate with diazonium salt) is a little slower than in cotton dyeing. In order to ensure that the rubbing fastness of the dyeing is not too low, the intermediate process must be carried out by rinsing with water, including salt, and two soaping steps must be carried out during the post-treatment.[2]
c. ***Vat (Indanthren) dyes:*** These dyes have good colour fastness and high resistance to light and bleaching. Their dissolution is a delicate process and can lead to uneven uptake

by the fibres.[4] In order to ensure satisfactory penetration, special dyeing methods are required for the application of vat dyes to flax yarns.[6] When dyeing flax with vat dyes, the pigment or vat acid methods[2] or the pad-jig method should be preferred.[7]

d. ***Sulphur dyes:*** It may be advisable to dye linen tents and tarpaulins with sulphur dyes.[2]
e. ***Reactive dyes:*** Reactive dyes are used to obtain bright colours. When dyeing linen fabrics, it should be noted that these fabrics shrink when exposed to high temperatures. Therefore, cold dyeing reactive dyes should be preferred.[4] The two-step method is generally recommended as dyeing method. Finishing treatments such as dirt repellency, softening and water repellency should be avoided before dyeing, as these are very difficult to remove and may cause unevenness in dyeing.[2] Dye selection should be made carefully for linen woven fabrics to be treated with wrinkle-free finishing. Caustic soda pre-treated linen woven fabrics generally give satisfactory results under normal conditions for cotton. It is recognised that pad-jig dyeing is the best approach for covering slubs or other surface irregularities.[8] In order to obtain reproducible dyeings with the best processing properties, the dye should be applied by the impregnation method, kept in the cold and then processed in an open-width washing machine.[4] The pad-batch process is particularly advantageous for many linen qualities due to excellent dye diffusion at very short soaking times.[8] Reactive dyes are readily diffused in the dyeing of woven linen fabrics by pad-batch method, but the impurities present can in some cases cause deviations in shades or a mottled appearances.[6] In bobbin dyeing, penetration of reactive dyes into linen yarns is easier compared to vat dyes. Dye-fibre reactivity and final fixing efficiency are closer to non-mercerised cotton than mercerised cotton and significantly less than viscose. Care must be taken during the critical levelling and diffusion phase just prior to alkali addition and during alkali addition.[8]

In the literature, a study on the dyeing of cotton, flax and hemp with reactive dyes showed that the dye uptake (%) values of the yarns decreased in the order of cotton > hemp > flax. The cellulose content decreases in the order of cotton > hemp > flax fibres, and it is quite normal for cotton with high cellulose content to absorb more dye than linen and hemp. However, when these results were compared with the colour yield values obtained in fabric dyeing, the opposite result was observed. Colour yield (K/S) of fabrics is listed as cotton < hemp < linen, from lightest to darkest. In other words, although cotton fibres take more dye than flax and hemp, cotton fabric is dyed lighter than linen and hemp fabric. To understand the reason for this, it is necessary to look at the fibre properties. Cotton fibres are finer than flax and hemp fibres. This results in the total surface area of cotton being higher than that of flax and hemp, which is one of the reasons why cotton fabric can be dyed in a lighter shade with the same amount of dye.[9] This actually explains the darker dyeing of bast fibres such as flax and hemp; in fact, flax and hemp fibres have a higher crystal structure than cotton as the fibril orientation is aligned along the fibre axis.[9,10]

Bleached jute fibres can be dyed with reactive dyes by the exhaust method with salt and careful addition of soda ash. Jute is somewhat sensitive to alkali, and it is important to control both alkaline pH and temperature conditions.[8] One of the most important problems encountered in jute fibres is colour change under light, that is, fading. With very little dye, jute fibres can reach the light fastness values obtained in cotton. Fastness values in jute fibres are generally 0.5–1.5 points lower than cotton and sometimes the decrease can reach

2–3 points. In the blend of jute with cotton, the problem is alleviated due to the cotton component; however, especially if the jute ratio is high, the blended fabric itself does not meet the properties expected from apparel and upholstery fabric. In order to obtain high light fastness in products containing jute, the use of 1:2 metal complex dyes, which provide very good light fastness in polyamide fibres, for dyeing jute and jute/cotton blend fabrics has been studied in the literature. It is stated that very high light fastness values such as 6 can be obtained in jute fibres when dyed with 1:2 metal complex dyes after pretreatment with polymer.[11] In limited quantities, ramie is dyed with reactive dyes by methods similar to cotton. Due to ramie's dark natural colour and high content of non-cellulosic impurities, the results obtained in other bast fibres are not very satisfactory.[8] Dyeing ramie fibre to dark shades is difficult due to its high degree of polymerisation, orientation and crystallinity. In order to improve the dyeing performance of ramie fibre, pretreatments such as cationic modification and mercerisation are usually applied to the fibre. Cationic modification of ramie is to provide a cationic group to the fibre. Thus, the fibres are given a positive charge, which reduces the repulsive force between the anionic reactive dye and the negative charge of the ramie fibre. As a result, dye uptake and fixation are increased. However, colour fastness to rubbing is problematic. Mercerisation of ramie fibre in caustic solution or liquid ammonia damages the crystalline region of the ramie fibre and increases the amorphous region. Therefore, dye uptake is increased. In addition, more free hydroxyl groups are formed, which are beneficial for dye fixation. There is also a study in the literature in which ramie fibre was treated with a solution containing a mixture of caustic and alcohol to improve its dyeing ability. It has been shown that isopropanol is the most suitable alcohol for modifying ramie fibres with a caustic solution mixture, resulting in higher dye uptake and fixation levels.[12]

11.2 Dyeing of Regenerated Cellulose Fibres

As it is known, the first regenerated cellulose fibre developed as an alternative to cotton is viscose. High-modulus modal fibres were produced as the second-generation regenerated fibre. The third-generation regenerated cellulose fibre is lyocell. In principle, these fibres are dyed with dyes (direct, reactive, etc.) and dyeing methods (exhaust, impregnation) used in cotton dyeing. Viscose fibres have found a wide range of applications due to their drapability and bright colours. However, the dyeing of fabrics made of viscose fibre is a little more difficult than cotton. Reasons for this can be listed as:

- Due to the low degree of polymerisation of regenerated cellulose fibres and the high amount of amorphous regions, the swelling ability of these fibres in aqueous solutions is high and their wet strength is low; and
- Rapid dye absorption and more difficult migration (levelling).[1]

Since the length (average degree of polymerisation) of cellulose macromolecules forming regenerated cellulose fibres is low and the amount of amorphous (easily penetrable) region in these fibres is higher than the crystalline part, the swelling ability of these fibres in aqueous solutions is high and their wet breaking strength is low. For this reason, during the dyeing of viscose and rayon, excessive stretching, squeezing, and compression of the fibres in the wet state should be avoided, and processing should be carried out carefully. For example, the use of conventional jiggers in the dyeing of products made of regenerated cellulose fibres can be dangerous. Special tensionless jiggers should be preferred. For the same reasons, care should be taken to ensure that the surfaces of the vessels and transport parts of the machines used are smooth and even.[2]

Since viscose and rayon fibres partially lose their softness and ability to bend easily at high temperatures, it is recommended that the temperature should not exceed 50–60 °C during their drying.[2]

The high swelling ability of viscose and rayon fibres in aqueous solutions (dyeing liquor) can lead to various difficulties. For example, it becomes difficult for the dye to penetrate into the yarn or fabric. Especially the dyeing of windings (such as cross bobbins) is difficult. To prevent this, the bobbins should be sparsely wound beforehand, as the winding will become very hard and tight as a result of swelling. However, since rayon yarns are slippery, it is not always possible to wind them loosely. Since regenerated cellulose fibres swell considerably more than cotton in aqueous media, the danger of permanent crease marks in rope dyeing is high. Like cotton fibres, regenerated cellulose fibres become more flexible by increasing the temperature, and the migration and diffusion of the dye are also increased. With the addition of anti-creasing agents, the formation of crease marks is relatively prevented. In order to prevent crease formation in dyeing, an effective anti-creasing agent should be used.[2]

Since rayon fibres are prepared with about 1% oil, this must be removed before dyeing, otherwise, irregularities may occur in the dyeing. The removal of the preparation is done by treatment with a suitable washing agent at 70–75 °C. Viscose woven fabrics must be desized before dyeing. Since knitted goods are treated with mineral oil or paraffin to facilitate the formation of loops, they should be subjected to a good cleaning before dyeing.[2]

As a result of the high content of amorphous regions in their structure, the affinity of regenerated cellulose fibres to dyes that dye cellulose fibres is higher than cotton fibres. Although their affinity to the dye is high, if the dye has a good ability of subsequent levelling, a uniform dyeing can be obtained despite a non-uniform uptake. For this purpose, the equilibrium between the dye taken up by the fibre and the dye remaining in the liquor should occur quickly, and sufficient time should be left for post levelling within the dyeing time. On the other hand, the wet fastness of dyes with good post-levelling ability is generally low. In many cases, although the wet fastness can be partially improved by post-dyeing processes, the cost increases, and in some cases, the washing fastness increases while the light fastness decreases.[2]

One of the reasons for the difficulty in the uniform dyeing of regenerated cellulose fibres is that these fibres do not always have the same properties, and various parts of the same batch may differ from each other. For example, during the production of viscose silk, the degree of ripening, the composition of the coagulation bath and the various tensions vary even by small amounts, which have an effect on the dyeing properties of the fibres obtained. For this reason, factories producing regenerated cellulose fibres assign batch numbers to their fibres. The dyeing properties of fibres belonging to the same batch number do not show great differences. However, if fibres with different batch numbers are used together during weaving or knitting, irregularities in dyeing (e.g. warp or weft streaks) may occur. In viscose fibres, the reason why changes in production conditions lead to uneven (striped) dyeing is the differences in the dense mantle layer outside the fibres. Therefore, in order to reduce the unevenness of products made of viscose fibres due to changes in the production conditions of the fibres, it is necessary to loosen this dense layer outside the fibres and thus reduce its inhibitory effect on the uptake of the dye. In practice, this can be done by treating the product with a 2–10 °Bé solution of caustic soda (it is useful to add an alkali-resistant wetting agent) for 15–20 minutes, or even longer if necessary, before dyeing.[2] Careful attention to dye selection and the determination of a suitable dyeing method are particularly important for obtaining optimum results in viscose. Viscose fibres are more easily affected by hot alkali than cotton. Therefore, it is preferable to use direct and reactive dyes for these fibres. However, to meet the highest fastness standards, significant quantities of vat dyes are successfully used in viscose fibres.[6] Some of the considerations for dyeing regenerated cellulose fibres with certain groups of dyes, different from cotton, are briefly described below.

a. ***Direct (substantive) dyes:*** If the direct dyeing process is extended too far, some of the available dye may degrade. A common cause of the problem is the reduction of azo bonds; degradation is accelerated in the presence of viscose fibres, which have a reducing effect under alkaline dyeing conditions. When dyes with low stability have to be used in viscose fibres, a stabilising effect is provided if the pH is buffered to 6 with ammonium sulphate. Weak oxidants such as sodium m-nitrobenzenesulphonate are sometimes useful, and stronger inorganic oxidants such as potassium chlorate or dichromate are most effective if used with care. An alternative approach is to use sodium perborate, but the dyes must be free of copper, which would cause catalytic degradation of cellulose, and the dyes used must be unaffected by sodium perborate.[13] Since the use of soda ash in the dyeing of viscose fibres with direct dyes will cause swelling of the fibres and thus make it difficult to obtain dye penetration into the fibres, dyeing is carried out without the use of soda ash as much as possible. Since the swelling of the fibres is less and the uptake of dye is slower at high temperatures, it may be useful to start dyeing at boiling temperature (especially in cuprammonium rayon) when necessary. There is no need to add salt in light shades; 2.5–10% salt addition is sufficient in dark shades. Salt should be added in portions after the dyeing progresses. It is recommended that levelling agents that can be used in the dyeing of cotton fibres should also be used in the dyeing of regenerated cellulose (especially in cuprammonium rayon). It is also useful to add an anionic softener to the liquor. This slightly reduces the sensitivity of the fibres to mechanical stress and improves the handle of the finished product. In order to obtain a uniform dyeing, the preparation substances in the fibres are removed before dyeing. Therefore, either by adding softener during dyeing or after dyeing, it is necessary to fill the place of the removed preparation material by passing it through an avivage. Wet fastness can be improved by post-dyeing treatments.[2]
b. ***Development (naphthol) dyes:*** Since naphthols have a higher substantivity towards regenerated cellulose fibres than cotton, lower concentrations are required to achieve the same depth. In dyeing by exhaust method, it is advisable to thoroughly pre-wet the material thoroughly in a wetting agent solution before the addition of naphthol to avoid uneven uptake due to insufficient wetting of the material.[14] Excess caustic soda

used during grounding with naphtholate is objectionable as it causes excessive swelling of the fibres. Either soda ash should be used during grounding or the amount of caustic soda should be kept as low as possible. Another reason for the suboptimal use of naphthol dyes in the dyeing of regenerated cellulose fibres is that they cannot be soaped to the desired degree (mechanical soaking is not possible), and therefore their rubbing fastness is low.[2]

c. ***Vat (Indanthren) dyes:*** Yarns made of regenerated cellulose fibres are dyed in bobbins or hanks according to the conventional leuco process or pre-pigmentation method. Dyeing at as high a temperature as possible minimises fibre swelling and promotes levelling.[5] Generally, the uptake of vat dyes up to 40 °C by regenerated cellulose fibres is similar to that of cotton fibres, but at higher temperatures, they are taken up faster and give darker colours compared to cotton. Some dyes give different nuances of colour in regenerated cellulose fibres than in cotton, and some of them cause a decrease in the brightness of the rayon. The addition of caustic soda and salt should be kept as low as possible. A protective colloid should be added to the liquor. Soaping should be carried out more carefully.[2] Upon completion of dyeing, the material should be subjected to acid neutralisation, as the removal of alkali from these fibres by rinsing alone is slow.[5] After dyeing, it is advisable to give the fibres an avivage to improve the handle.[2]

d. ***Vat-leucoester (Indigosol, Anthrasol) dyes:*** Since strong alkalis are not used during dyeing with them, they are advantageous compared to other dyes that give dyeings having good fastness. They are used in knitwear and for dyeings requiring high fastness in pale shades. Working conditions are the same as for cotton.[2]

e. ***Sulphur dyes:*** Conventional sulphur dyes are not recommended for dyeing regenerated cellulose fibres because the soda ash and sodium sulphide used in conventional sulphur dyes swell the regenerated cellulose fibres too much and cause the fibres to become stiff. Since sodium sulfhydrate can be used instead of sodium sulphide in water-soluble sulphur dyes, regenerated cellulose fibres can be dyed with them.[2] The pad-jig method is particularly advantageous for dyeing viscose fabrics where uniform penetration of the dye in the next step is difficult due to rapid uptake onto the fibres.[15]

f. ***Reactive dyes:*** Many reactive dyes give significantly higher fastness to light on viscose than on cotton dyed at equivalent depths. The pad jig dyeing method used in cotton also gives satisfactory results in woven viscose fabrics. Under alkaline conditions in closed machines, viscose tends to create reduction problems with some sensitive azo reactive dyes. It is therefore important to add a reduction inhibitor from the beginning of dyeing.[8] When dyeing regenerated cellulose fibres with reactive dyes, fewer and weaker alkalis are used compared to dyeing cotton fibres.[2] For this reason, the soda ash method is preferred in the dyeing process, and soda ash is used less than in normal cotton. For this purpose, optimisation should be made using the appropriate tables given in the catalogues. In viscose dyeing, since the fibres already absorb the dye fast, the salt ratio given for cotton in the catalogues should be reduced by 30%. In addition, Glauber's salt should be preferred as the salt as it provides more uniform dyeing. The outer part of viscose fibres has a mantle layer which is denser than the inner part. It will be more advantageous to dye the fibres with hot-dyeing reactive dyes in order to loosen this mantle layer and reduce the inhibiting effect of this layer on dye uptake. In addition, in order to prevent the fibres from swelling too much and thus making it

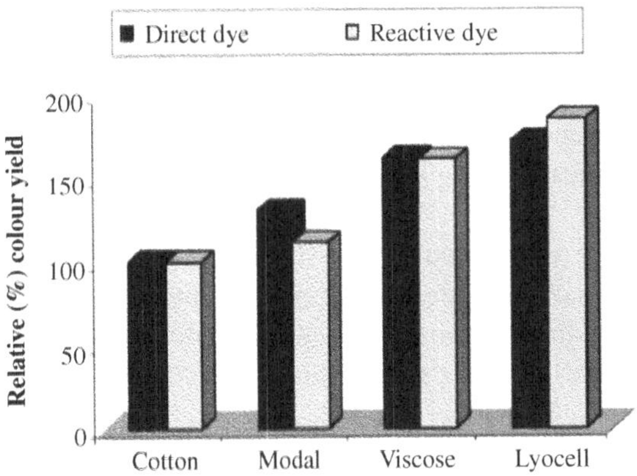

Figure 11.1 Comparison of dyeing properties of cotton and regenerated cellulose fibres. Adapted from Ref. [1].

difficult to obtain dyeing that has penetrated into the material, it is recommended to start dyeing at high temperature and then decrease to normal dyeing temperature or to start at normal temperature and then increase to high temperature (e.g. 80 °C instead of 60 °C or 95 °C instead of 80 °C) and then decrease to normal dyeing temperature (migration method).[1]

Although cuprammonium rayon, viscose, modal and lyocell are all cellulose-based fibres, it should be noted that there will be marked differences in their dyeing behaviour. Generally, the dye uptake is faster in cuprammonium fibres than in viscose fibres. Therefore, the equilibrium between the dye taken up by the fibres and the dye remaining in the liquor is faster. However, if the dye used has good post-levelling properties, the covering of dyeing differences that occur during the production of the fibres is better in this type of regenerated cellulose than in normal viscose.[2]

Unlike viscose, the absence of a core/sheath layer in modal and lyocell, along with the amorphous region, pore structure and morphological differences between fibres all change the dyeing properties of these fibres. If the dyeing properties of cotton and regenerated cellulose fibres with direct and reactive dyes are compared, it can be said that the colour depth generally increases in the order of cotton → modal → viscose → lyocell. Figure 11.1 gives an example of the relative colour yield of viscose, modal and lyocell fibres when the colour yield of cotton is assumed to be 100 in a dyeing process with a direct and reactive dye.[1]

References

1 Yurdakul, A., Atav, R. Boya baskı esasları. Bornova, İzmir: Ege Üniversitesi Tekstil ve Konfeksiyon Araştırma Uygulama Merkezi Yayını, 2006.
2 Tarakçıoğlu, I. Tekstil Boyacılığı I Teksiri. İzmir, Türkiye: Ege Üniversitesi Mühendislik Fakültesi Çoğaltma Yayınları, 1982.
3 Nevell, T.P. Cellulose: structure, properties and behaviour in the dyeing process. In Cellulosics Dyeing (Ed. J. Shore, Chapter 1, 1–73). Bradford, West Yorkshire, UK: Society of Dyers and Colourists, 1995.

4 Atav, R., Namlıgöz, E.S. Keten ve jüt liflerinin boyanması ve bu konudaki yenilikler. Tekstil Teknolojileri Elektronik Dergisi, 2009; 3(3), 65–69.
5 Latham, F.R. Dyeing with vat dyes. In Cellulosics Dyeing (Ed. J. Shore, Chapter 5, 246–278). Bradford, West Yorkshire, UK: Society of Dyers and Colourists, 1995.
6 Shore, J. Selection dyes for cellulosic fibers. In Cellulosics Dyeing (Ed. J. Shore, Chapter 8, 352–395). Bradford, West Yorkshire, UK: Society of Dyers and Colourists, 1995.
7 Mwaikambo, L. Review of the history, properties and application of plant fibres. African Journal of Science and Technology, 2006; 7(2), 121.
8 Shore, J. Dyeing with reactive dyes. In Cellulosics Dyeing (Ed. J. Shore, Chapter 4, 189–242). Bradford, West Yorkshire, UK: Society of Dyers and Colourists, 1995.
9 Atav, R., Büşra Dilden, D., Keskin S., Ergünay, U. Investigation of the dyeability and various performance properties of fabrics produced from flax and hemp fibres and their blends with cotton in comparison with cotton. Coloration Technology, 2024; 140(3), 440–450.
10 Cardomone, J.M., Marmer, W.N. The whitening of textiles. In Chemistry of the Textiles Industry (Ed. C.M. Carr). Blackie Academic & Professional, UK, 1995.
11 Cai, Y., David, S.K., Pailthorpe, M.T. Dyeing of jute and jute/cotton blend fabrics with 2:1 pre-metallised dyes. Dyes and Pigments, 2000; 45(2),161–168.
12 Cai, Y., Huang, G., Huang, X., Pervez, Md N., Lin, L., Xiong, X., Naddeo, V., Wang, L. Modification of ramie fibre with a mixture of NaOH solution and iso-propanol solvent for enhancing its dyeing performance, Journal of Materials Research and Technology, 2024; 30, 1172–1182.
13 Shore, J. Dyeing with direct dyes. In Cellulosics Dyeing (Ed. J. Shore, Chapter 3, 152–186). Bradford, West Yorkshire, UK: Society of Dyers and Colourists, 1995.
14 Shore, J. Dyeing with azoic components. In Cellulosics Dyeing (Ed. J. Shore, Chapter 7, 321–350). Bradford, West Yorkshire, UK: Soceity of Dyers and Colourists, 1995.
15 Senior, C. Dyeing with sulphur dyes. In Cellulosics Dyeing (Ed. J. Shore, Chapter 6, 280–319). Bradford, West Yorkshire, UK: Society of Dyers and Colourists, 1995.

12

Correction of Faulty Dyeings

No matter how carefully the dyeing process is carried out, defects will inevitably occur. Correction of these errors is an additional process and reduces the value of the goods. Sometimes it can even render the goods completely useless. Each nuancing process and repetition of dyeing not only increases the cost of dyeing but also leads to a decrease in production and delays in deadlines.[1] It is possible to categorise the errors occurring in dyeing into two groups:

a. Colouring that is uniform but lighter, darker or of a different nuance than it should be.
b. Uneven colouring (barrê, moirê, abrage, edge-middle-edge difference, tailing, etc.).

In the first group, if the material is dyed in a lighter colour than it should be or if the nuance is not matched, the desired colour is attempted to be achieved by over-dyeing, or if the error is noticed before the dye bath is drained, appropriate additions are made before the dyeing is finished. When addition is to be made, the liquor is first cooled and then the dye is added and the dyeing temperature is raised again and dyeing is continued for a while. If the colour is darker than it should be, then it is necessary to lighten the colour with the appropriate method. In order to do this, it is first necessary to consider which dye the fabric is dyed with. Uneven dyeings can be divided into two categories:

1. ***Extensive unevenness throughout the material***: This type of unevenness is primarily related to the dyeing equipment or process. The material is generally uniform in both chemical and physical properties.
2. ***Local unevenness***: This type of unevenness is primarily due to the material being physically and/or chemically non-uniform.[2]

Uneven dyeings can have various sources. These are:

- ***Faults due to raw material***: Differences in fibre diameters and degree of damage, etc. and differences in pretreatment
- ***Faults of chemical origin***: Incompatibilities between auxiliary and dye, low solubility of dye, etc.
- ***Machine-induced faults***: Improper use of machines, insufficient circulation/movement of liquor and/or material, etc.

In this group of faults, it is necessary to carry out a levelling process or completely strip the dye from the material. However, although some dyes can be easily stripped from the material, some of them are quite difficult to strip. Considering that the fibres will also be damaged during the stripping of the dye, it can be said that in some cases it would be more profitable to dye the goods black or, if possible, to put them on the market as second quality rather than attempting to strip the dye. Dye stripping can be done, in principle:

- either by breaking the bonds of the dye with the fibre and allowing it to bleed into the liquor and retaining the bleeding part in the liquor with auxiliaries
- or the degradation (usually by reduction or oxidation) of the dye molecule into colourless degradation products.

It is also possible to work as a combination of both possibilities.[3] It should be noted that some dyes are affected by oxidation but not by reduction, some are affected by reduction but not by oxidation, and some are affected by both.[4] In the following, information is given on the correction of slight irregularities in cotton materials, the lightening of darker than desired colours and the stripping of faulty dyeings.

12.1 Levelling and Colour Lightening Processes

12.1.1 Direct Dyes

In order to lighten the colour of the dyeing, the dyed material is subjected to a boiling process for 30–40 minutes. with a liquor containing suitable levelling agent and, if necessary soda ash. If 0.5–2 g/L levelling agent and 0.5–2 g/L soda ash are used together, the colour lightening is enhanced.[3]

12.1.2 Vat Dyes

With a few exceptions, it is impossible to completely remove vat dyes. It is only possible to lighten the colour. For this purpose, material is treated with a liquor containing caustic soda and hydrosulphite. With 12–20 mL/L caustic soda (38 °Bé) and 5–7 g/L hydrosulphite-containing liquor, it is possible to lighten the shade of the dye by up to half at most by treating at 70–90 °C for 1 hour.[3]

12.1.3 Sulphur Dyes

For powdered sulphur dye, sodium sulphide, and for soluble sulphur dye, sodium hydrosulphite and sodium hydroxide mixture can be used at boiling temperature to partially remove dyes and correct faulty dyeing processes.[5] It is possible to work with a liquor containing 2–5 g/L calcined soda ash and 5–10 g/L sodium sulphide, or some hydrosulphite can be added to the liquor, or it can be worked directly according to the recipe described for vat dyes.[3]

12.1.4 Reactive Dyes

In order to lighten the colour of the reactive dyed material, the material can be treated with a liquor containing 2 mL/L hydrogen peroxide (30%) and 4 mL/L water glass at 80–90 °C for 60 minutes. It should be noted that materials dyed with dyes containing copper in the molecule should not be processed according to this method. Otherwise, fibres may be damaged as a result of catalytic decomposition.[3]

12.2 Dye Stripping Processes

Colour removal from dyed fabrics is required to correct faulty dyeing or to re-dye the excess fabric in different shades and colours for reuse.[6] The problem of uneven and faulty dyeing in fabrics is usually solved by a chemical stripping process to remove loose dyes from the surface of dyed fabrics.[7] Traditional chemical stripping involves reducing agents and auxiliaries.[8] Chemical stripping leads to the danger of reducing the bursting strength of the fabric and may also cause different chemicals to be released into the water.[9] In recent years, many studies focusing on the use of microorganisms that can degrade dyes suggest that biotechnological dye colour removal is an environmentally friendly and cost-effective alternative. Both bacteria and fungi and their mixed cultures have been reported to remove the colour of textile dyes.[10] When the literature is examined, it is seen that there are various studies on the microbial removal of dyes.[11–14] In addition, there are studies in the literature on colour removal with ozone,[15] and hydroxy methane sulphinic acid and potassium persulphate,[16] which can be an environmentally friendly alternative to colour removal with hydrosulphite.

12.2.1 Direct Dyes

Unless aftertreated, the stripping of direct dyes from dyed fabric is relatively easy, especially in the case of Class A dyes. The dyes may be removed by boiling with a soap solution, or the dyes may be stripped by the reduction and solubilisation of the dyes using sodium hydrosulphite in the presence of caustic soda or by bleaching with a suitable bleaching agent.[17]

The stripping of direct dyes from the fabric is carried out with hydrosulphite in alkaline environment. Stripping can be achieved by treating the material in a liquor containing 2–5 g/L hydrosulphite and 2–5 g/L calcined soda ash or 2–3 mL/L caustic soda (38 °Bé) at 80–95 °C for 30–40 minutes. Most of the dyes that cannot be stripped by reduction in this way, can be stripped by oxidation. For this purpose, it is necessary, to process the dyed material either with a liquor containing 1–1.5 g/L active chlorine (5–10 mL/L NaOCl solution) and 1–2 g/L calcined soda ash at 20–25 °C for 20–40 minutes or with a liquor containing 0.5–1 g/L sodium chlorite, 1 g/L sodium nitrate and 0.4–0.6 mL/L formic acid (85%) at 80–85 °C for 30–45 minutes.[3]

After dyeing with direct dyes, it is more difficult to decolourise or strip the colours of the dyes whose wet fastness has been improved by post-treatment with cation-active auxiliaries. Therefore, it is recommended to treat them with acids before decolourisation. For this purpose, after 30 minutes treatment at 40–60 °C with a liquor containing 2–5 mL/L conc. hydrochloric acid and 1 g/L levelling agent, they are well rinsed and neutralised if necessary.[3]

In order to remove special direct dyes which are copperised or chromated after dyeing and have improved fastness, the metal ion must first be removed, and the material must subsequently be treated with a liquor containing 1–5 g/L sequestering agent at 90–95 °C for

30–60 minutes or with a liquor containing 3–5 mL/L hydrochloric acid (20 °Bé) at 40–50 °C for 30 minutes. Then, as with normal direct dyes, it is treated with a liquor containing 2–5 g/L hydrosulphite and 1–2 mL/L caustic soda (38 °Bé) for 30 minutes at 80–95 °C.[3]

12.2.2 Vat Dyes

When uneven dyeing occurs or the desired colour shade is not achieved, it is necessary to correct it by partial or complete stripping. Vat dyes give very specific colours, and complete stripping of the dye is not practical due to the strong affinity of the dye to cotton. Dyeings are treated with $Na_2S_2O_4$ and NaOH at a certain reduction temperature for 30 minutes for shade correction and partial stripping. The use of non-ionic surfactant increases the levelling rate. For complete stripping, the method used for partial stripping can be carried out by adding a cationic surfactant.[18] Although the stripping effect can be slightly increased by adding 5 g/L magnesium sulphate to the liquor half an hour after starting the process at 70–90 °C with a liquor containing 12–20 mL/L caustic soda (38 °Bé) and 5–7 g/L hydrosulphite, better results can be obtained by using levelling agents (cationic and some non-ionic levelling agents) that have a retarder effect. In case of addition of cationic products to the liquor, since these products form complex with the dye bleeding through the fabric after being vatted under the effect of hydrosulphite and caustic soda, the following equilibrium is constantly disturbed and more and more dye bleeds.

$$\text{Dye in liquor} \rightleftarrows \text{Dye on fibres}$$

It should be noted that the solubility in water of the addition product formed by the sodium leuco form of the dye and the cationic auxiliary is low, and if precautions are not taken, tar-like precipitates may form when stripping dark shades, especially at short liquor ratio. Once these precipitates are formed, it is not possible to dissolve them. Therefore, attention should be paid to the following points:

- Liquor ratio should be increased in parallel with the depth of the colour to be stripped. Liquor ratio should not fall below 1:5 in light shades, 1:10 in medium shades and 1:20 in dark shades.
- The caustic soda (38 °Bé) concentration should not exceed 10–12 mL/L.
- The temperature should not exceed 90–95 °C.
- The processing time should not be too long.
- Auxiliaries should not be used more than recommended.[3]

Polyvinylpyrrolidone (PVP) can be used as a stripping agent. This complexes more readily with the leuco dye than the levelling agent, rapidly forming soluble leuco dye polymer complexes in the liquor. The concentration of PVP should not exceed 1 g/L.[19] Most vat dyes are soluble in hot dimethyl sulphoxide (DMSO) and can therefore be stripped, but this is not of commercial significance from a cost perspective.[18]

12.2.3 Sulphur Dyes

Hot soda ash and sodium sulphide solution can be used to remove sulphur dyes.[6] Most sulphur dyeings are not fast to chlorine. Therefore, faulty dyeings can be stripped by treatment with sodium hypochlorite.[20] If complete removal of dyes made with sulphur dyes is desired, a hypochlorite or chlorite bleaching is used. Sulphur dyes (except for black dyes with good chlorine fastness) decompose during this bleaching. However, since some of the decomposition products remain on the product, they may interfere with a subsequent dyeing and prevent dyeing, especially in vivid tones.[3] Colour fading is less severe when treated with hydrogen peroxide. The loss of colour value is thought to be due to excessive oxidation and to the action of hypochlorous acid (the fall in pH) formed by the decomposition of hypochlorites in the open air. Potassium permanganate also causes decolourisation of sulphur dyes in the same way as hypochlorites.[5]

12.2.4 Reactive Dyes

Removal of reactive dyes can be done according to different methods.

a. ***Neutral hydrosulphite:*** Treatment is done with 5 g/L hydrosulphite containing liquor at 60–80 °C for 30 minutes and the material is rinsed with hot water.
b. ***Alkali hydrosulphite:*** Material is treated with a liquor containing 5 g/L hydrosulphite, 15 mL/L caustic soda (38 °Bé) or 15 g/L calcined soda ash at 60–90 °C for 60 minutes and rinsed with hot water.
c. ***Hypochlorite bleaching:*** Material is treated in a liquor containing 3 g/L active chlorine (NaOCl) and 2 g/L calcined soda ash (pH 10–11) at 20–30 °C for 60 minutes and then chlorine residues are removed from the material with the help of sodium thiosulphate.
d. ***Chlorite bleaching:*** Material is treated in a liquor containing 3 g/L sodium chlorite and formic acid (pH 3.5–4) at 85 °C for 60 minutes, then it is rinsed with hot water.
e. ***Potassium permanganate:*** After treatment at 30 °C for 30 minutes with a liquor containing 1 g/L potassium permanganate and 15 mL/L caustic soda (38 °Bé), material is rinsed with a liquor containing 2 g/L stabilised hydrosulphite. It is a mild method. It can be used especially for the stripping of dyes from regenerated cellulose.
f. ***Alkali hydrosulphite-hypochlorite combination:*** After treatment with a liquor containing 5 g/L hydrosulphite and 15 mL/L caustic soda (38 °Bé) at 60–80 °C for 30 minutes, the material is rinsed with hot water and then treated with a liquor containing 2 g/L active chlorine (NaOCl) and 1–2 g/L calcined soda ash (pH 10) at 20 °C for 30 minutes. The chlorine residues are then removed first with acetic acid and then with sodium bisulphite liquor.
g. ***Hypochlorite-alkali hydrosulphite combination:*** After treatment with a liquor containing 5 g/L active chlorine and 2 g/L calcined soda ash (pH 10–11) at 20 °C for 30 minutes, the material is rinsed with a liquor containing acetic acid and then sodium sulphite to remove chlorine residues. Then it is treated with a liquor containing 5 g/L hydrosulphite and 15 mL/L caustic soda (38 °Bé) at 60–80 °C for 30 minutes and rinsed.
h. ***Sodium chlorite-alkali hydrosulphite:*** After treatment with a liquor containing 3 g/L sodium chlorite and formic acid (pH 3.5–4) at 70–80 °C for 60 minutes, the material is rinsed with hot water and then treated with a liquor containing 5 g/L hydrosulphite and 15 mL/L caustic soda (38 °Bé) at 60–80 °C for 30 minutes and rinsed.[3]

References

1 Yurdakul, A., Orzel, S., Atav, R. Numune ve üretim ölçeklerinde yapılan boyamalar arasındaki farklılıkların ve nedenlerinin incelenmesi. Proje No: TÜBİTAK-TAM 2003-01, Aralık 2004.

2 Baldwinson, T.M. Classification of dyeing and printing auxiliaries by function. In Colorants and Auxiliaries, Organic Chemistry and Application Properties, Second Edition, Volume 2 – Auxiliaries (Ed. J. Shore). Manchester, UK: Society of Dyers and Colourists, 2002.

3 Tarakçıoğlu, I. Tekstil Boyacılığı I Teksiri. İzmir, Türkiye: Ege Üniversitesi Mühendislik Fakültesi Çoğaltma Yayınları, 1982.

4 Jowett, H. Stripping of dyed woollens and other textile materials. Journal of the Society of Dyers and Colourists, 2008; 60(8), 197–200.

5 Chakraborty, J.N. Sulphur dyes. In Handbook of Textile and Industrial Dyeing, Volume 1: Principles, Processes and Types of Dyes (Ed. M. Clark, Chapter 14, 466–485). Cambridge, Birleşik Krallık: Woodhead Publishing Limited, 2011.

6 Chatha, S.A.S., Asgher, M., Ali, S., Hussain, A.I. Biological color stripping: a novel technology for removal of dye from cellulose fibers. Carbohydrate Polymers, 2012; 87(2), 1476–1481.

7 Chatha, S.A.S., Mallhi, A.I., Hussain, A.I., Asgher, M., Nigam, P.S. A biological approach for color-stripping of cotton fabric dyed with C.I. Reactive Black 5 using fungal enzymes from solid state fermentation. Current Biotechnology, 2014; 3(2), 166–173.

8 Chatha, S.A.S., Asgher, M., Iqbal, H.M.N., Ali, S. A novel enzymatic stripping of CI Reactive Black B dyed knitted textiles fabric as an environmentally responsible technology. Electronic Journal of Biology, 2016; 12(3), 276–281.

9 Asgher, M., Muzammil, M., Qamar, S.A., Khalid, N., Bilal, M. Environmentally friendly color stripping of Solar Golden Yellow R dyed cotton fabric by ligninolytic consortia from Ganoderma lucidum IBL-05. Case Studies in Chemical and Environmental Engineering, 2020; 100031.

10 Chatha, S.A.S., Ali, S., Asgher, M., Bhatti, H.N. Investigation of the potential of microbial stripping of dyed cotton fabric using white rot fungi. Textile Research Journal, 2011; 81(17), 1762–1771.

11 Balan, D.S.L. Monteiro, R.T.R. Decolorization of textile indigo dye by ligninolytic fungi. Journal of Biotechnolology, 2001; 89, 141–145.

12 Chander, M., Arora, D.S., Bath, H.K. Biodecolorization of some industrial dyes by white-rot fungi. Journal of Industrial Microbiololgy and Biotechnololgy, 2004; 31, 94–97.

13 Eichlerova, I., Homolka, L., Nerud, F. Synthetic dye decolorization capacity of white rot fungus Dichomitus squalens. Bioresource Technolology, 2006; 97, 2153–2159.

14 Asgher, M., Kousar, S., Bhatti, H.N., Shah, S.H.A., Ali, M. Optimization of medium for decoloriozation of Solar golden yellow R direct textile dye by Schizophyllum commune IBL-06. International Biodeterioration and Biodegradation, 2008; 61, 189–193.

15 Atav, R., Yüksel M.F., Özkaya S., Buğdaycı, B. The use of ozone technology for color stripping and obtaining vintage effect on reactive dyed cotton fabrics. Journal of Natural Fibers, 2022; 19(13), 6648–6658.

16 Atav, R., Çağman F.N., Şahin N.H., Çolakoğlu Ö. Determination of color removal and fading performance of environmentally friendly chemical alternatives to hydrosulfite in reactive dyed cotton knitted fabrics depending on dye chromophore. Journal of Natural Fibers, 2022; 19(17), 15882–15890.

17 Chattopadhyay, D.P. Chemistry of dyeing. In Handbook of Textile and Industrial Dyeing, Volume 1: Principles, Processes and Types of Dyes (Ed. M. Clark, Chapter 4, 150–183). Cambridge, Birleşik Krallık: Woodhead Publishing Limited, 2011.

18 Chakraborty, J.N. Fundamentals and Practices in Colouration of Textiles. Woodhead Publishing India Pvt Ltd, 2010.

19 Latham, F.R. Dyeing with vat dyes. In Cellulosics Dyeing (Ed. J. Shore, Chapter 5, 246–278). Bradford, West Yorkshire, UK: Society of Dyers and Colourists, 1995.

20 Leube, H. Textile dyeing. In Industrial Dyes: Chemistry, Properties, Applications (Ed. Dr. K. Hunger, Chapter 4, 339–426). Wiley-VCH Verlag GmbH & Co. KGaA, 2003.

13

Laboratory Applications Related to Dyeing of Cellulose Fibres

13.1 Dyeing with Direct Dyes

13.1.1 The Effect of Salt and Soda Ash in Dyeing with Direct Dyes

Dyeing liquors containing dye and soda ash, if to be used, are prepared as summarised in Table 13.1. Bleached cotton fabrics are placed in the liquor and the dyeing process is started at 40 °C. After 10 minutes the temperature is increased to 100 °C with a heating rate of 2 °C/min. After this 10 minutes, 1/3 of the salt is added and after 20 minutes of dyeing, the remaining portion (2/3) is added and dyeing is continued for 30 minutes and then the temperature is reduced to 80 °C at a rate of 1 °C/min and the liquor is drained after 20 minutes of operation. After dyeing, the fabric sample is rinsed and dried.

Table 13.1 Effect of salt and soda ash on dyeing with direct dyes.

	1	2	3	4
Direct dye (%)	2	2	2	2
Salt (g/L)	—	15	—	15
Soda ash (%)	—	—	2	2
Liquor ratio	1:20	1:20	1:20	1:20

Evaluation of Results

1. Measure the colour of the fabrics and evaluate how salt and soda ash affect direct dye uptake.
2. Explain for what purpose salt and soda ash are used in dyeing with direct dyes.

13.1.2 The Effect of Pretreatment Process and Cellulosic Fibre Type on Dyeing with Direct Dyes

Dyeing liquors are prepared as summarised in Table 13.2. Five cotton fabrics with varying pretreatments (raw, desized, scoured, bleached, and mercerised) and mercerised five cotton fabrics and bleached linen and viscose fabrics are placed into the liquor. The dyeing process is started at 40 °C and after 10 minutes the temperature is increased to 100 °C with a heating rate of 2 °C/min. After this initial 10 minutes, 1/3 of the salt is added and after 20 minutes of dyeing, the remainder (2/3) is added and dyeing is continued for 30 minutes and then the temperature is reduced to 80 °C at a rate of 1 °C/min and the liquor is drained after 20 minutes. After dyeing, the fabric sample is rinsed and dried.

Table 13.2 Effect of pretreatment and fibre type in dyeing with direct dyes.

	5
Direct dye (%)	2
Salt (g/L)	15
Liquor ratio	1:20

Evaluation of Results

1. Using the colour yield (K/S) values of different pretreated cotton samples, evaluate how the pretreatment process affects direct dye uptake.
2. Using the colour yield (K/S) values of the samples of each fibre type, evaluate how the fibre type affects direct dye uptake.

13.1.3 The Effect of Salt Amount, Soda ash Amount and Liquor Ratio on Dyeing Result in Dyeing with Direct Dyes

Dyeing liquors are prepared as summarised in Table 13.3. The bleached cotton fabric is placed in the liquor and the dyeing process is started at 40 °C and after 10 minutes the temperature is increased to 100 °C with a heating rate of 2 °C/min. After this initial 10 minutes, 1/3 of the salt is added and after 20 minutes of dyeing, the remainder (2/3) is added and dyeing is continued for 30 minutes and then the temperature is reduced to 80 °C at a rate of 1 °C/min and the liquor is drained after 20 minutes of operation. After dyeing, the fabric sample is rinsed and dried.

Table 13.3 Effect of salt and soda ash amount and liquor ratio on dyeing with direct dyes.

	6	7	8	9
Direct dye (%)	2	2	2	2
NaCl (g/L)	15	15	25	15
Soda ash (%)	5	5	5	2
Liquor ratio	1:10	1:20	1:20	1:20

Evaluation of Results

1. Measure the colour yield (K/S) values of the fabric samples with a spectrophotometer and interpret how salt concentration, soda ash concentration and liquor ratio affect the direct dye uptake of the fibres.

13.1.4 Effect of Water Quality on Dyeing Result in Dyeing with Direct Dyes

Dyeing liquors are prepared as summarised in Table 13.4. The bleached cotton fabric is placed in the liquor and the dyeing process is started at 40 °C. After 10 minutes the temperature is increased to 100 °C with a heating rate of 2 °C/min. After this initial 10 minutes, 1/3 of the salt is added and after 20 minutes of dyeing, the remainder (2/3) is added. Dyeing is continued for 30 minutes and then the temperature is reduced to 80 °C at a rate of 1 °C/min and the liquor is drained after 20 minutes of operation. After dyeing, the fabric sample rinsed and dried.

Table 13.4 Effect of water quality on dyeing with direct dyes.

	10	11	12
Direct dye (%)	2	2	2
Salt (g/L)	15	15	15
Liquor ratio	1:20	1:20	1:20
Water type	Pure water	Water containing 50 ppm Ca^{++}	Water containing 5 ppm Fe^{+++}

Evaluation of Results

1. Using the fabric sample from experiment 10 as a reference, calculate the total colour difference (ΔE) of the other samples and discuss the effect of metal ions on the colour obtained in dyeing.

13.1.5 Testing the HT Stability of Direct Dyes

Dyeing liquors are prepared as summarised in Table 13.5. The bleached cotton fabric is placed in the liquor and the dyeing process is started at 40 °C. After 10 minutes the temperature is increased to 100 °C for experiment 13 and 130 °C for experiment 14 with a heating rate of 2 °C/min. After this initial 10 minutes, 1/3 of the salt is added and after 20 minutes of dyeing, the remainder (2/3) is added. Dyeing is continued for 30 minutes in experiment 13 and 90 minutes in experiment 14. Then the temperature is reduced to 80 °C at a rate of 1 °C/min and the liquor is drained after 20 minutes of operation. After dyeing, the fabric sample is rinsed and dried.

Table 13.5 Determination of the HT stability of direct dyes.

	13	14
Direct dye (%)	2	2
Salt (g/L)	15	15
pH adjustment	—	4–5 (with acetic acid/sodium acetate)
Liquor ratio	1:20	1:20

Evaluation of Results

1. Calculate the colour yields of fabric samples 13 and 14 by colour measurements with a spectrophotometer. Based on the difference between the colour yields of the samples and also on the total colour difference values of the two samples with respect to each other, comment on whether the dye you have is resistant to high-temperature HT conditions.

13.1.6 Dyeing with Direct Dyes in Standing Bath

The waste liquor of experiment 4 is collected 3/4 of the chemicals used in experiment 4 are added to this waste liquor and the bleached cotton fabric is dyed. The dyeing process is executed similarly to experiment 4.

Evaluation of Results

1. By comparing the colour of the fabric sample from this experiment with the colour of fabric sample 4, discuss the suitability of working in a standing bath for dyeing with direct dyes.

13.1.7 Determination of Migration Capabilities of Different Types of Direct Dyes

Dyeing liquors are prepared as summarised in Table 13.6. 5 g of bleached cotton fabrics are placed into the liquors and the dyeing process is started at 40 °C. After 10 minutes the temperature is increased to 100 °C with a heating rate of 2 °C/min. After 10 this initial minutes, 1/3 of the salt is added and after 20 minutes of dyeing, the remainder (2/3) is added and dyeing is continued for 30 minutes and then the temperature is reduced to 80 °C at a rate of 1 °C/min and the liquor is drained after 20 minutes of operation. After dyeing, the fabric is rinsed and dried.

Migration test: An equal weight (5 g) accompanying fabric is sewn on one side of the dyed fabrics and treated with four different liquors containing 10 g/L Na_2SO_4 with a liquor ratio of 1:10 at 98 °C for 20–40–60–80 minutes. At the end of the test, the migration ability of the dyes is determined by evaluating the dye migration to the accompanying fabrics with a spectrophotometer.

Table 13.6 Determination of migration ability of different types of direct dyes.

	15	16	17
Direct dye (%)	2	2	2
Salt (g/L)	15	15	15
Liquor ratio	1:20	1:20	1:20
Dye type	Class A	Class B	Class C

Evaluation of Results

1. Measure the colour yields (K/S) of all samples by spectrophotometer. Calculate the relative (%) colour yields of dyed fabrics and accompanying fabrics subjected to migration for 20, 40, 60 and 80 minutes, assuming an initial colour yield of 100 for each dyed sample. Using these values, plot a time-dependent migration graph for each dye. Compare the migration capabilities of the dyes based on your results. With which of these dyes is it easier to obtain a uniform colour? Why?

13.1.8 Combination Dyeing with Direct Dyes Suitable and Unsuitable for Combination

For this experimental group, a yellow-coloured dye is selected (Dye 1) and combination dyeings are made with two different blue dyes from the dye catalogue, one which is deemed suitable to be combined with this dye (Dye 2) and one which is not (Dye 3). Three bleached cotton fabrics of equal weight are placed in each liquor: the first sample is removed before the first portion of salt is added, the second sample is removed before the second portion is added and the third sample is removed at the end. Dyeing liquors are prepared as summarised in Table 13.7. The fabrics are placed into the liquors and the dyeing process is started at 40 °C. After 10 minutes the temperature is increased to 100 °C with a heating rate of 2 °C/min. After this initial 10 minutes, 1/3 of the salt is added and after 20 minutes of dyeing, the remainder (2/3) is added. Dyeing is continued for 30 minutes and then the temperature is reduced to 80 °C at a rate of 1 °C/min and the liquor is drained after 20 minutes of operation. After dyeing, 3 cold rinsings are applied to the fabrics.

Table 13.7 Combination of dyes suitable and unsuitable for combination.

	18	19
Direct dye (%)	1% Dye 1 + 1% Dye 2	1% Dye 1 + 1% Dye 3
Salt (g/L)	15	15
Liquor ratio	1:20	1:20

Evaluation of Results

1. How does the uptake of dyes occur when dyes that are and are not suitable for the combination are combined? Comment on the nuance changes that occur when non-combinable dyes are used together.

13.1.9 Fastness Improvement Processes After Dyeing with Direct Dyes

In this experiment, the previously dyed fabric sample 1 is divided into four pieces. One piece is treated with a cationic fixing agent, another piece is treated with a reactive cationic aftertreatment agent and the third one is rinsed with salt to see how the washing fastness values change compared to the untreated sample. The treatment recipes are given in Table 13.8.

Table 13.8 Fastness improvement processes after dyeing with direct dyes.

	20	21	22
Cationic fixing agent (%)	2	—	—
Reactive cationic fixing agent (%)	—	4	—
Salt (g/L)	—	—	10
Acetic acid	pH 4–5	—	—
Caustic soda (38 °Bé) (mL/L)	—	5	—
Liquor ratio	1:20	1:20	1:20
Processing time (min)	20	35*	15
Temperature (°C)	40	40	100

*Firstly, the fabric is placed in the bath and the reactive cationic fixing agent is dosed within 5 minutes. After 10 minutes of processing, caustic soda is dosed within 10 minutes and the aftertreatment is finished by processing for an additional 10 minutes. The fabric is subjected to cold rinsing and neutralisation (with acetic acid).

Evaluation of Results

1. Compare the fastness values of aftertreated and untreated fabric samples separately by applying the washing fastness test. Explain the possible reasons for the differences.

13.1.10 The Effects of Various Parameters on Dyeing Result in Dyeing with Direct Dyes by Pad-batch Method

Firstly, 1 L of dye solution (10 g/L) is prepared and divided into four flasks, each of which is used as a flask for one impregnation experiment. The chemicals required for that experiment are added to each flask in the amounts specified in Table 13.9. The bleached cotton woven fabrics are then impregnated with these liquors under the specified temperature and squeezing pressure (pick-up), wrapped with polyethylene foil and kept at room temperature for the specified time. After unwrapping, the fabrics are rinsed and dried.

Table 13.9 Effect of various parameters on pad-batch dyeing with direct dyes.

	23	24	25	26
Direct dye (g/L)	10	10	10	10
Wetting agent (g/L)	2	2	2	2
Urea (g/L)	100	—	100	100
Pick-up (%)	80	80	60	80
Impregnation temperature (°C)	25–30	25–30	25–30	25–30
Batching time (h)	16	16	16	8

Evaluation of Results

1. Comment on the effect of urea usage by comparing the colour of fabric samples 23 and 24, the effect of pick-up by comparing the colour of fabric samples 23 and 25, and the effect of dyeing time by comparing the colour of fabric samples 23 and 26.

13.2 Dyeing with Vat Dyes

13.2.1 Effect of Physical Form and Particle Size of Vat Dye on Dyeing Result

A vat dye is selected from the catalogue and a 2% dyeing is performed with powder and liquid forms of this dye (Table 13.10). The dyeing recipe and graph for both dyeing processes to be carried out by vatting in long liquor are found from the catalogue. In both experiments, the process temperature and duration are the same. After dyeing, the rinsed fabric sample is subjected to oxidation with hydrogen peroxide (15 minutes at 50–60 °C with 0.7% peroxide). The fabric sample is then washed (with 2 g/L washing agent) and dried.

Table 13.10 Effect of physical form and particle size of the vat dye on the dyeing result.

	27	28
Dye form	Powder	Liquid
Dyeing depth (%)	2	2
Liquor ratio	1:20	1:20

Evaluation of Results

1. Measure the colour of both fabric samples with a spectrophotometer and compare the colour yields. State which form of dye gives better colour yields and give the reason for this.
2. Considering the difficulty in vatting of powder dyes compared to liquid brands, which method do you recommend for powder brands, vatting in main liquor or vatting in long liquor? Why?

13.2.2 Dyeing with IK, IW and IN Type Vat Dyes

One IK, one IW and one IN class vat dye are selected from the catalogue. The chemicals specified in Table 13.11 are added to the liquors, and after 10 minutes, the bleached cotton

Table 13.11 Dyeing with IK, IW and IN type vat dyes.

	29	30	31*
Dye class	IK	IW	IN
Dyeing depth (%)	2	2	2
NaOH (38 °Bé) (mL/L)	5.6	7	12.5
$Na_2S_2O_4$ (g/L)	2	2.4	2.8
Na_2SO_4 (g/L)	22	13.5	—
Dyeing temperature (°C)	25	40	60
Dyeing time (min)	60	60	60
Liquor ratio	1:20	1:20	1:20

*Fabric sample no. 31 will be used in the experimental group no. 13.2.6

knitted fabric is added. The dyeing process is carried out at the specified temperatures and times. After dyeing, the rinsed fabric sample is subjected to oxidation with hydrogen peroxide (15 minutes at 50–60 °C with 0.7% peroxide). The fabric sample is then washed (with 2 g/L washing agent) and dried.

Evaluation of Results

1. Find the molecular structures of the dyes you use in the dyeing process from the Colour Index (CI) catalogue and calculate their molecular weights (g/mol). Compare the values you have found for IK, IW and IN class vat dyes.
2. What is the other name of the hydrosulphite you used in dyeing? Explain the function of hydrosulphite in the dyeing process.
3. Write and balance a general redox reaction for an anthraquinone-based vat dye.
4. In IN class vat dyes, salt is not used to avoid increasing the already high aggregation. On the other hand, although the degree of association is low in IK-type vat dyes, caustic soda (although less is used compared to other dye classes) is added to the dyeing liquor to reduce the degree of association. What could be the reason for this?

13.2.3 Dyeing with Vat Dyes in Open and Closed Systems

A vat dye belonging to the IN class is selected from the catalogue and the liquor prepared with the chemicals specified in Table 13.12 is divided into two parts. The dyeing process is started with one of these liquors in the open dyeing system and with the other one in the closed dyeing system. After 10 minutes, bleached cotton knitted fabrics are added and the dyeing process is carried out at the specified temperature and time. After dyeing, the rinsed fabric sample is subjected to oxidation with hydrogen peroxide (15 minutes at 50–60 °C with 0.7% peroxide). The fabric sample is then washed (with a 2 g/L washing agent) and dried.

Table 13.12 Dyeing with vat dyes in open and closed systems.

	32	33
Dye class	IN	IN
Dyeing depth (%)	2	2
NaOH (38 °Bé) (mL/L)	12.5	12.5
$Na_2S_2O_4$ (g/L)	2.8	2.8
Dyeing system	Open	Closed
Liquor ratio	1:20	1:20

Evaluation of Results

1. Measure the colour of the dyed fabric samples by spectrophotometer and compare the colour yields. Comment on the reason for the difference.

13.2.4 The Effect of Vatting Method on Dyeing Result in Dyeing with Vat Dyes

Two vat dyes are selected from the catalogue, one of which is suitable for both vat dyeing methods (Dye 1) and the other one is suitable for vat or dyeing only in the main vat or long liquor (Dye 2). And 2% dyeing is done with these dyes as summarised in Table 13.13. The recipe and process graph of the dyeing process are found from the catalogue. After dyeing, the rinsed fabric samples are subjected to oxidation with hydrogen peroxide (15 minutes at 50–60 °C with 0.7% peroxide). The fabric sample is then washed (with a 2 g/L washing agent) and dried.

Table 13.13 Effect of dicing method on the dyeing result in dyeing with vat dyes.

	34	35	36	37
Dye	Dye 1	Dye 1	Dye 2	Dye 2
Dyeing depth (%)	2	2	2	2
Vatting method	Main vat	Long liquor	Main vat	Long liquor
Liquor ratio	1:20	1:20	1:20	1:20

Evaluation of Results

1. Calculate the colour yields of fabric samples 34 and 35 by colour measurements with a spectrophotometer. Indicate which vatting method gives the highest colour yield and explain the reason.
2. By comparing fabric samples 36 and 37, comment on the drawbacks that may arise if a vatting method that is not recommended for a dye is used.

13.2.5 The Effect of Hydrosulphite, Alkali and Salt Amount and Dyeing Temperature on Dyeing Result in Dyeing with Vat Dyes

Dyeing processes are carried out by making the changes specified in Table 13.14 in the dyeing recipe and temperature with a vat dye belonging to the IW class selected from

Table 13.14 The effect of various factors in dyeing with vat dyes.

38	39	40
No hydrosulphite	20% less hydrosulphite	20% more hydrosulphite
41	**42**	**43**
No caustic soda	20% less caustic soda	20% more caustic soda
44	**45**	**46**
No salt	20% less salt	20% more salt
47	**48**	
10 °C low-temperature dyeing	10 °C high-temperature dyeing	

the catalogue for experiment 31. After dyeing, the rinsed fabric samples are subjected to oxidation with hydrogen peroxide (15 minutes at 50–60 °C with 0.7% peroxide). The fabric sample is then washed (with a 2 g/L washing agent) and dried.

Evaluation of Results

1. Compare the colour yields and dyeing uniformity of test sample 30 with test samples 38, 39 and 40 and comment on the effect of the amount of hydrosulphite on dyeing with vat dyes.
2. Compare the colour yields and dyeing uniformity of sample 30 with samples 41, 42 and 43 and comment on the effect of alkali content on dyeing with vat dyes.
3. Compare the colour yields and dyeing uniformity of test sample 30 with test samples 44, 45 and 46 and comment on the effect of salt content on dyeing with vat dyes.
4. Comment on the effect of dyeing temperature on dyeing with vat dyes by comparing the colour yields and dyeing uniformity of test sample 30 with test samples 47 and 48.

13.2.6 Post-treatments After Dyeing with Vat Dyes

The previously dyed fabric sample 31 is divided into eight equal parts. One of them (49) is oxidised with hydrogen peroxide without rinsing. The other fabric samples are oxidised with air (50), water (51), hydrogen peroxide (52), sodium perborate (53), potassium bichromate/acetic acid (54), sodium chloride/acetic acid (55) and sodium hypochlorite (56) after rinsing. All fabric samples are then subjected to soaping (with 2 g/L washing agent).

Evaluation of Results

1. Measure the colour of the fabric samples by spectrophotometer and compare the differences of the samples.
2. Compare the rubbing fastness of samples 49 and 52. Interpret the results you have found.

13.2.7 Pad-batch Dyeing with Vat Dyes

Impregnation with dye dispersion → intermediate drying (100°C) → impregnation with vatting liquor → batching (Table 13.15) → rinsing → oxidation → soaping → rinsing

Table 13.15 Pad-batch dyeing with vat dyes.

57	58	59
2 hours of batching	4 hours of batching	2 hours of batching without covering with foil

Evaluation of Results

1. Measure the colour of samples no. 57 and 58 with a spectrophotometer and compare the colour yields. State the effect of prolonging the batching time.
2. By comparing test samples no. 57 and 59, state the disadvantages of not covering the fabric with polyethylene foil during batching, and explain the reason.

13.3 Dyeing with Indigo Dye

For the dyeing process, the main vat is prepared first. For this, 4 g of indigo dye is made into a paste with 2 mL of alcohol and 40 mL of hot water, 7.8 mL of caustic soda (38 °Bé) and 4 g of hydrosulphite are added and the mixture is vatted at 55–60 °C for 30 minutes. Then, the main vat is poured into the liquor containing 0.5 mL/L caustic soda (38 °Bé), 0.5 g/L hydrosulphite, 1 g/L auxiliary and 20 g/L NaCl and mixed thoroughly. The dyeing process is carried out in three passes:

Passage I: After dyeing at 20–25 °C for 10 minutes, squeezed and oxidised for 15 minutes **(60)**.

Passage II: After dyeing at 20–25 °C for 5 minutes, squeezed and oxidised for 15 minutes **(61)**.

Passage III: After dyeing at 20–25 °C for 5 minutes, squeezed and oxidised for 15 minutes **(62)**.

The number of passages can be increased depending on the depth of the dyeing. After dyeing, the fabric sample is rinsed thoroughly with cold water. In the dyeing process, three fabrics of equal weight are used. One of the pieces is separated at the end of each passage.

Evaluation of Results

1. Measure the colour of the fabric samples with a spectrophotometer and examine the effect of the number of passages on the colour yield. State why single-pass dyeing does not give sufficient colour yield.
2. Can the vatting in long liquor method be used instead of the main vatting method for dyeing warp yarns with indigo dye? Why not?
3. Investigate the causes and potential solutions for the two main problems encountered in indigo-dyed goods, back staining and yellowing, and give brief information.
4. What are the ecological drawbacks of hydrosulphite, which has found the most widespread use in the vatting of indigo dyes? Discuss the possibilities of using electrochemical reduction technology instead.
5. What do you think could be the reason for adding NaCl to the liquor in indigo dyeing? Explain it in relation to the structural properties of the dye.

13.4 Dyeing with Sulphur Dyes

13.4.1 Dyeing with Conventional Sulphur Dye

The dyeing liquor is prepared as summarised in Table 13.16, containing dye, soda ash, sodium sulphide and a complexing agent. The bleached cotton fabric is placed in the liquor and the dyeing process is started at 40 °C and after 10 minutes the temperature is increased to 90 °C with a heating rate of 2 °C/min. After 60 minutes of dyeing, salt is added and the temperature is reduced to 80 °C at a rate of 1 °C/min and the liquor is drained. After dyeing, the bleached cotton fabric is rinsed and oxidised and rinsed again.

Table 13.16 Dyeing with conventional sulphur dye.

	63
Conventional sulphur dye (%)	2
Sodium sulphate (g/L)	10
Soda ash (%)	2
Complexing agent (g/L)	1
Sodium sulphide (%)	4
Liquor ratio	1:20
Oxidation	at 50–60 °C with 0.7% peroxide for 15 min

Evaluation of Results

1. Evaluate the brightness of the colour of the fabric sample visually.
2. Apply the washing and rubbing fastness test to the fabric sample and evaluate with the grey scale.

13.4.2 Comparison of the Performance of Reducing Agent Alternatives for Dyeing with Sulphur Dyes

Dyeing liquors are prepared as summarised in Table 13.17, containing dye, soda ash, reducing agent and complexing agent. Bleached cotton fabric is placed in the liquor and the dyeing process is started at 40 °C. After 10 minutes the temperature is increased to 90 °C with a heating rate of 2 °C/min and 1/2 of the salt is added. After 30 minutes of dyeing, the remainder part of the salt is added and the temperature is reduced to 80 °C at a rate of 1 °C/min and the liquor is drained. After dyeing, the fabric is rinsed and subjected to oxidation with 1% H_2O_2 in a liquor adjusted to pH 5–6 with acetic acid at 70 °C for 10 minutes and then rinsed again.

Table 13.17 Comparison of the performance of reducing agent alternatives that can be used in dyeing with sulphur dyes.

	64	65	66	67	68
Conventional sulphur dye (%)	2	2	2	2	2
Sodium sulphate (g/L)	10	10	10	10	10
Soda ash (%)	2	—	6	6	6
Caustic soda (38 °Bé) (mL/L)	—	1	8	8	8
Sodium sulphide (%)	4	—	—	—	—
Glucose (g/L)	—	—	10	20	—
Sodium hydrogen sulphide (mL/L)	—	5	—	—	—
Hydrosulphite (g/L)	—	—	6	—	10
Stabilisol S (mL/L)	—	3	—	—	—
Liquor ratio	1:20	1:20	1:20	1:20	1:20

Evaluation of Results

1. Measure the colour of the test samples and compare the colour yields.
2. Check whether the samples are dyed uniformly. If there is any sample that is not dyed uniformly, discuss what the cause might be.
3. Write the chemical formulae of the reducing agents used in the experiments. Compare their redox potentials.
4. Compare the unit prices of the reducing agents used in the experiments.

13.4.3 Dyeing with Water-soluble Sulphur Dyes

The dyeing liquor is prepared as summarised in Table 13.18, containing dye, soda ash and sodium hydrogen sulphide. Bleached and mercerised cotton fabrics and viscose fabrics are placed into the liquor and the dyeing process is started at 40 °C. After 10 minutes the temperature is increased to 90 °C with a heating rate of 2 °C/min. After 60 minutes of dyeing, salt is added and the temperature is reduced to 80 °C at a rate of 1 °C/min and the liquor is drained. After dyeing, the fabrics are rinsed and oxidised with 0.7% peroxide in a liquor adjusted to pH 5–6 with acetic acid at 50–60 °C for 15 minutes and then rinsed again.

Table 13.18 Dyeing with water-soluble sulphur dyes.

	69
Water-soluble sulphur dye (%)	2
Sodium sulphate (g/L)	10
Soda ash (%)	2
Sodium hydrogen sulphide (g/L)	4
Liquor ratio	1:20

Evaluation of Results

1. Evaluate the dry and wet rubbing fastness of the bleached cotton fabrics from experiments 63 and 69 on a grey scale. Which fabric has better rubbing fastness? Why?

13.4.4 Dyeing with Sulphur Dyes by Impregnation Method

The dyeing liquor is prepared as summarised in Table 13.19. Bleached cotton woven fabric is subjected to impregnation (50 °C) – intermediate drying – air passage – impregnation (30 °C) – steaming (30–60 seconds at 102–105 °C) – air passage – rinsing – drying processes, respectively.

Table 13.19 Dyeing with sulphur dyes by impregnation method.

First bath	70
Dye (g/L)	30
Wetting agent (mL/L)	1
Pick-up (%)	70
Second bath	
Sulfhydrate F (mL/L)	70
Soda ash (g/L)	17.5
Na_2SO_4 (g/L)	20
From first bath liquor (mL/L)	20
Pick-up (%)	80

Evaluation of Results

1. Investigate the uses of pad-steam dyeing with sulphur dyes. Evaluate whether the dyeing obtained is uniform or not.

13.5 Dyeing with Reactive Dyes

13.5.1 Dyeing with Reactive Dyes By Exhaust Method

13.5.1.1 The Effect of Dyeing Method on Dyeing Result in Dyeing with Reactive Dyes

Two-step method (71): The dyeing process is started in a salt-containing liquor is started at 30 °C and the dye is added after 10 minutes (Table 13.20). After 25 minutes of dyeing, the temperature is increased to 60 °C with a heating rate of 1 °C/min and after 10 minutes of dyeing at this temperature, alkali is added and dyeing is finished after 45 minutes.

The method of adding some alkali from the beginning (72): Dyeing is started with a salt-containing liquor at 30 °C and the dye is added after 10 minutes. After 15 minutes of dyeing, 1/3 of the alkali is added. After 10 minutes, the temperature is increased to 60 °C with a heating rate of 1 °C/min and after 10 minutes of dyeing at this temperature, the remainder (2/3) of the alkali is added and the dyeing is finished after 45 minutes.

All-in method (73): Dyeing is started with a liquor containing salt, alkali and dye at 30 °C. After 25 minutes of dyeing, the temperature is increased to 60 °C with a heating rate of 1 °C/min and the dyeing is finished after 55 minutes at this temperature.

Washing steps: Overflow cold rinsing → neutralisation at 60 °C (10 minutes with 0.5 g/L acetic acid) → rinsing at 95 °C (15 minutes) → rinsing at 80 °C (10 minutes) → cold rinsing (5 minutes).

Table 13.20 Exhaust Dyeing with vinylsulphone-based reactive dyes.

	71	72	73
Vinylsulphone-based reactive dye (%)	2	2	2
Na_2SO_4 (g/L)	50	50	50
Na_2CO_3 (g/L)	12	12	12
Liquor ratio	1:20	1:20	1:20

Evaluation of Results

1. Measure the colour of the fabric samples by spectrophotometer and compare the colour yields.
2. Compare the levelness of the fabric samples and indicate whether there is any difference between the dyeing methods for the dye you used.
3. Do you think the all-in method is suitable for every colour? Why? State which of the hot and cold dye groups it is more suitable for and why.

13.5.1.2 The Effect of Alkali Type Used in Dyeing with Reactive Dyes

The dyeing process is started in a salt-containing liquor is started at 30 °C and the dye is added after 10 minutes (Table 13.21). After 15 minutes of dyeing, the first alkali is added. After 10 minutes, the temperature is increased to 60 °C with a heating rate of 1 °C/min. After 10 minutes of dyeing at this temperature, second alkali is added and dyeing is finished after 45 minutes.

Washing steps: Overflow cold rinsing → neutralisation at 60 °C (10 minutes with 0.5 g/L acetic acid) → rinsing at 95 °C (15 minutes) → rinsing at 80 °C (10 minutes) → cold rinsing (5 minutes).

Table 13.21 Effect of alkali type in exhaust dyeing with reactive dye.

	74	75	76	77
Vinylsulphone-based reactive dye (%)	2	2	2	2
NaCl (g/L)	50	50	50	50
First alkali addition	4 g/L Na_2CO_3	5 g/L Na_2CO_3	0.4 mL/L NaOH*	4 g/L Na_3PO_4
Second alkali addition	8 g/L Na_2CO_3	0.7 mL/L NaOH*	0.8 mL/L NaOH*	8 g/L Na_3PO_4
Liquor ratio	1:20	1:20	1:20	1:20

*50%

Evaluation of Results

1. Measure the colour of the fabric samples by spectrophotometer and compare the colour yields. Comment on the effect of alkali type.
2. Which parameters should be considered in the selection of alkali for dyeing? Why?

3. Compare the levelness of the fabric samples and indicate whether there is any difference in terms of the type of alkali for the dye you used.
4. For each experiment, measure the pH of the liquor at the time of addition of all alkali with a pH-meter.
5. Write the reactions for the covalent binding of vinylsulphone and heterocyclic ring reactive dyes to cellulose fibre and show the necessity of alkali presence in the medium.

13.5.1.3 The Effect of Salt Type in Dyeing with Reactive Dyes

Dyeing with vinylsulphone-based dye: The dyeing process is started with salt-containing liquor at 30°C and the dye is added after 10 minutes (Table 13.22). After 15 minutes of dyeing, 1/3 of the alkali is added. After 10 minutes, the temperature is increased to 60°C with a heating rate of 1°C/min and after 10 minutes of dyeing at this temperature, the remainder (2/3) of the alkali is added and dyeing is finished after 45 minutes. Then, the following steps are applied: overflow cold rinsing → neutralisation at 60°C (10 minutes with 0.5 g/L acetic acid) → rinsing at 95°C (15 minutes) → rinsing at 80°C (10 minutes) → cold rinsing (5 minutes) is applied, respectively.

Dyeing with heterocyclic ring-based dye: The dyeing process is started with a salt-containing liquor at 50 °C and the dye is added after 10 minutes. After 20 minutes of dyeing, the temperature is increased to 85 °C with a heating rate of 1 °C/min and after 10 minutes of dyeing at this temperature, alkali is added and dyeing is finished after 45 minutes. Then, the following steps are applied: overflow cold rinsing → rinsing at 70 °C (10 minutes) → rinsing at 70 °C (10 minutes) → rinsing at 95 °C (15 minutes) → rinsing at 70 °C (10 minutes) → cold rinsing (5 minutes) is applied, respectively.

Table 13.22 Effect of salt type in exhaust dyeing with reactive dye.

	78	79	80	81
Vinylsulphone-based reactive dye (%)	2	2	—	—
Heterocyclic ring-based reactive dye (%)	—	—	2	2
NaCl (g/L)	50	—	60	—
Na_2SO_4 (g/L)	—	50	—	60
Na_2CO_3 (g/L)	12	12	15	15
Liquor ratio	1:20	1:20	1:20	1:20

Evaluation of Results

1. Measure the colour of the fabric samples by spectrophotometer and compare the colour yields. Comment on the effect of salt type.
2. Did you observe any difference in the effect of salt type on the dyeing result depending on the chemical structure of the dye? If there is a difference, explain what this difference is due to.
3. Compare the levelness of the fabric samples and indicate whether there is any difference in terms of the type of dye used.

4. Investigate the properties of sodium chloride and sodium sulphate salts for wastewater treatment and indicate which type of salt would be more advantageous from an ecological perspective.

13.5.1.4 Dyeing with Reactive Dyes Requiring Low Salt Usage

The dyeing process is started in a salt-containing liquor is started at 30 °C and the dye is added after 10 minutes (Table 13.23). After 15 minutes of dyeing, alkali is added. After 10 minutes, the temperature is increased to 60 °C with a heating rate of 1 °C/min and dyeing is performed for 60 minutes at this temperature.

Washing steps: Overflow cold rinsing → neutralisation at 60 °C (10 minutes with 0.5 g/L acetic acid) → rinsing at 95 °C (15 minutes) → rinsing at 80 °C (10 minutes) → cold rinsing (5 minutes)

Table 13.23 Dyeing with reactive dye requiring low salt.

	82
Reactive dye requiring low salt (%)	2
Na_2SO_4 (g/L)	20
Na_2CO_3 (g/L)	8
Liquor ratio	1:20

Evaluation of Results

1. Which characteristic of the dye you used in the dyeing experiment is suitable for low-salt dyeing? Find the chemical composition of the dye you used in the experiment in the CI catalogue and comment on its structure.
2. As is well known, the aim of washing is to remove the unfixed dye (hydrolysed with water) fraction. From this point of view, would you expect the washing of low-salt dyeable reactive dyes to be easier or more difficult than for normal reactive dyes? Why?

13.5.1.5 The Effect of Fabric Knit Structure in Dyeing with Reactive Dyes

Bleached cotton single jersey, 1X1 rib, interlock, purl and three thread fleece (futter) fabrics are placed in the dye liquor and the dyeing process is started at 60 °C and salt is added after 10 minutes (Table 13.24). After 15 minutes of dyeing, alkali is added. After 60 minute of dyeing at this temperature, the fabric sample is subjected to rinsing at 50 °C → neutralisation at 50 °C (10 minutes with acetic acid) → rinsing at 90 °C (15 minutes) → rinsing at 50 °C (10 minutes) → cold rinsing (10 minutes).

Table 13.24 The effect of fabric knit structure in exhaust dyeing with reactive dye.

	83
Vinylsulphone-based reactive dye (%)	2
Na_2SO_4 (g/L)	50
Na_2CO_3 (g/L)	12
Liquor ratio	1:20

Evaluation of Results

1. Measure the colour of the fabric samples with a spectrophotometer and compare the colour yields. Discuss whether the difference is due to the difference in dye uptake or to the different light reflection properties due to the knit structure of the fabric.
2. What are porosity and pore area? What is the relationship between fabric knit structure and porosity? What effect does the change in fabric porosity and pore area have on the ability of the fabric to uptake dye? Explain.

13.5.1.6 Bleaching of Cotton Fabrics in The Same Bath After Dyeing with Reactive Dyes

The dyeing process is started in a salt-containing liquor is started at 40°C and dye, salt, alkali and cotton fabric are added to the liquor (Table 13.25). After 20 minutes of dyeing, the temperature is increased to 80°C with a heating rate of 1°C/min and after 45 minutes of dyeing at this temperature, hydrogen peroxide is added and the dyeing process is finished after 20–30 minutes. Then overflow cold rinsing → rinsing at 70°C (10 minutes) → rinsing at 70°C (10 minutes) → rinsing at 95°C (15 minutes) → rinsing at 70°C (10 minutes) → cold rinsing (5 minutes) is applied, respectively.

Table 13.25 Bleaching in the same bath after dyeing with reactive dye.

	84	85
Hot-dyeing reactive dye (%)	0.5	0.5
Sequestering agent (g/L)	1	1
Wetting agent (g/L)	1	1
Na_2SO_4 (g/L)	30	30
Na_2CO_3 (g/L)	5	5
NaOH (38 °Bé) (mL/L)	1	1
H_2O_2 (35%) (mL/L)	4	—
Liquor ratio	1:20	1:20

Evaluation of Results

1. Measure the colour of the fabric samples by spectrophotometer and compare the colour yields.
2. Perform a washing fastness test on the fabric samples and compare the fastness values.

13.5.1.7 Use of Sequestering Agent in Dyeing with Reactive Dyes

The dyeing process is started in a salt-containing liquor is started at 30°C and the dye is added after 10 minutes (Table 13.26). After 15 minutes of dyeing, 1/3 of the alkali is added. After 10 minutes, the temperature is increased to 60°C with a heating rate of 1°C/min and after 10 minutes of dyeing at this temperature, the remainder (2/3) of the alkali is added and the dyeing is finished after 45 minutes.

Washing regime: Overflow cold rinsing → Neutralisation at 60°C (10 minutes with 0.5 g/L acetic acid) → rinsing at 80°C (10 minutes) → rinsing at 95°C (15 minutes) → rinsing at 80°C (10 minutes) → cold rinsing (5 minutes)

Table 13.26 Effect of the use of sequestering agent in exhaust dyeing with reactive dye.

	86	87	88	89
Metal-containing reactive dye (%)	2	2	2	2
Na_2SO_4 (g/L)	50	50	50	50
Na_2CO_3 (g/L)	12	12	12	12
Sequestering agent type	—	—	Polyamino carboxylic acid	Acrylate
Sequestering agent (g/L)	—	—	1	1
Water to be used	Pure	Hard	Hard	Hard
Liquor ratio	1:20	1:20	1:20	1:20
	90	91	92	93
Metal-free reactive dye (%)	2	2	2	2
Na_2SO_4 (g/L)	50	50	50	50
Na_2CO_3 (g/L)	12	12	12	12
Sequestering agent type	—	—	Polyamino carboxylic acid	Acrylate
Sequestering agent (g/L)	—	—	1	1
Water to be used	Pure water	Hard water	Hard water	Hard water
Liquor ratio	1:20	1:20	1:20	1:20

Evaluation of Results

1. Measure the colour of the fabric samples with a spectrophotometer and compare the colour yields. Also measure the total colour difference values of other fabric samples with reference to the dyeing with pure water. Based on your results, comment on the effect of water hardness on metal-containing and metal-free reactive dyes and the most suitable type of sequestering agent for them.
2. What could be the reason why dyeing with a metal-containing reactive dye using a polyamino carboxylic acid-based sequestering agent (e.g. EDTA) gives a worse result than dyeing with hard water? Explain.

3. Determine the ion-binding capacity according to the boiling method for one of the sequestering agents used in the experiments. Calculate the optimum amount of sequestering agent to be used by taking into account the ion-binding capacity and hardness values of your water.

13.5.1.8 Investigation of the Effect of Various Parameters on Washing Processes After Dyeing with Reactive Dyes

In this series of experiments, dye hydrolysates were first formed with two different vinylsulphone-based reactive dyes with low and high substantivity. For this purpose, the dye solution prepared with 2.5 g/L vinylsulphone-based reactive dye and 10 g/L soda ash is boiled at 95 °C for 4 hours to form hydrolysate and cooled to room temperature. The prepared hydrolysate solutions were applied to 100% cotton woven fabric with a pick-up value of 80% at room temperature and the fabric samples were dried at 120 °C for 5 minutes. These fabric samples are then washed at 95 °C for the time specified in Table 13.27.

Table 13.27 Investigation of the effect of various parameters on washing processes after dyeing with reactive dyes (H: Hard, S: Soft).

	Fabric dyed with low-substantivity reactive dye						
	94	**95**	**96***	**97**	**98**	**99**	**100**
Water type	S	S	S	S	H	S	S
Washing agent ionity	—	—	—	—	—	Anionic	Non-ionic
Washing agent (g/L)	—	—	—	—	—	2	2
Temperature (°C)	95	60	95	95	95	95	95
Duration (min)	15	15	15 + 15	15	15	15	15
Liquor ratio	1:10	1:10	1:10	1:20	1:10	1:10	1:10
	Fabric dyed with high-substantivity reactive dyes						
	101	**102**	**103***	**104**	**105**	**106**	**107**
Water type	S	S	S	S	H	S	S
Washing agent ionity	—	—	—	—	—	Anionic	Non-ionic
Washing agent (g/L)	—	—	—	—	—	2	2
Temperature (°C)	95	60	95	95	95	95	95
Duration (min)	15	15	15 + 15	15	15	15	15
Liquor ratio	1:10	1:10	1:10	1:20	1:10	1:10	1:10

*Two pieces of fabric of equal weight are placed in the washing liquor and one of the fabric samples is taken after washing (96-a and 103-a) and the other one is washed for the second time in the same way (96-b and 103-b).

Evaluation of Results

1. Measure the colour of the fabric samples by spectrophotometer and calculate the relative (%) colour yields of the other samples by accepting the colour yield of the unwashed dyed fabric sample as 100 for experiments 94–100 and the colour yield of the unwashed dyed fabric sample as 100 for experiments 101–107. Using these values, interpret the effect of water type, washing temperature, washing time, washing agent type and liquor ratio on the hydrolysate removal efficiency for both dyes. Are the effects you found parallel for both dyes? Why?
2. Calculate the difference between the relative colour yields of the fabric samples after application and after washing with soft water for both dyes and interpret the effect of substantivity on the washing process.
3. Compare the relative colour yields of fabric samples (96-b and 103-b) washed twice at a liquor ratio of 1:10 and fabric samples washed once at a liquor ratio of 1:20 and indicate which washing has the better hydrolysate removal efficiency.

13.5.1.9 The Effect of Post-treatments on Fastness After Dyeing with Reactive Dyes

Firstly, cotton fabric is dyed with a vinylsulphone-based reactive dye by the two-step method as specified in Section 13.5.1.1. After dyeing, the fabric sample is divided into five equal parts and one of the parts is not post-treated with a softener or fixing agent, while each of the other four fabric parts is post-treated separately at 40°C for 20 minutes with the recipes given in Table 13.28.

Table 13.28 Effect of post-treatments on fastness after dyeing with reactive dyes.

	108 (a)	108 (b)	108 (c)	108 (d)	108 (e)
Softener type	—	Non-ionic	Cationic	Silicone	—
Softener amount (%)	—	2	2	2	—
Fixing agent amount (%)	—	—	—	—	2
pH (with acetic acid)	5	5	5	5	5
Liquor ratio	1:20	1:20	1:20	1:20	1:20

Evaluation of Results

1. Calculate the total colour difference values of the fabric samples in the spectrophotometer with reference to fabric sample 108 (a). On the basis of your results, indicate whether softeners and fixing agents cause colour changes.
2. Rank the handle of the fabric samples from best to worst.
3. Apply the hydrophilicity test to fabric samples 108 (a-b-c-d) according to the drop test method and analyse the effect of fabric softener type on fabric hydrophilicity.
4. Evaluate the effect of post-treatment with fixing agent on fastness by applying washing, water, dry and wet rubbing and light fastness tests to samples 108 (a) and 108 (e) samples.

13.5.1.10 Comparison of Dyeing Properties of Linen and Regenerated Cellulose Fibres with Cotton Fibres by Reactive Dyes

Fabric samples of similar structure (density, weight, etc.) made of cotton **(109)**, linen **(110)**, viscose **(111)**, modal **(112)** and Lyocell **(113)** fibres are dyed with a vinylsulphone-based reactive dye by the two-step method as specified in Section 13.5.1.1.

Evaluation of Results

1. Compare the dyeing uniformity of the fabric samples and their colour yields based on the results of spectrophotometer measurements. Comment on the reason for the difference.
2. Would the results have been different if the dyeing had been done in the same bath instead of separate baths? Why?
3. If the fabric is a blend of cotton or viscose containing elastane, it needs to be heat-set before dyeing. What is the purpose of this and what are the optimal temperature/time values for each blend? Research.

13.5.1.11 Effect of Cellulase Enzymatic Treatment on the Dyeability of Cellulosic Fibres with Reactive Dyes

Cellulase-treated and untreated fabrics are added to the dye liquor and the dyeing process is started at 60 °C and salt is added after 10 minutes (Table 13.29). After 15 minutes of dyeing, alkali is added. After 60 minutes of dyeing at this temperature, the fabric sample is subjected to rinsing at 50 °C → neutralisation at 50 °C (10 minutes with acetic acid) → rinsing at 90 °C (15 minutes) → rinsing at 50 °C (10 minutes) → cold rinsing (10 minutes). After the dyeing process, the fabric sample without enzymatic treatment is divided into two equal parts and one of them is subjected to enzymatic treatment with cellulase after dyeing.

Enzymatic treatment with cellulase: Fabric is subjected to treatment in a liquor containing 2% enzyme with pH adjusted to 4.5–5.5 (with acetic acid) at a ratio of 1:20 at 60 °C for 30 minutes.

Table 13.29 Effect of enzymatic treatment with cellulase in dyeing with reactive dyes.

	114
Vinylsulphone-based reactive dye (%)	2
Na_2SO_4 (g/L)	45
Na_2CO_3 (g/L)	11
Liquor ratio	1:20

Evaluation of Results

1. Measure the colour of samples which were not enzymatically treated, enzymatically treated before dyeing and enzymatically treated after dyeing by spectrophotometer. Compare and interpret the colour yield values.
2. Perform a tensile strength test on fabric samples and interpret the effect of enzymatic treatment on the strength.

3. Perform a pilling test on samples which were not enzymatically treated, enzymatically treated before dyeing and enzymatically treated after dyeing and interpret the effect of enzymatic treatment and its step (before or after dyeing) on the pilling tendency of the fabric.

13.5.2 Dyeing by Impregnation Method with Reactive Dyes

13.5.2.1 Pad-batch Dyeing

13.5.2.1.1 Single Bath Pad-batch Method

Dyeing liquors containing the chemicals given in Table 13.30 are prepared. Cotton woven fabrics are impregnated with these liquors at room temperature, wrapped in foil and kept at the temperatures and times specified in Table 13.30. After batching, the fabric samples are unwrapped and subjected to rinsing at 50 °C → rinsing at 50 °C → rinsing at 80 °C → rinsing at 95 °C → rinsing at 95 °C → rinsing at 80 °C → rinsing at 50 °C (pH is adjusted to 5–6 with acetic acid).

Table 13.30 One-bath pad-batch dyeing with reactive dyes.

	115	116	117	118	119	120	121*
Vinylsulphone-based reactive dye (g/L)	50	50	50	50	50	50	50
Wetting agent (g/L)	2	2	—	2	2	2	2
Urea (g/L)	—	—	—	100	—	—	—
Sodium silicate (1:3.3) (37–40 °Bé) (mL/L)	50	50	50	50	50	50	50
Caustic soda (50%) (mL/L)	14	14	14	14	14	14	14
Batching temperature (°C)	25	25	25	25	25	50	25
Batching time (hours)	12	12	12	12	48	12	12
Pick-up (%)	60	80	60	60	60	60	60

*The fabric sample is kept without being wrapped in foil.

Evaluation of Results

1. Measure the colour yield and CIE L*a*b* values of the fabric samples by spectrophotometer. Explain the effects of pick-up value, impregnation time, use of wetting agent, use of urea, batching time, batching temperature and batching method on the colour obtained.
2. Investigate the errors that may occur in pad-batch dyeing and their reasons.
3. How is it tested whether there is an edge-centre-edge difference in the fabric during pad-batch dyeing? Investigate and apply this test to fabric sample 115 and evaluate the result.
4. What are powder and carbon paper tests? For what purpose and how are they done? Apply these tests on the padding mangle you use in your laboratory and check whether squeezing pressure is uniform across the width.
5. Examine the pad-batch dyeing recipe and indicate the differences between it and the recipes used in exhaust dyeing and the reasons for these differences.

13.5.2.1.2 Two Bath Pad-batch Method

Conduct of the experiment (122): The fabric is impregnated with a dyeing liquor containing 50 g/L dye and 2 g/L wetting agent, squeezed to 60% pick-up and dried. Then, the fabric sample is impregnated with 14 mL/L NaOH (50%) (Pick-up 80%), wrapped in foil and kept for six hours at room temperature. After batching the fabric sample is unwrapped and post-treatments given in Section 13.5.2.1.1 are applied.

Evaluation of Results

1. Measure the colour yield and CIE L*a*b* values of the fabric sample with a spectrophotometer. Compare the results with sample 115.
2. Evaluate the fabric sample with the grey scale by applying washing fastness and rubbing fastness tests.
3. Explain how to prepare a 50% caustic soda solution. How much 38 °Bé caustic (mL/L) should be used instead of 50% caustic soda?

13.5.2.2 Pad-steam Dyeing

13.5.2.2.1 Single Bath Pad-steam (Pad-dry-steam) Method

The dyeing liquor containing the chemicals given in Table 13.31 is prepared. The fabric is impregnated with this liquor at room temperature, dried (100–120 °C) and then steamed with saturated steam at 101–105 °C for 60 seconds. After steaming, the fabric sample is subjected to the post-treatments given in Section 13.5.2.1.1.

Table 13.31 One-bath pad-steam dyeing with reactive dyes.

	123
Vinylsulphone-based reactive dye (g/L)	50
Wetting agent (g/L)	2
Antimigration agent (g/L)	10
Weak oxidising agent (g/L)	10
Urea (g/L)	100
Soda ash (g/L)	20
Pick-up (%)	60

Evaluation of Results

1. Measure the colour yield and CIE L*a*b* values of the fabric sample with a spectrophotometer.
2. Evaluate the fabric sample with the grey scale by applying washing fastness and rubbing fastness tests.
3. How does the addition of salt to the impregnation liquor prevent migration? Comment.
4. Where can reducing effects occur during pad-steam dyeing? Which chemicals can be added to the liquor to prevent this?

13.5.2.2.2 Single Bath Wet (without Intermediate Drying) Pad-steam Method

The dyeing liquor containing the chemicals given in Table 13.32 is prepared. The fabric is impregnated with this liquor at room temperature and then steamed at 101–105 °C with saturated steam for 60 seconds. After steaming, post-treatments given in Section 13.5.2.1.1 are applied to the fabric sample.

Table 13.32 One-bath wet pad-steam dyeing with reactive dyes.

	124
Vinylsulphone-based reactive dye (g/L)	50
Wetting agent (g/L)	2
Salt (g/L)	30
Weak oxidising agent (g/L)	10
Soda ash (g/L)	20
Caustic soda (50%) (mL/L)	6
Pick-up (%)	60

Evaluation of Results

1. Measure the colour yield and CIE L*a*b* values of the fabric sample with a spectrophotometer.
2. Apply washing fastness and rubbing fastness tests to the fabric samples and evaluate them with the grey scale.
3. Why is salt added to the impregnation liquor?
4. Find out why intermediate drying is required for viscose fabrics, cord fabric and towels.

13.5.2.2.3 Two-bath Pad-steam (Pad-dry-pad-steam and Pad-pad-steam) Method

Dyeing liquors containing the chemicals given in Table 13.33 are prepared. The fabrics are impregnated with this liquor at room temperature and only sample 125 is dried at 110–120 °C. Then the samples are impregnated with the chemical liquor containing the substances given in Table 13.33 and steamed at 101–105 °C with saturated steam for 60 seconds. After steaming, the fabric sample is subjected to the post-treatments given in Section 13.5.2.1.1.

Table 13.33 Two-bath pad-steam dyeing with reactive dyes.

	125	126
Dye liquor		
Vinylsulphone-based reactive dye (g/L)	50	50
Wetting agent (g/L)	2	2
Antimigration agent (g/L)	10	—
Weak oxidising agent (g/L)	10	10
Pick-up (%)	60	60

	125	126
Chemical liquor		
Salt (g/L)	250	250
Soda ash (g/L)	20	20
Caustic soda (50%) (mL/L)	6	6
Pick-up (%)	60	60

Evaluation of Results

1. Measure the colour yield and CIE L*a*b* values of the fabric samples with a spectrophotometer. Compare the results obtained and discuss the effects of various pad-steam methods on the colour to be obtained in dyeing.
2. Apply washing fastness and rubbing fastness tests to the fabric samples and evaluate them with the grey scale.
3. State the reason for adding salt to the impregnation liquor in pad-dry-pad-steam and pad-pad-steam method dyeings.

13.5.2.3 Dyeing by Thermofixing (Pad-dry-thermofix) Method

Dyeing liquors containing the chemicals given in Table 13.34 are prepared. The fabrics are impregnated with this liquor at room temperature and subjected to an IR pre-drying process. Then, they are dried at 100–120 °C and fixed at 140°C for 1 minute. After the fixation process, the post-treatments given in Section 13.5.2.1.1 are applied to the fabric samples.

Table 13.34 Pad-dry-thermofix dyeing with reactive dyes.

	127	128
Vinylsulphone-based reactive dye (g/L)	50	50
Wetting agent (g/L)	2	2
Antimigration agent (g/L)	10	10
Weak oxidising agent (g/L)	10	10
Urea (g/L)	100	—
Soda ash (g/L)	30	30
Pick-up (%)	60	60

Evaluation of Results

1. Measure the colour yield and CIE L*a*b* values of the fabric samples with a spectrophotometer. Based on your results, indicate how the use of urea affects the colour obtained.
2. Apply washing fastness and rubbing fastness tests to the fabric samples and evaluate them with the grey scale. Compare the results and indicate whether the use of urea has an effect on the fastness.
3. Investigate the chemical structure of borax and its effects during the dyeing process.

4. How does the variation in temperature and duration of the thermofixing process affect the dyeing result? Which parameters should be taken into account when determining the optimum thermofixing temperature and time?

13.5.2.4 Pad-dry (Impregnation-drying) Dyeing

The dyeing liquor containing the chemicals given in Table 13.35 is prepared. The fabric is impregnated with this dyeing liquor at room temperature and then dried at 100 °C for 8 minutes. The fabric sample is then subjected to the post-treatments given in Section 13.5.2.1.1.

Table 13.35 Pad-dry dyeing with reactive dyes.

	129
Vinylsulphone-based reactive dye (g/L)	50
Wetting agent (g/L)	2
Antimigration agent (g/L)	10
Weak oxidising agent (g/L)	10
Urea (g/L)	100
Soda ash (g/L)	28
Pick-up (%)	60

Evaluation of Results

1. Measure the colour yield and CIE L*a*b* values of the fabric sample with a spectrophotometer.
2. Apply washing fastness and rubbing fastness tests to the fabric sample and evaluate with the grey scale.
3. Examine new dye catalogues to determine whether this method is still in use.

13.5.2.5 Dyeing by Wet Fixation Method

The dyeing liquor containing the chemicals given in Table 13.36 is prepared. The fabric sample is impregnated with this liquor at room temperature and dried at 100–120 °C. Then the fabric is passed through the chemical liquor at 95 °C containing the substances given in Table 13.36 within 3–5 seconds. Then the post-treatments given in Section 13.5.2.1.1 are applied to the fabric sample.

Table 13.36 Wet fixation dyeing with reactive dyes.

	130
Dye liquor	
Vinylsulphone-based reactive dye (g/L)	50
Wetting agent (g/L)	2
Antimigration agent (g/L)	10
Weak oxidising agent (g/L)	10
Pick-up (%)	60

(Continued)

Table 13.36 (Continued)

	130
Chemical liquor	
Salt (g/L)	250
Caustic soda (38 °Bé) (mL/L)	50
Pick-up (%)	60

Evaluation of Results

1. Measure the colour yield and CIE L*a*b* values of the fabric sample with a spectrophotometer.
2. Apply washing fastness and rubbing fastness tests to the fabric sample and evaluate with grey scale.
3. Discuss the effect of the method used in dyeing by impregnation method on the dyeing result by comparing the colours, dyeing levelness and fastness of samples 115, 122, 123, 124, 125, 127, 129 and 130.

13.6 Dyeing with Other Dyes

13.6.1 Dyeing with Indosol Dyes

Dyeing liquors are prepared to contain dye as given in Table 13.37. Bleached cotton woven fabric is placed in the liquor and the dyeing process is started at 40 °C and after 10 minutes the temperature is increased to 100 °C within 30 minutes. After 10 minutes, 1/3 of the salt is added and after 10 minutes of dyeing, the remainder (2/3) is added and dyeing is continued for 45 minutes and then the levelness of the dyeing is checked. If sufficient uniformity is achieved, the dyeing is finished by cooling the dyeing liquor to 80–90 °C within 15 minutes. After dyeing, the fabric sample is thoroughly rinsed and dried and then the fabric is divided into three parts. One part is kept as is. One of the other parts is impregnated with reactive fixing agent only and the other with a reactive fixing agent + softener, dried at 100–120 °C and condensed at 160 °C for 3 minutes. The post-treatment recipe is given in Table 13.38.

Table 13.37 Dyeing with indosol dye.

	131
Indosol dye (%)	2
Salt (g/L)	15
Liquor ratio	1:20

Table 13.38 Post-treatment after indosol dyeing.

	Post-treatment with reactive fixing agent	Post-treatment with reactive fixing agent + softener
Reactive fixing agent (g/L)	100	100
Cationic softener (g/L)	—	20
Non-ionic wetting agent (g/L)	1	1
Acetic acid (g/L)	0.5–1	0.5–1
Pick-up (%)	80	80

Evaluation of Results

1. Perform a washing fastness test on samples with and without post-treatment with reactive fixing agent and evaluate the results with the grey scale.
2. Subjectively evaluate the handle of fabric samples that have not been post-treated with reactive fixing agent, treated only with reactive fixing agent and treated with reactive fixing agent + softener. Is it possible to evaluate the handle of fabric samples objectively (numerically)? Investigate.
3. What could be the reason for choosing woven fabric for dyeing? What difficulties can be encountered when dyeing knitted materials with this dye?
4. Do you think it is appropriate to use this dye class for a product to be used in the production of sportswear? Why do you think so?

13.6.2 Dyeing with Naphthol Dyes

After 1 kg Naphthol dye is pasted with 1.5 L ethyl alcohol, 0.4 L caustic soda (38 °Bé) is added and diluted with 2 L water at 40 °C and dissolved by stirring. 0.75 L formaldehyde (33%) is added and the mixture is left to stand for 5–15 minutes. Using this naphtholate solution, a grounding liquor containing the chemicals given in Table 13.39 is prepared and the fabric samples are grounded with this liquor at 30 °C for 30 minutes. Then the fabric samples are squeezed and the excess naphtholate is removed.

To obtain diazonium salt, an amine compound (Base) suitable for diazotisation is taken. 1 kg. base is made into a pasted with 0.1 L dispersing agent and 2 L hydrochloric acid (20 °Bé). The solution prepared with 0.43 kg of sodium nitrite and 2 L of cold water is added

Table 13.39 Dyeing with naphthol dye.

	132
Naphthol dye (%)	2
NaOH (38 °Bé) (mL/L)	6
Protective colloid (g/L)	3
NaCl (g/L)	15
Liquor ratio	1:20

while stirring. The diazotisation process takes 30 minutes at 5–15 °C (can be cooled with ice). Then the grounded fabric samples are coupled with diazonium salt at 20–25 °C for 30 minutes. 20–50 g/L sodium chloride can be added to the development liquor.

Fabric sample 132 is divided into two parts (132 (a) and (b)). All fabric samples are treated with liquor containing 2 g/L dispersing agent, 2 g/L soda ash, 2 g/L sequestering agent at 60 °C for 10 minutes. Then rinsed with hot and cold water. Afterwards, only fabric sample 132 (a) is soaped again with a liquor containing 1 g/L dispersing agent, 1 g/L soda ash and 1 g/L sequestering agent. It is rinsed with hot and cold water and dried.

Evaluation of Results

1. Perform washing, rubbing and light fastness tests on the fabric samples and evaluate the results. How did you observe a difference between the fastness to rubbing of fabric samples 131 (a) and (b)? Explain why.
2. Why is formaldehyde added to the grounding liquor? Explain the mechanism of action of formaldehyde by writing the reaction.
3. Why have naphthol dyes lost their importance today? Discuss.

13.6.3 Dyeing with Vat-leucoester (Indigosol, Anthrasol) Dyes

13.6.3.1 Dyeing with Vat-leucoester Dyes by One-bath Nitrite Method

(133): 1 g/L wetting-washing agent, 1–2 g/L dispersing agent, 1 g/L calcined soda ash are added to the flask heated up to 40–50 °C and dye (1 g/L) is added in portions within 10–20 minutes. Again, 20 g/L sodium sulphate, which was previously dissolved, is added in portions within 20 minutes and then 0.5 g/L sodium nitrite is added. In the meantime, the temperature of the liquor should have decreased to 35 °C. 5 mL/L sulphuric acid (66 °Bé), which is diluted by pouring four times its volume in cold water, is slowly poured and allowed to react at 30–35 °C for 15 minutes. The fabric sample is then rinsed, neutralised (2–5 g/L soda ash), washed at boiling temperature (1 g/L washing agent) and rinsed, respectively.

Evaluation of Results

1. Why should dyes with high substantivity be preferred for dyeing with vat-leucoester dyes by the one-bath nitrite method? Explain.

13.6.3.2 Dyeing with Vat-leucoester Dyes by Two-bath Nitrite Method

(134): When working with dyes whose substantivity to fibres is good in cold, after the temperature of the liquor containing the wetting agent is adjusted to 30 °C, dissolved dye and 1 g/L calcined soda ash are added and grounding is carried out isothermally at this temperature. After treatment with this liquor for 15–20 minutes, 20 g/L calcined sodium sulphate is added in two portions and the grounding is continued for 30 minutes. After grounding, the product is squeezed before treatment with the development liquor. The development liquor contains 5–20 mL/L sulphuric acid (96%), 0.5–1 g/L sodium nitrite. This liquor is treated at 25 °C for 15 minutes. The fabric sample is then rinsed, neutralised (2–5 g/L soda ash), washed at boiling temperature (1 g/L washing agent) and rinsed, respectively.

Evaluation of Results

1. Measure the colour of the fabric sample with a spectrophotometer and compare the results of dyeing by one- and two-bath nitrite methods.

13.6.3.3 Dyeing by Impregnation (Semi-continuous and Continuous) Method with Vat-leucoester Dyes

(135): 10 g/L dye, 2 g/L wetting agent and 1 g/L calcined soda ash are dissolved in boiling water and added to the liquor, water is added and 10 g/L $NaNO_2$ dissolved in a separate pot is added. Then the volume of the liquor is adjusted with water. After impregnation, the fabric is passed through an air passage and impregnated with the development liquor. The development liquor contains 20 mL/L sulphuric acid (66 °Bé). The temperature of the development liquor is 70 °C and the residence time in the liquor is 8–10 seconds. The subsequent processes are overflow rinsing with cold water → soaping with boiling soda ash and detergent liquor (2.5 g/L soda ash and 1 g/L detergent) → rinsing for 10–15 minutes.

Evaluation of Results

1. Apply washing and rubbing fastness tests to the fabric sample and evaluate with the grey scale.

13.6.4 Dyeing with Cationic (Basic) Dyes

***Pretreatment*:** The cotton fabric is treated with 3% tannin at 60 °C for 15 minutes at a liquor ratio of 1:20. Then the liquor is drained, the fabric is squeezed and treated with 1.5% tartrate (potassium antimonyl tartrate) at a liquor ratio of 1:20 for 15 minutes with cold liquor.

***Dyeing*:** The dyeing process is started at 40 °C with the liquor containing the chemicals given in Table 13.40 and the temperature is increased to 50 °C. After 60 minutes of dyeing at this temperature, the liquor is cooled and drained. Then the fabric is rinsed and dried.

Table 13.40 Dyeing of cotton with cationic dye.

	136	137
Material to be dyed	Pretreated cotton	Untreated cotton
Cationic Dye (%)	2	2
CH_3COOH (%)	2	2
Liquor Ratio	1:20	1:20

Evaluation of Results

1. Calculate the "% Dye Uptake" values for both samples by measuring the absorbance values of the dyeing liquors before and after the dyeing process. Interpret the reason for the results you obtained.

2. Measure the colour yields and CIE L*a*b* values of the fabric samples by spectrophotometer. Based on the results of the colour measurements, interpret how pretreatment with tannin/tartrate affects the dyeing result.
3. Apply washing and light fastness tests to fabric samples and discuss the reasons for the results.

13.6.5 Dyeing with Pigment

13.6.5.1 Dyeing with Pigment by Impregnation Method

Dyeing liquors containing the chemicals given in Table 13.41 are prepared. Fabric samples are impregnated with these liquors at room temperature and dried at 100–120°C. Then, they are subjected to fixation at the temperatures and times specified in Table 13.41.

Table 13.41 Dyeing with pigment by impregnation method (A: Acrylic, B: Butadiene based binder).

	138	139	140	141	142	143	144
Pigment (g/L)	1	10	30	10	10	10	10
Non-ionic wetting agent (g/L)	5	5	5	5	5	5	5
Antimigration agent (g/L)	20	20	20	20	20	20	20
Binder (g/L)	3	30	90	15	30	—	30
Binder type	A	A	A	A	B	A	A
Fixation temperature (°C)	150	150	150	150	150	150	170
Fixation time (min)	4	4	4	4	4	4	2
pH (with NH_3)	8–9	8–9	8–9	8–9	8–9	8–9	8–9
Pick-up (%)	60	60	60	60	60	60	60

Evaluation of Results

1. Measure the colour yield and CIE L*a*b* values of the fabric samples by spectrophotometer. Based on your results, discuss the effects of dye concentration, binder concentration and type, and fixing temperature and time on the colour and dyeing uniformity obtained in dyeing.
2. Perform washing and rubbing fastness tests on fabric samples 138, 139, 140, 141 and 142 and interpret how dyeing depth, binder concentration and type affect the wet fastness.
3. Subjectively evaluate the handle of fabric samples 138, 139, 140, 141 and 142 and indicate how the dyeing depth, binder concentration and type affect the fabric handle.
4. Investigate what types of pigments exist in terms of chemical structure and what their properties are.
5. Examine whether the use of pigments in viscose fabrics would be appropriate.

13.6.5.2 Combination of Dyeing with Pigment by Impregnation and Wrinkle-free Finishing

Dyeing liquors containing the chemicals given in Table 13.42 are prepared. Fabric samples are impregnated with these liquors at room temperature and dried at 100–120 °C. Then they are subjected to fixation at the temperatures and times specified in Table 13.42.

Table 13.42 Combination of pigment dyeing by impregnation and wrinkle-free finishing.

	145
Pigment (g/L)	10
Non-ionic wetting agent (g/L)	5
Antimigration agent (g/L)	20
Binder (acrylic-based) (g/L)	30
Anti-wrinkling agent (g/L)	100
$MgCl_2$ (g/L)	10
Fixation temperature (°C)	180
Fixation time (min)	1
pH (with ammonia)	8–9
Pick-up (%)	60

Evaluation of Results

1. Measure the colour yield and CIE L*a*b* values of the fabric sample by spectrophotometer. Compare with sample 137.
2. Test the handle, dry and wet wrinkle recovery angles and strength of fabric samples 139 and 145. Comment on the reasons for the results.
3. Which type of anti-wrinkling agent did you use? Why did you prefer it?

13.6.5.3 Dyeing with Pigment by Exhaust Method

Two-step method (146): Four cotton fabrics (146 (a), (b), (c) and (d)) are treated with 5% cationising agent and 4.5% acrylic-based binder at 40 °C for 20 minutes, then 0.5 g/L soda ash is added to the liquor and treated at this temperature for another 20 minutes. The fabrics are thoroughly rinsed after the process. Then, they are treated at 65 °C for 30 minutes in a new liquor containing the chemicals given in Table 13.43. After the process, the fabric is rinsed with cold water.

Three-step method (147): After the cotton fabric is treated with 5% cationising agent at 40 °C for 20 minutes, 0.5 g/L soda ash is added to the liquor to reach pH 9 and the liquor is treated at this temperature for 20 minutes more. After the process, the fabric is thoroughly rinsed. Then, it is treated at 65 °C for 30 minutes in a new liquor containing the chemicals given in Table 13.43. After the process, the fabric is rinsed with cold water. Finally, it is treated with 4.5% acrylic-based binder at 40 °C for 20 minutes and then rinsed and dried.

Table 13.43 Pigment dyeing by exhaust method.

	146	147
Pigment (%)	3	3
Anti-crease agent (g/L)	4	4
Soda ash (g/L)	0.5*	0.5*
Liquor ratio	1:20	1:20

Evaluation of Results

1. Compare the colour, fastness, handle and appearance properties of samples 146 and 147 and determine the differences between the two- and three-step methods.
2. Investigate the mechanism of action of the alkali added to the liquor in the application of the cationising agent.

13.6.5.4 Obtaining Bleeding Dyeing with Pigment

Cotton fabrics are placed in liquors containing the chemicals given in Table 13.44 and treated at 65 °C for 30 minutes. After the process, they are rinsed with cold water and dried.

Table 13.44 Obtaining bleeding dyeing in pigment dyeing.

	148	149	150
Material to be dyed	Cotton	Cotton	Cationised cotton**
Pigment (%)	3	3	3
Binder (%)	6	3	3
Anti-crease agent (g/L)	4	4	4
Soda ash (g/L)	0.5*	0.5*	0.5*
Liquor ratio	1:20	1:20	1:20

**pH is 9*

**The *fabric* is treated with 5% cationising agent at 40 °C for 20 minutes *before dyeing*.

Evaluation of Results

1. Carry out three consecutive washings of the fabric samples at a liquor ratio of 1:20. Take samples of each washing water and compare the dye bleeding during repeated washings. Comment on the reason for the differences.
2. Compare the colour, fastness, handle and appearance properties of the fabric samples.
3. Analyse the bleeding dyeing process in terms of environmental ecology.

Index

Printed and bound by CPI Group (UK) Ltd, Croydon, CR0 4YY
17/09/2026
14964270-0003